International Conference on Environmental Resources Management in Global Region (ICERM 2017)

IOP Conference Series: Earth and Environmental Science
Volume 148

Bali, Indonesia
25 November 2017

ISBN: 978-1-5108-6280-7
ISSN: 1755-1307

Printed from e-media with permission by:

Curran Associates, Inc.
57 Morehouse Lane
Red Hook, NY 12571

Printed by Curran Associates, Inc. (2018)

For permission requests, please contact the Institute of Physics at the address below.

Institute of Physics
Dirac House, Temple Back
Bristol BS1 6BE UK

Phone: 44 1 17 929 7481
Fax: 44 1 17 920 0979

techtracking@iop.org

Additional copies of this publication are available from:

Curran Associates, Inc.
57 Morehouse Lane
Red Hook, NY 12571 USA
Phone: 845-758-0400
Fax: 845-758-2633
Email: curran@proceedings.com
Web: www.proceedings.com

Table of contents

Volume 148

International Conference on Environmental Resources Management in Global Region (ICERM 2017)

25 November 2017, Bali, Indonesia

Accepted papers received: 19 April 2018
Published online: 11 May 2018

Geomaritime

Land Degradation

Urban Climate and Energy

Water and Waste Management

There are no words that can describe our respect for all of the guests and participants of the International Conference on Environmental Resources Management on Global Region 2017 or sorted as ICERM 2017.

ICERM 2017 is our first year international conference which directly organized under the Departement of Environmental Geography, Faculty of Geography, Universitas Gadjah Mada Yogyakarta. In this year, we brought six topics: Water and Waste Management, Degradation, Efficient Urban Resources, Disaster Risk Reduction, Geomaritime, and Urban Climate and Energy Management.

This year, we are proudly presents six (6) keynote and invited speakers, they are:

- Dr. Sadjuga, Ministry of Research and Technology for Higher Education, Indonesia
- Prof. Dr. Lars Ribbe, Institute for Technology and Resources Management in the Tropics and Subtropics, Cologne University of Applied Sciences, Germany
- Dr. I Nyoman Radiarta, Head of Institute for Marine Research and Observation (BROL), Bali, Indonesia
- Prof. Makoto Takahashi, Department of Social and Human Environment, Graduate School of Environmental Studies, Nagoya University, Japan
- Prof. Marwa Khalifa, Planning & Urban Design Department, Faculty of Engineering, Ain Shams University, Egypt
- Prof. Aly El Bahrawy, Faculty of Engineering, Ain Shams University, Egypt

All efforts that we have done here, were not only for fulfill our task as commitee of this conference, but we realized that environmental issues has to be mentioned in our daily life. Because what we have received today, are the deposit from our descendant, so we have an obligation to keep all of the resources in a balance, as we campaign for our mission.

In the end, by this conferece we wish our mission will be spreaded to all over the world and will always be remembered, and delivered without end.

Denpasar, 25 November 2017
Chairman of the organizing Committee,

Dr. Muhammad Anggri Setiawan, M.Si.

Peer review statement

All papers published in this volume of *IOP Conference Series: Earth and Environmental Science* have been peer reviewed through processes administered by the proceedings Editors. Reviews were conducted by expert referees to the professional and scientific standards expected of a proceedings journal published by IOP Publishing.

IOP Conf. Series: Earth and Environmental Science **148** (2018) 012001 doi:10.1088/1755-1315/148/1/012001

An investigation of Digital Elevation Model (DEM) structure influence on flood modelling

Sahid[1], A W Nurrohman[2] and M P Hadi[3]

[1] Graduated School, Geo-Information for Spatial Planning and Disaster Risk Management, Environmental Science, Universitas Gadjah Mada, Indonesia
[2] Master Program on Planning and Management of Coastal Area and Watershed, Faculty of Geography, Universitas Gadjah Mada, Indonesia
[3] Centre of Environmental Studies, Universitas Gadjah Mada, Indonesia

maani.sahid@gmail.com

Abstract. Flood is one of the natural calamities that cause huge losses and damages. Flood hazard zonation has been widely produced to face the impact of the disaster. DEM as the primary data to construct the earth surface has been developed from rough to fine resolution. Aster GDEM v.2 within 1arc spatial resolution has an ability to derived DEM and TIN data as bases river geometrics data. Maximum daily peak discharges used to calculate flood peak discharge. Furthermore, steady flow analysis has been used to produce flood inundation model based on four scenarios with return periods 5yr, 10yr, 50yr, and 100yr. The model results have been validated using UAV flood map in 2016 by means of pixel by pixel operation and the result shows that the vertical variance between grid DEM and TIN data about 0.3 m.

1. Introduction

Flood is one of the most disaster commonly ensued in a year as well as damaged catastrophe [19] Generally it is not only influence on human life but also economics [30] . Anthropogenic activities and climate change are common factors that bring about flood event, namely agricultural activities, land-use change, extreme rainfall which result in reducing river capacity to receive the discharge in when the intense rainfall happen [9] [14].

In addition, the reason Cimanuk River upper part was flooded in September 2016 can be indirectly explained by extreme rain fall that struck seven districts in Garut along Cimanuk River. Approximately 6.361 people were injured and homeless, 33 died, 20 individuals went missing, and 1766 houses broke because of the flood disaster [3]. Flood-prone zone mapping based on return period or probability is important to reduce flood impacts. Flood modelling has been used by many researchers to examine the flood event in order to face its negative impacts [27] [31] [36].

Recently, HEC-RAS provided from U.S. Army Corps of Engineers (USACE) has become popular analysing and routing flood modelling. It is a freeware software that can be downloading freely from USACE site and it has helped to create flood mapping [16]. Moreover, HEC-RAS as a hydrologic analysis and GIS within HEC-GeoRAS as software to derive river profile can be combined to visualize 2D of the flood inundated area. The combination of HEC-RAS and HEC-GeoRAS can be accurate to visualize the flood extent and it can widely simulate flood disaster [11] [24].

On the other hand, in order to gain reliable hydrologic model, it needs more data, for instance, hydrologic data, DEM, land-use (LU), soil, river bathymetry [12]. Nowadays, only a few data can be accessed freely, for example, the DEM data with high resolution may not accessible for everyone also the data is restricted. On the contrary, the level precise topographic data as an input data for terrain models is a key element of the accuracy of hydraulic flood modelling [5]. Consequently, many researchers have developed DEM as an alternative data to obtain reliable flood modelling. This research attempts to compare two different basis data (DEM) between regular Grid DEM and TIN data derived from ASTER GDEM V2. The value between structure Grid DEM and TIN are differents, even though used in similar region. Grid DEM data is suitable of plain area and TIN for rough area along man-made object [37] [38]. What of those data that can correctly result in flood inundation model.

Simulation of inundation or extent area will use HEC-RAS and GIS based on the return period of the flood. Models validation use flood event in 2016. The operations processed pixel by pixel and using raster calculator in ArcGIS. The purposes of this paper are to identify the bias vertical of the ASTER GDEM v.2, construct river geometries, identify appropriate ASTER GDEM v.2 to build flood modelling, develop reliable flood inundation model both from Grid DEM and TIN data.

2. The Methods and Data

2.1. Geographic setting

Cimanuk river is the main river in Cimanuk Watershed which has total drainage approximately 3.636 km^2 [25]. The upper part of this river from Mt. Papandayan (±1805m above sea level) to flow through to the lower part of the river in Indramayu Regency. The main river has distance about 337.67 km [32] [33]. Lying on administrative boundaries, Cimanuk Watershed is included in 4 regencies which are Garut Regency, Sumedang Regency, Majalengka Regency, and Indramayu Regency.

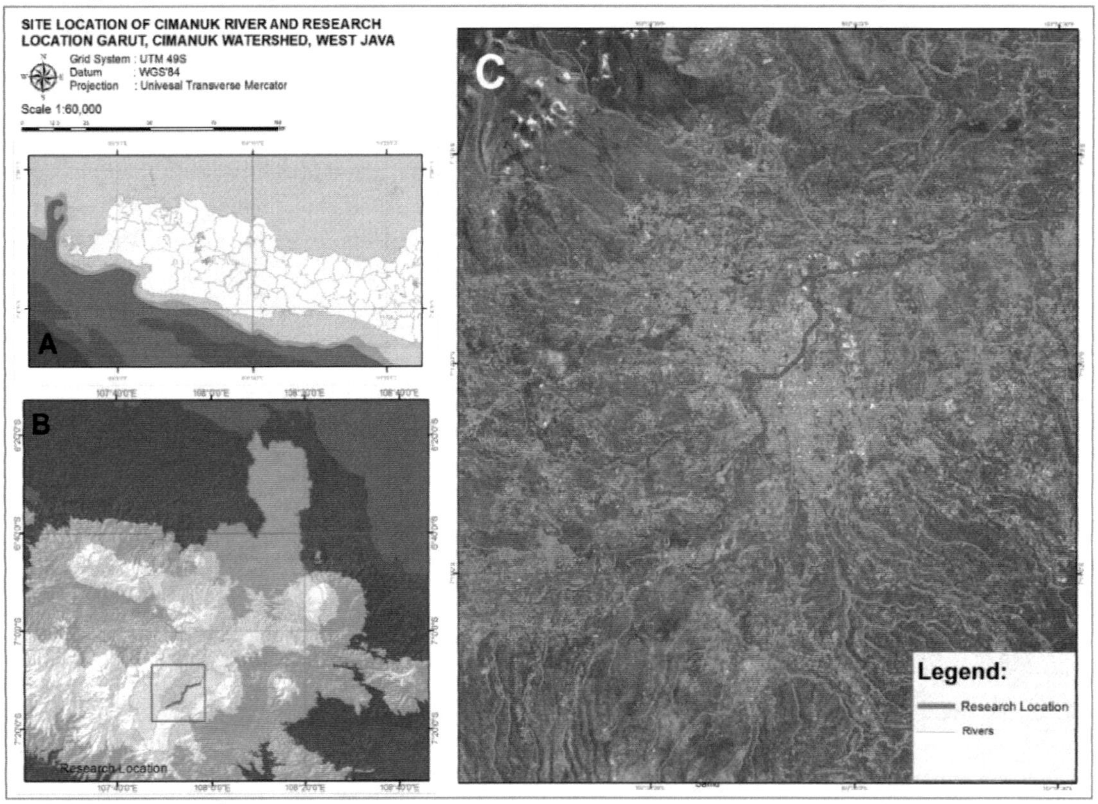

Figure 1. Map A the location of the Cimanuk Watershed in West Java Province. Image B shows the location of the research area around the upper part of the Cimanuk River. Map C illustrate the position of the research area.

The upper part of the Cimanuk River was chosen for this study which has distance around 25.5 km (figure1). Geographically, the upper part of Cimanuk Watershed is located among 107° 49' 57,74"E –

107° 57' 47,77"E and 7° 8' 20,43"S – 7° 17' 37,58"S. Administratively, the site area is located in Garut Regency contained by 6 districts, consists of Bayongbong, Tarogong Kidul, Cilawu, Garut Kota, Banyuresmi, and Karangpawitan from the upper to lower part respectively. The average annual rainfall is about 4000mm and maximum peak discharge is approximately 1222 m^3/s over 10 years from 2007-2016. Similarly, the highest maximum discharge that has ever recorded was 1470 m^3/s in 1978 [1]. The complexity of the hydrological condition at the upper part of the catchment is influenced by various land-use (LU) changes. As many researchers believe that the changing of the LU function have significant contribution of the runoff pattern [18] [21] [35].

2.2. Data and Analysis

2.2.1. ASTER GDEM v.2 characteristics. The advanced space-borne thermal emission and reflection radiometer global digital elevation model (ASTER GDEM) v.2 released on 17 October 2011 was freely downloaded from USGS (*www.earthexplorer.usgs.gov*). It is developed by Joining space program between NASA and the Ministry of Economy, Trade, and Industry (METI) of Japan [17] [29]. This version has an improvement from previous ASTER GDEM v1 launched onboard NASA's Terra spacecraft in December 1999 [2] in which several benefits from the inclusion of additional scenes to reduce artifacts, higher horizontal resolution using a smaller correlation kernel (5 x 5 versus 9 x 9 used for GDEM 1.0). In line with this, the data are delivered in tiles of 1° by 1° referenced to the WGS84 ellipsoid and EGM96 geoid. An improved water mask can be more accurate to define the water surface. Furthermore, a 5-m overall bias observed in GDEM 1.0 was removed in the newer version [22]. It provides the first complete land surface model over the whole globe [17].

2.2.2. Regular grid DEM and TIN data. Digital Elevation Model (DEM) is a 3D digital representation of terrain surface which has an ability to represent the topographic features such as river system [8] [34]. A DEM has important role to modulating the Earth surface that can be represented as a raster (grid) or a vector-based like Triangular Irregular Network (TIN) [6]. Both of them are the most important shape to represent the landscape which has own characteristics and different usages [34]. The benefit of the regular Grid-based model are to minimize data storage and also due to simplicity of the data structure [37]. The data is efficient to represent terrain form with small slope variability. However, Triangulated Irregular Network (TIN) is a vector based which commonly used to represent any continuous and complex terrain [26]. Consequently, the data storage is much larger and more complex than more complex than a grid-based structure. Therefore, an efficient solution for terrain representation can be achieved by applying a TIN-based model to represent rough areas and man-made objects and a Grid-based model to represent relatively plain areas [37]. In this research, we used both regular grid DEM and TIN data that derived from ASTER GDEMv2 which have 1arc (30m) spatial resolution.

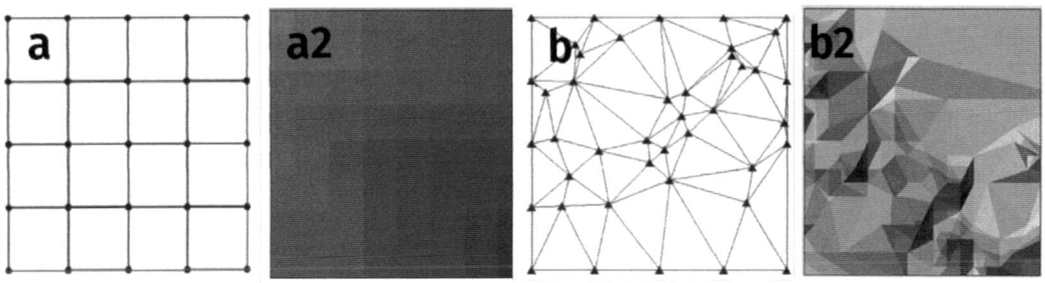

Figure 2. Image a and a2 are a simple representation of grid DEM that comprises a rectangular shape. The image b and b2 are representation of an irregular DEM. Image b2 depicts simple form of TIN data.

2.2.3. Bias vertical accuracy assessment. Prior to build the geometric data, height information needs to be validated, as DEM from imagery satellite catches the canopy height of the forest or tree rather than record bare earth elevation [10]. The topographic map with scale 1:25.000 within vertical accuracy about 6.25 m released by Geospatial Information Agency (BIG) has been applied as a referenced map. The datum is based on local sea water level in Tanjung Priok with horizontal accuracy approximately 0.3 mm and National Geodetic datum 1995 (DGN-95) or similarly with the WGS-84 (BSN, 2010). The height's information is calculated in metric format, thus transformation of the datum to metric coordinate (UTM) has been operated by means of project transformation in ArcGIS. In addition, the referenced map should has higher accuracy than the calculated map [6]. Hence, the assumption of the research is the topographic map more accurate. Root mean square error (*RMSE*) has been calculated to get the vertical bias correction in each (200) random point around the study area.

$$RMSE = \sqrt{\frac{\sum_{i=1}^{N}\left(H^{i}_{DEM} - H^{i}_{ref}\right)^{2}}{N}} \tag{1}$$

Where $RMSE$ = root mean square error, H^{i}_{DEM} = DEM elevation at the point i, H^{i}_{ref} = reference elevation at the point i, and N = the number of ground check points.

2.2.4. Hydrological analysis. The calculation of flood frequency has been computed based on daily maximum peak discharges from Cibatu station which is collected from BBWS Cimanuk-Cisanggarung. The Cibatu station was chosen to get maximum daily discharge due to the location within the outlet and the availability of the data around 10 years from 2007-2016. Statistically, log normal distribution has been selected due to the fit result of the reoccurrence statistic computation compared with other distributions method. The four scenarios return periods 5yr, 10yr, 50yr and 100yr were used to produce the flood's peak discharge as an input data to simulate flood modelling. Landuse types are also obtained from BBWS Cimanuk-Cisanggarung which will be used as input data to get the value of manning's coefficient by performing hydraulic modelling. Manning coefficient is important data to get the friction value when the water flows through to downstream area [15].

2.2.5. Hydraulic Analysis. The combination of GIS application ArcGIS v.10.4, HEC-GeoRAS v. 10.4 and hydraulic application HEC-RAS v.5.0.3 have capability to perform the 2D flood modelling [23]. Both HEC-RAS and HEC-GeoRAS are freely available software which can be downloaded from US Army Corps of Engineers-Hydrologic Engineering Center (*http://www.hec.usace.army.mil*). HEC-GeoRAS is a GIS tool to create river geometries and attributing them as an input data [20]. Either TIN or Grid DEM data is used as a terrain basis data to digitize the river center line, banks, flow direction, and the start and end stations of a river reach [10].

The 2D steady flow analysis has been applied to produce inundated area in which this model is applicable to accomplish in the flat region [7]. The main data to process the simulation are (1) flood peak discharges from 4 return periods and (2) DEM data to derive the river geometries. The derivative of river geometric from both DEM requires ArcGIS and HEC-GeoRAS which is utilized to create RAS theme in HEC-RAS, namely stream center line, main channel bank, flow path center line, and cross-sectional (Cutline). Moreover, the validation process has been computed by comparing the model results with UAV flood hazard map on 23 September 2016 Garut whom published by BNPB 2016 (*http://geospasial.bnpb.go.id*).

Table 1. Land-use types in relation to manning values.

Land-use	N Values
Cultivated Crops	0.035
Developed, High Intensity	0.15
Mix Garden/Pasture/Hay	0.1
Open Space	0.04
Open Water	0.04
Pasture/Hay	0.1
Shrub/Scrub	0.1

3. Results and Discussions

3.1. River Geometry

Different types of DEMs obtained from ASTER GDEM v.2 have been utilized to construct a river geometric data. Although Grid DEM originally acquired from the USGS data, TIN data was constructed by means of 3D analyst in ArcGIS v. 10.4. The flood inundation model was performed using river geometrics data from both DEMs, namely river center line, cross section, bank, and flow path. In case constructing of the data, an assessment of the vertical accuracy has done to reduce elevation error value (figure.3).

The assessment of vertical accuracy has been calculated from 200 random points distributed around the research area by using map algebra prior to use the origin data. The results show that total RMSE is 9.09 m (table 2). Although the calculation result is higher than total bias of [6] assessment, the RMSE is close in [28] study. Both of them work out to assess the vertical accuracy from ASTER GDEM v.2 and the conclusion show ±7.97 m and ±9.27 m respectively. The factors that influence of the different elevation are depend on surface variability and geographic location [6] [28]. In addition, this research also assessing the height accuracy based on the land-use types and the study shows that the higher bias vertical is 10.9 m located in open space area and the lower value is 6.5 m in developed area (table 3). Another factor that might render the variety are resolution and level accuracy of the referenced map. In tandem with this, the forest area is ignored because there is no forest area in research location.

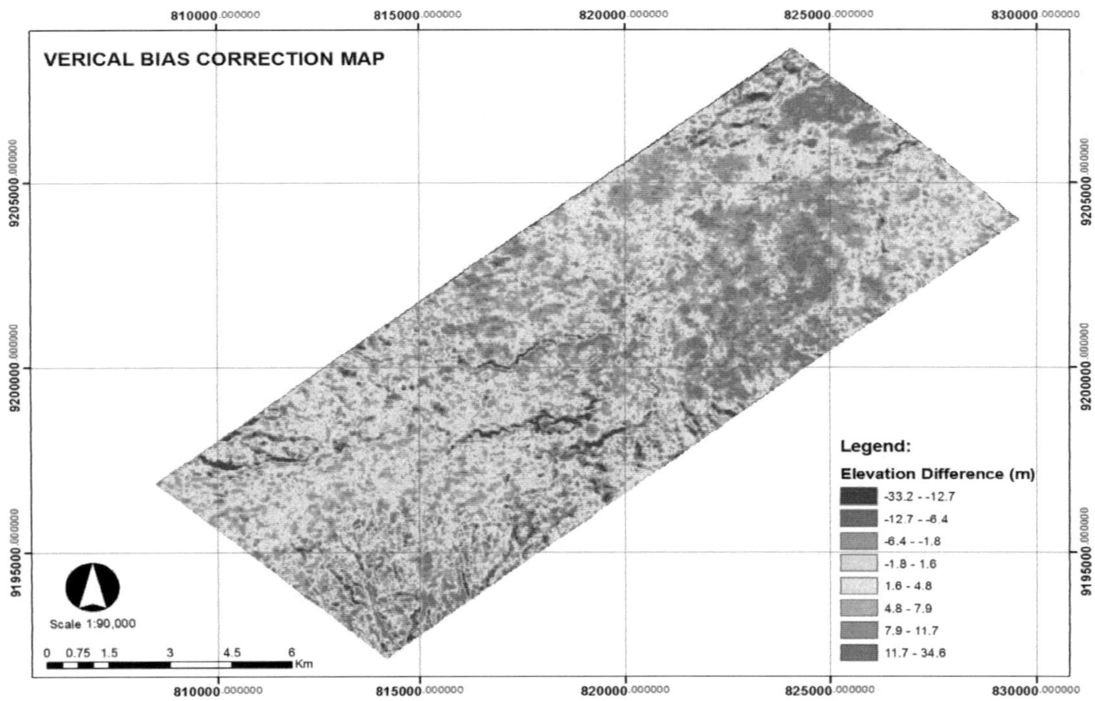

Figure 3. Vertical Bias Correction Map of referenced DEM and ASTER GDEM v.2. The red colours in the map show that elevation from referenced DEM higher than ASTER GDEM v.2. On the contrary, green colours depict elevation value from referenced DEM lower than other.

Table 2. Basic statistics calculation from comparing of grid and TIN with referenced DEM.

Statistics	Grid DEM Value (m)	TIN Value (m)
Maximum (m)	41.47	7.99
Minimum (m)	-31.80	-9.27
Mean (m)	3.67	-0.15
Standard Deviation (m)	8.32	3.10
RMSE (m)	9.09	3.10

Moreover, the corrected DEM which has original form as a Grid DEM has been applied to construct a TIN data. A transformation needed to gain TIN data by using 3D analyst in ArcGIS. Aforementioned, the Grid DEM has rectangular shapes, on the other hand TIN data is the vector base developed from triangular shapes. As mentioned before that transforming the Grid DEM to TIN data needs recalculation of the vertical value. Clearly shows in table 2 that data transformation has reduce the RMSE about ±6 m comparing to the referenced map. In other word, it means that reconstruction Grid DEM becoming TIN data has changed the vertical value.

Table 3. Vertical bias correction based on the land-cover types.

Land-use Types	RMSE (m)
Cultivated Crops	6.9
Developed, High Intensity	6.5
Mix Garden/Pasture/Hay	7.8
Open Space	10.9
Pasture/Hay	7.9
Shrub/Scrub	8.9

In addition, the river's geometries have been collected after the assessment of the vertical value, for instance cross sections, banks and flow paths. They have been manually digitized by means of HEC-GeoRAS v.10.4 software. Besides then, the river centerline has been performed using spatial analyst in GIS to get the precious location of the center stream. At least 209 cross sections have been built manually from the each DEM data in the same location (figure 4.). Although automatically cross section built is available, it require fine DEM's resolution to apply this function due to the limitation of the software. A DEM within spatial resolution near 30 m is not adequately distributed well all the river when creating automatically cross section. Sometimes the elevation of the banks is higher than the streamline [10] . The lack detail of the cross section have been built to simulate the flood inundation model. Selecting cross sections before performing the model needed because sometimes meeting point of the bank information is settled in right location (figure 5).

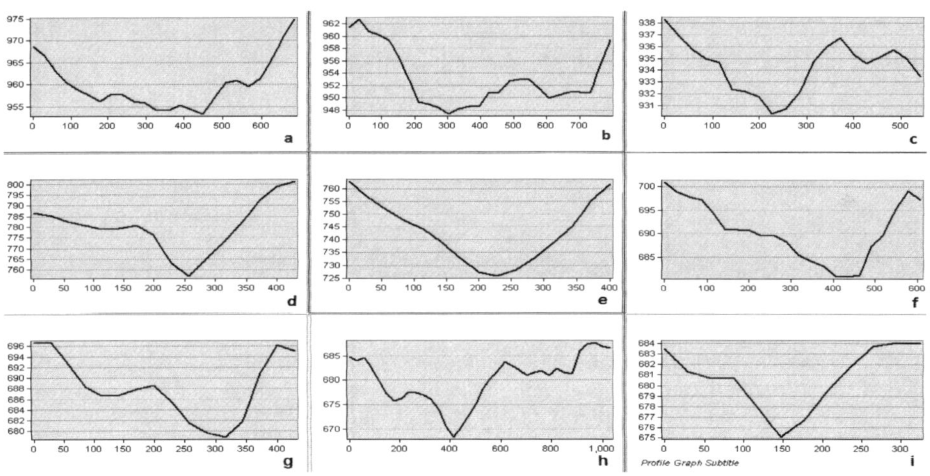

Figure 4. Examples of the cross sections' profile graphs derived from the ASTER GDEM v.2 from upper (a) to downstream (i) respectively.

3.2. Flood inundation model

Underlying on the daily maximum peak discharges over 10 years starting from 2007-2016, these data have been used to calculate the flood frequency in Cimanuk River. As mentioned before, the log normal distribution has been selected due to the fit statistical analysis using chi-square. It shows that the calculation value of the statistical analysis less than basic chi-square table, namely 12 and 14.067 respectively. Thus the log normal analysis can be performed to analyse the flood frequency analysis. Furthermore, the calculation of the 4 return periods shows that flood peak discharges for 5yr, 10yr,

50yr and 100yr return periods are 152.18 m^3/s, 207.40 m^3/s, 463.44 m^3/s and 619.06 m^3/s respectively. Likewise, the cross sections have been evaluated since the results of the simulation over the cross areas.

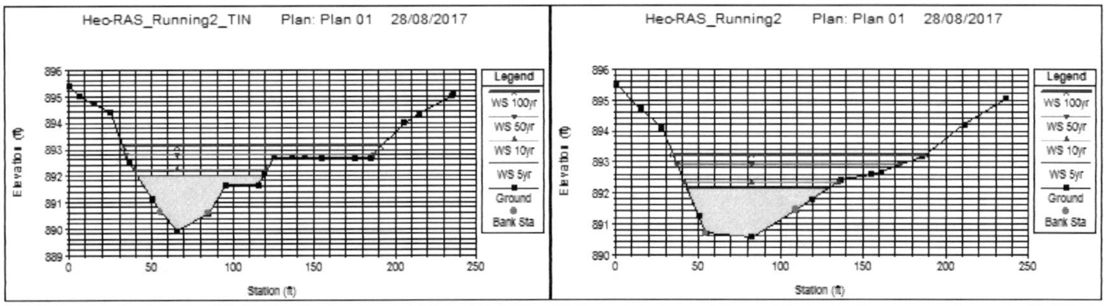

Figure 5. Two examples of the hydraulic simulation from two kind DEM as the input data. Figure left shows the simulation with Grid DEM while figure right depicts the output proceeding with TIN data at the same cross-section.

Moreover, a steady flow simulation using HEC-RAS v. 5 has been utilized in order to get inundated area. The simulation has been performed based on both of grid DEM and TIN data. The results show that the total area of the flood inundation from both DEM have difference. For example, in the 5 return period calculated within input data as the Grid DEM and TIN data is about 461.17 ha and 445.76 ha subsequently. To sum up, the total calculation shows that there are different area approximately 15.41 ha or equal to 154.100 m^2 when the flood inundation model perform both of DEMs. The total flooded area from Grid DEM tend to higher than TIN. All of the calculation of the return periods shows in table 3. As mention before that there are about 3.10 m the elevation difference between both DEM. As a consequence, the depth flood from both DEM also has difference. It is ± 0.4 m the deep variance.

Table 4. Comparison of the Flood Extend and flood depth based on Grid DEM and TIN data for each return periods.

Data	Return Period			
Flood Extend	5yr	10yr	50yr	100yr
Grid DEM-Extend Flood Inundation (ha)	461.17	470.64	547.63	570.34
TIN Data-Extend Flood Inundation (ha)	445.76	467.01	532.04	564.79
Total Area Differences (ha)	15.41	3.63	15.59	5.55
Total Area Differences (%)	3.34	0.77	2.85	0.97
Flood Depth				
RMSE Flood Depth (m)	0.38	0.58	0.38	0.38
Mean flood depth differences (m)	0.43			

In addition, the UAV flood hazard map Garut on 23 September 2016 has been used to validate the flood modelling. It has been utilized due to the minimum information of all regions about the flood map. The validation has been calculated by means of map algebra analysis with comparing the results of the modelling with the original flood extent. It is compared to examinethe total pixel of the model within the total pixel from referenced flood map. Moreover, we have chose the 5yr return period to validate the result because of the extent of the flood area similar to the UAV map. Both of the flood extents from return period based on Grid and TIN data has been evaluated. The results show that the total area of the grid DEM and TIN data within referenced map about 60% and 57.5% respectively. Approximately the Grid DEM data has 3.2% higher value included in the referenced map than TIN.

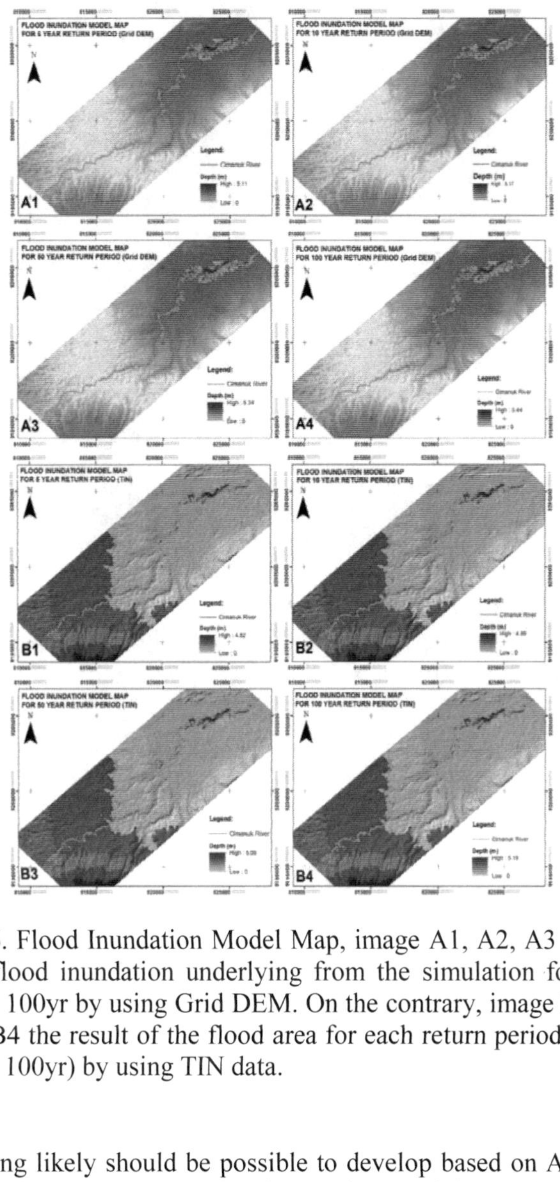

Figure 6. Flood Inundation Model Map, image A1, A2, A3 and A4 are the flood inundation underlying from the simulation for 10yr, 50yr and 100yr by using Grid DEM. On the contrary, image B1, B2, B3 and B4 the result of the flood area for each return periods (10yr, 50yr and 100yr) by using TIN data.

4. Conclusions

Flood inundation modeling likely should be possible to develop based on ASTER GDEM v.2. Since several processing should be done to cope with the roughness of the resolution. Assessing of vertical accuracy should be performed before extract the height information. Based on the referenced DEM from topographic map with scale 1:25.000, approximately 9.09 m the RMSE value. In case of the assessment of the vertical accuracy, the referenced data should be more accurate than others in order to get the closest value (bare earth). In addition, ASTER GDEM v.2 also can be used to derive the river geometries data after the bias correction has been analyzed. In case, this is the finest resolution that can be freely downloaded for everyone. Constructing the river geometries data namely streamline, bank, cross section, and flow path should be produced by means of the DEM with collaboration HEC-GeoRAS and ArcGIS. Evaluating and rechecking the river geometries after the flood simulation needs owing to the miss widely of the cross sections also banks location. Transforming Grid DEM into TIN data might be changed of the vertical elevation about 3.10 m (RMSE) in case of the ASTER GDEM v.2.

Underlying from the flood simulation results. Transforming the data from Grid DEM to TIN should influence the total flood area as well as depth of the water surface. Moreover, the total covering areas from Grid DEM tend to higher than flood covering area from the TIN data. Hence, it is wise to choose what types of the DEM data and also evaluates the vertical accuracy when converting Grid DEM into TIN data. Moreover, the flood frequency analysis is useful to develop flood hazard zonation.

5. Acknowledgments

Authors express their sincere gratitude to USGS and BIG for making this work possible by processing and distributing free the DEMs as well as topographic data to the scientific community. Also, the authors would like to thank three anonymous reviewers for their helpful comments. The authors are grateful to the supervisors, also to BBWS Cimanuk-Cisanggarung, and PSDA West Java.

6. References

[1] Anwar S 2010 Flood Forecasting using a Transfer Hydrograph *The Journal for Technology and Science* 21 33–39

[2] Baral S S, Das J, Saraf A K, Borgohain S, and Singh G 2016 Comparison of Cartosat ASTER and SRTM DEMs of Different Terrains *Asian Journal of Geoinformatics* 16 1–7

[3] BNPB (National Agency for Disaster Management) 2016 Banjir Kab. Garut-Jawa Barat (per26 September 2016) East Jakarta: BNPB

[4] BSN (National Standardization of Indonesia) 2010 Spesifikasi penyajian peta rupa bumi – Bagian 2: Skala 1:25.000 (Jakarta: National Standardization of Indonesia - BSN

[5] Casas A Benito G Thorndycraft V R and Rico M 2006 The topographic data source of digital terrain models as a key element in the accuracy of hydraulic flood modelling *Earth Surface Processes and Landforms* 31 444–456

[6] Elkhrachy I 2017 Vertical accuracy assessment for SRTM and ASTER Digital Elevation Models : A case study of Najran city Saudi Arabia *Ain Shams Engineering Journal* 2 1–11

[7] FEMA 2009 Guidelines and Specifications for Flood Hazard Mapping Partners New York

[8] Gain A K Mojtahed V Biscaro C Balbi S and Giupponi C 2015 An integrated approach of flood risk assessment in the eastern part of Dhaka City *Natural Hazards* 79 1499–1530

[9] Gao P and Hartnett J J 2016 Exploring the causes of an extreme flood event in Central New York, USA *Physical Geography* 37 38–55

[10] Gichamo T Z Popescu I Jonoski A and Solomatine D 2012 Environmental Modelling & Software River cross-section extraction from the ASTER global DEM for flood modeling. *Environmental Modelling and Software* 31 37–46

[11] Huo A D Guan W K Dang J Wu T Z Shantai H Wang W and Liew M W V 2016 Submerged area of typical torrential flood and debris-flow disasters in Mengzong Gully China *Geomatics, Natural Hazards and Risk* 7 18–24

[12] Jafarzadegan K and Merwade V 2017 A DEM-based approach for large-scale floodplain mapping in ungauged watersheds *Journal of Hydrology* 550 650–662

[13] Janses C and SCE 2016 Manning's Values for Various Land Covers To Use for Dam Breach Analyses by NRCS in Kansas

[14] Jena P P Chatterjee C Pradhan G and Mishra A 2014 Are recent frequent high floods in Mahanadi basin in eastern India due to increase in extreme rainfalls? *Journal of Hydrology* 517 847–862

[15] Khattak S M Anwar F Saeed T U Sharif M Sheraz K and Ahmed, A 2016 Floodplain Mapping Using HEC-RAS and ArcGIS : A Case Study of Kabul River *Arabian Journal for Science and Engineering* 41 1375–1390

[16] Knebl M R Yang Z L Hutchison K and Maidment D R 2005 Regional scale flood modeling using NEXRAD rainfall, GIS, and HEC-HMS/ RAS: A case study for the San Antonio River Basin Summer 2002 storm event *Journal of Environmental Management* 75 325–336
[17] Li P Shi C Li Z Muller J and Drummond J 2013 Evaluation of ASTER GDEM using GPS benchmarks and SRTM in China *International Journal of Remote Sensing* 34 1744–1771
[18] Li Z Liu W Zhang X and Zheng F 2009 Impacts of land use change and climate variability on hydrology in an agricultural catchment on the Loess Plateau of China *Journal of Hydrology* 377 35–42
[19] Lund J R 2012 Flood management in California *Water (Switzerland)* 4 157–169
[20] Marfai M A 2003 GIS Modelling of River and Tidal Flood Hazards in a Waterfront City case study: Semarang City Central java Indonesia Enschede: ITC
[21] Moiwo J P Lu W Zhao Y Yang Y and Yang Y 2010 Impact of land use on distributed hydrological processes in the semi-arid wetland ecosystem of Western Jilin. *Hydrological Processes* 503 492–503
[22] NASA METI ERSDAC and Caltecth U 2011 Advanced Spaceborne Thermal Emission and Reflection Radiometer (ASTER) Global Digital Elevation Model (GDEM) Version 2
[23] Quiroga V M Kure S Udo K and Mano A 2016 Application of 2D numerical simulation for the analysis of the February 2014 Bolivian Amazonia flood : Application of the new HEC-RAS version 5 *RIBAGUA - Revista Iberoamericana Del Agu* 3 25–33
[24] Rahmati O Zeinivand H and Besharat M 2017 Flood hazard zoning in Yasooj region, Iran, using GIS and multi-criteria decision analysis *Geomatics, Natural Hazards and Risk* 7 1000–1017
[25] Rau M I Samodra G Pachri H Irwansyah E and Subair M 2016 Physical Vulnerability Modelling Based On Flood *Junral Teknik Sipil Dan Lingkungan* 1 137–145
[26] Rolf de B Huisman O Knippers R Kraak M and Stein A 2012 Spatial data modelling, collection and management. *The Core of GIScience a systems-based approach* (1st ed., p. 246) Enschede: The International Institute for Geo-information Science and Earth Observation (ITC)
[27] Saini S S Kaushik S P and Jangra R 2016 Flood-risk assessment in urban environment by geospatial approach: a case study of Ambala City, India *Applied Geomatics* 8 163–190
[28] Santillan J R and Makinano-Santilan M 2016 Vertical Accuracy Assessment of 30-M Resolution ALOS, ASTER, and SRTM Global DEMS Over Northeastern Mindanao, Philippines *The International Archives of the Photogrammetry, Remote Sensing and Spatial Information Sciences* vol XLI-B4 (Prague) pp. 149–156)
[29] Satgé F Bonnet M P Timouk F Calmant S Pillco R Molina J Arsen A Crétaux J F Garnier J 2015 Accuracy assessment of SRTM v4 and ASTER GDEM v2 over the Altiplano watershed using ICESat / GLAS data *International Journal of Remote Sensing* 36 465–488
[30] Schumann G J P and Moller D K 2015 Microwave remote sensing of sflood inundation. *Physics and Chemistry of the Earth, Parts A/B/C* 83 84–95
[31] Shahiriparsa A Noori M Heydari M and Rashidi M 2016 Floodplain Zoning Simulation by Using HEC-RAS and CCHE2D Models in the Sungai Maka River *Air, Soil and Water Research* 9 55–62
[32] Sulaksana N Sukiyah E Sjafrudin A and Haryanto E T 2013 Karakteristik Geomorfologi DAS Cimanuk Bagian Hulu Dan Implikasinya Terhadap Intensitas Erosi Serta pendangkalan Waduk Jatigede *Ilmu-Ilmu Hayati Dan Fisik* 15 100–106
[33] Susetyaningsih A 2012 Pengaturan Penggunaan Lahan Di Daerah Hulu DAS Cimanuk Sebagai

Upaya Optimalisasi Pemanfaatan Sumberdaya Air *Jurnal Konstruksi* 10 1–8

[34] Tan L Wan G Li F Chen X and Du W 2017 GPU based contouring method on grid DEM data. *Computers and Geosciences* 105 129–138

[35] Tarigan S D 2016 Land cover change and its impact on flooding frequency of Batanghari Watershed , Jambi Province , Indonesia *Procedia Environmental Sciences* 33 386–392

[36] Teng J Jakeman A J Vaze J Croke B F W Dutta D and Kim S 2017 Environmental Modelling & Software Flood inundation modelling : A review of methods , recent advances and uncertainty analysis *Environmental Modelling and Software* 90 201–216

[37] Yang B Shi W and Li Q 2005 An integrated TIN and Grid method for constructing multi - resolution digital terrain models *International Journal of Geographical Information Science* 19 1019–1038

[38] Zhao W Zhao J Ma L Zhou W Tian J and Qian J 2015 Can Grid and TIN coexist ? *Proceedings of the 13th International Conference on GeoComputation Geospatial Information Sciences* (Texas: GeoComputation Conference series) pp. 223–229)

IOP Conf. Series: Earth and Environmental Science **148** (2018) 012002 doi:10.1088/1755-1315/148/1/012002

Vulnerability assessment to frost disaster in dieng volcanic highland using spatial multi-criteria evaluation

A Pradana[1], Y A Rahmanu[1], I Prabaningrum[1], I Nurafifa[1] and D R Hizbaron[1]

[1]Environmental Geography Department, Faculty of Geography, Universitas Gadjah Mada

adityapradana@mail.ugm.ac.id

Abstract. Dieng Volcanic Highland is one of frost disaster prone area which is very unique phenomenon in tropical region. Frost indicated by appearance of frozen dew or ice layer on the ground or vegetation surface due air inversion and cold temperatures during midnight in dry season. Appearance of frost significantly causes plant damage and losses on agricultural land, while the impacts were strongly influenced by level of vulnerability within agricultural communities. This study aims to analyze the impact of frost on agricultural land in Dieng, to identify characteristics of physical, social, economic vulnerability and coping capacity of agricultural communities to frost disaster in Dieng, and to estimate total vulnerability of frost disasters in Dieng through SMCE scenario. Research was conducted in Dieng Village, Wonosobo and Dieng Kulon Village, Banjarnegara. Method to assess vulnerability level is performed by Spatial Multi Criteria Evaluation (SMCE) method using ILWIS software through a combination of physical, social, and economic vulnerability regarding frost hazard, as well as coping capacity of farmers. Data collected by interview within different agricultural plots using questionnaire and in-depth interview method on frost affected agricultural land. Impact of frost mostly causes damage on potato agricultural land than any other types of commodities, such as carrot, leek or cabbage. Losses varies in range of 0 million to 55 million rupiah, at most events in range of 10 million to 15 million rupiah during frost season on July-August-September. Main factors determining vulnerability comes from crop losses, preparedness effort, and type of commodity. Agricultural land dominated by high level physical vulnerability (95.37 percent), high level social vulnerability (70.79 percent), moderate level economic vulnerability (79.23 percent) and moderate level coping capacity (73.18 percent). All five scenarios indicated that level of total vulnerability vary only from moderate level up to high level.

1. Introduction

Indonesia as a wet tropical climate region characterized by relatively high air humidity, heavy annual rainfall and high annual temperatures above 25°C [1]. Tropical climate conditions not possible to allow the emergence of natural ice phenomenon in Indonesia, but the emergence of natural ice occurred in the Dieng Volcanic Highland, Central Java. This condition characterized by appearance of "*embun upas*" in local languange or well known as frost or freezing dew in the surface of thin leaves or soil during dry season. Theoretically, frost only can be found in sub-tropical regions [2]. The emergence of frost in Dieng Volcanic Highland became a unique phenomenon because it usually occurs only at the peak of the dry season between July, August and September at midnight. Frost is not only unique but causes losses on agricultural plants. Dieng Volcanic Highland is at an elevation between 1,200 meters to 2,000 meters above sea level. Physical characteristic causes Dieng has fertile soil and cool climate that is very suitable for agricultural activities mainly potato farming. Nowadays, Dieng became national potato cultivation center with harvest frequency two to three times a year [3]. Frost leads to freezing of outer leaf tissue in thin-leaved plants so that potato plants experience death

[4]. Highest damage and losses occur on thin-leaved plants up to tens of millions rupiah, mostly on potato as main commodities.

Distribution of frost occurs unevenly strongly determined by the condition of farming management, so there is a need for a frost disaster vulnerability mapping assessment as one of the foundations of disaster risk reduction policy in frost affected areas. Vulnerability assessment combines the physical, social and economic conditions of agricultural areas, so as to assist in the efficient delivery of disaster management directions. Frost disaster is very dynamic, while level of farmer's vulnerability is highly determined by many unstructured multi-level factors. In order to forecast future losses, the Spatial Multi Criteria Evaluation (SMCE) modeling is capable of generating various vulnerability scenarios for disaster management decision-making for future possibilities [5]. This study aims to analyze the impact of frost disaster on agricultural land in Dieng, to identify the characteristics of physical, social, economic vulnerability and coping capacity of community to frost disaster in Dieng, and to estimate the total vulnerability of frost disasters in agricultural land in Dieng through the SMCE scenario.

2. Methods

Research was conducted in Dieng, Kejajar, Wonosobo and Dieng Kulon, Batur, Banjarnegara. Research was conducted during May to September 2017. Vulnerability assessment was conducted on frost prone areas that have been mapped and located on 125.59 hectares Tulis river alluvial plain basin surrounded by four volcanic cone (Prau, Binem, Pangonan and Kendil) in Dieng Volcanic Highland wihich is very suitable the suitable location for frost formation [6] (figure 1). Vulnerability assessment undertaken on 110.15 hectars agricultural land located on frost prone area. During dry season (July until September), temperatures can reach 0°C in the mid night and create radiative frost that known as *embun upas* by locals. Frost causes damage and economic losses to potatoes agricultural crops as main commodities.

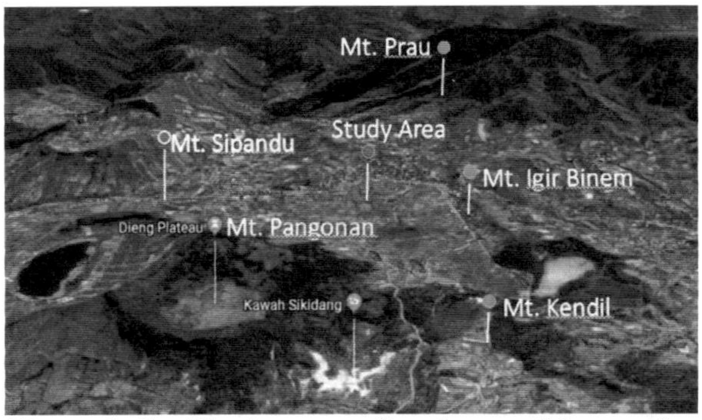

Figure 1. Satellite imagery of study area (blue area).

2.1. Material and equipment

Research was collecting sets of geospatial data, includes; agricultural land owner's boundary, administrative boundary, rivers, and land use in shp format. Data obtained from land observation, RBI Indonesia sheet of Kejajar and Batur as well as from satellite imagery. Equipment used among others, include; questionnaire, interview guidance, and Global Positioning System (GPS). Data operation was conducted using ILWIS 3.3 and ArcGIS 10.3.

2.2. Procedure

Research aims to develop a method by using Spatial Multi Criteria Evaluation (SMCE) for determining vulnerability scenario. There were four main activities in this research, namely data preparation, data collecting, data processing and data analysis (figure 3). Preparation stages consist of literature study, geospatial database preparation and defining unit of analysis. Vulnerability

IOP Conf. Series: Earth and Environmental Science **148** (2018) 012002 doi:10.1088/1755-1315/148/1/012002

assessment undertaken based on land owner units analysis on agricultural land affected by frost. Site selection was performed only in frost affected areas precisely in the 125.59 hectare alluvial plains between Dieng and Dieng Kulon village. Data collection was done through structured interview techniques with questionnaires to identify vulnerability parameters as well as in-depth interviews to understand the conditions of frost loss and community preparedness. Determination of vulnerability parameters based on preliminary survey and literature study. Questionnaire interviews conducted on 40 different plots in frost affected area. In-depth interviews were conducted with key informants who were village administrators and heads of farmer groups.

Data processing was performed by ILWIS 3.3 and ArcGIS 10.3 software. Interview results are grouped into parameters of vulnerability either physical, social and economic, as well as coping capacity. Multi-scenario vulnerability assessment using Spatial Multi Criteria Evaluation (SMCE) in ILWIS is selected for several important reasons, such as to process vulnerability data into spatial data results and to allow weighting of vulnerability parameters through different scenarios to reduce bias compared to if parameters of frost vulnerability are processed by one scenario and multiple different scenarios can be used to accommodate decision making in disaster management for various future possibilities. Spatial Multi Criteria Evaluation (SMCE) possible to work with multi-level and diverse input data to explain unstructured future conditions based on mathematical logics using problem tree analysis, standardization, weighting scenarios and map generation to generate multiple scenarios, so that corresponds to real conditions in the research area since data are common only available in multiple proxies [5]. Data processing in SMCE was done by making model of through setting up problem tree and parameters (figure 2).

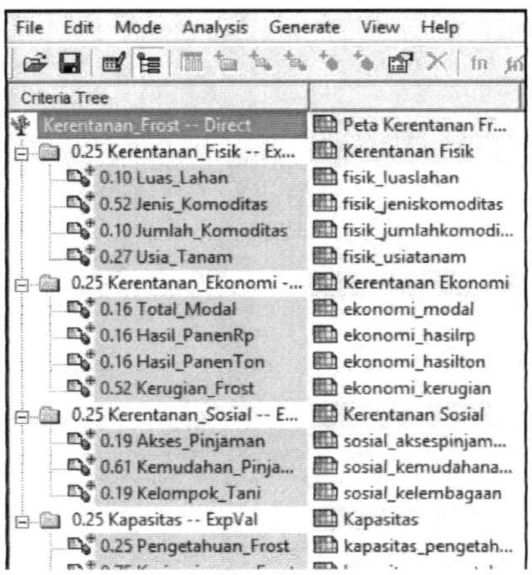

Figure 2. Problem Tree Criteria in ILWIS.

Each parameter must be standardized based on its effect on the level of vulnerability. Increase in the values of parameters (+) indicates that parameters have a positive linear relationship to increase level of vulnerability, while value (-) indicates that the parameter has a negative linear relationship with vulnerability. Weighting process for each vulnerability and coping capacity parameter uses a rank-order system by assessing primary, secondary and tertiary vulnerability determinants so resulting different value of weights and standard (figure 3b). Weighting process for total vulnerability parameters using pairwase system with five scenarios (figure 3a). Data analysis used descriptive analysis technique toward vulnerability assessment results. Descriptive analysis is used to describe the distribution of frost vulnerability based on the percentage of vulnerability levels. Vulnerability is classified into three criteria, i.e low vulnerability (0-0.33), moderate susceptibility (0.34-0.66), and high susceptibility (0.64-1). Analysis also emphasizes most influential factors on the five vulnerability scenarios whether physical, social, and economic vulnerabilities, as well as coping capacity.

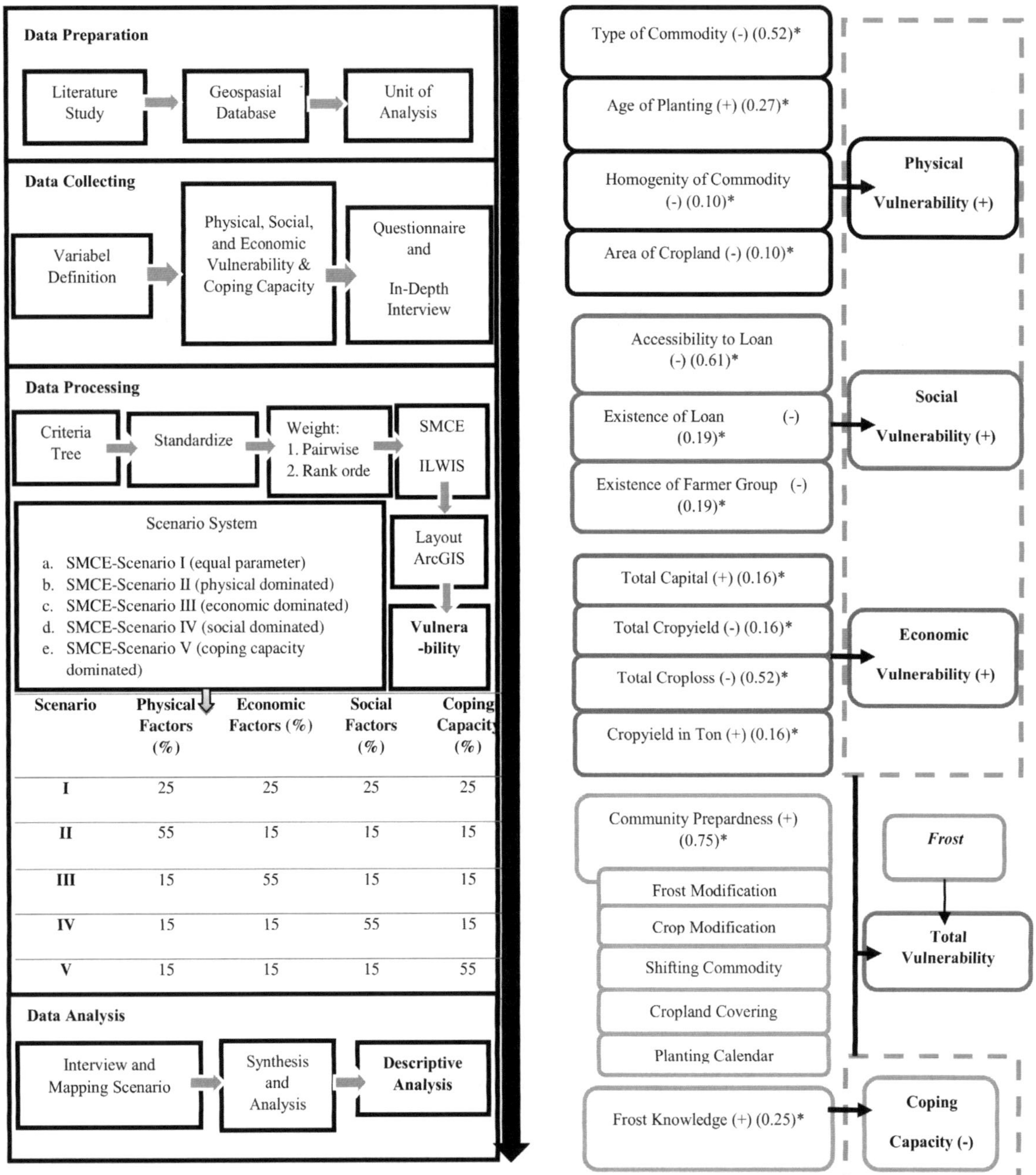

(a) (b)

Figure 3. (a) Research Procedure and (b) Weighting and Standardized System of Parameters.

3. Result and Discussion

3.1. Frost impact on agricultural land

Frost causing harm on potato farmland during dry season. Loss determined by whether or not there is an element at risk or can be interpreted as all assets both physical and nonphysical vulnerable exposed to losses [7]. Dieng Volcanic Highland has 125.59 ha frost hazard areas with varying magnitude [6]. Element at risk affected by frost disaster is agricultural land, so that all forms of non-vegetative land use and unproductive vegetation cover, i.e. grassland, swamp and shrub had no element at risk. Damage of frost can vary between one plots of farmland with others. Losses only occur on potatoes farming system, while other commodity i.e. carrot, leeks and cabbage in study area tend to be more resistant to frost occurrence. Plant damage strongly determined by the level of plant resistance to cold temperature conditions [2]. Based on table 1 potato has the highest level of vulnerability due to potato's critical temperature are at warmer degree than other commodities so that more easily damaged when exposed even to low-intensity frost [6].

Table 1. Critical temperature of agriculture commodities.

Commodities	Binomial Name	Critical Temperature (°C)
Carrot	*Daucus carota*	-7
Cabbage	*Brassica oleracea*	-5
Leek	*Allium porrum*	-6
Potato	*Solanum tuberosum*	-0,8

All frost hazard zones distributed in alluvial plains with dominance land use among others farmland, swamp and grass. Frost is formed at night in the dry season due to plains cooling faster than slopes. At night during dry season, ground and air above it loose radiation offset by earth radiation releases into space rapidly without any cloud cover as an insulating atmospheric blanket that prevents radiation escape into space [8]. Plains have better radiator properties than the air layer above it, so it tends to cooling faster [2]. Study area affected by the topographical effects which is surrounded by volcanic cone then generates gravitational cold air drainage from upper slope to the plains, so that the plains are cooling faster and causes frost appearance as can be seen in figure 4 [6]. Cold air drainage forms a warmer air layer above the soil surface called radiation inversion. The cold air descending from the slopes into the plains is accelerated by the absence of wind (calm) at night. Incidence of radiation inversion causes the alluvial plains to experience more intensive frost formation than the slopes.

Figure 4. Cold air drainage mechanism.

Frost recorded on August 14, 2015, destroying about 100 hectares farmland in Dieng and Dieng Kulon villages with losses reaching 15 million rupiah per hectare [9]. On August 24, 2015, frost resulted losses reaching 10 million rupiah per hectare, while on 3 August 2016 area affected almost 25 hectares and losses occurred in potato fields aged less than 60 days [10]. Newest in early September 2017 frost resulted in leaf wilting on a two-hectare potato field in Dieng Kulon [11]. Potato plants damaged by frost are characterized by wilting, twig and dry condition even blackened (figure 5). Losses could reach tens of millions of rupiah, depending on age of planting (phenological stage) when frost and mitigation efforts undertaken in order to reduce losses. For potatoes at young age (1-2 month), losses can reach up to 100%. At adult age (3 months), after the frost one to two days, the potatoes should be quickly harvested so losses are only in 50% range. Potatoes at the age of 4 months (ready to harvest) have smallest losses i.e. 20% because the potatoes more resistance to combat cold air and big enough, so then potatoes still can be sold.

Based on figure 6, losses caused by frost at most are in the range of 10 million to 15 million rupiah (37.1 percent). Cost of losses incurred by farmers includes the cost of seed, fertilizer, medicines and daily labor. Most small losses can reach 0 rupiah (zero losses) due to the success of mitigation on the land, such as shifting commodity into carrots or leek and granting of mulch on the soil surface to prevent frost. The highest losses suffered farmers can reach up to 55 million rupiah, generally owned by land with a fairly high capital input and large farmland located in Dieng Kulon. No losses (0% loss) only appear on plots of farmland who do frost disaster preparedness in a manner not to plant potatoes on its land and replace them with other commodities or conducting active preparedness like using mulch, sprinkle irrigation during night, and furnace.

Frost: August 2015 Frost: September 2017

(A) (B)

Figure 5. (A) Strong Frost (2015): Thickness More Than 3cm, Potato Experienced Death and Rot, Leek Not Damaged, (B) Low Frost (2017): Thickness Less Than 1cm, Potato Withered Leaves, Leek Does Not Wilt.

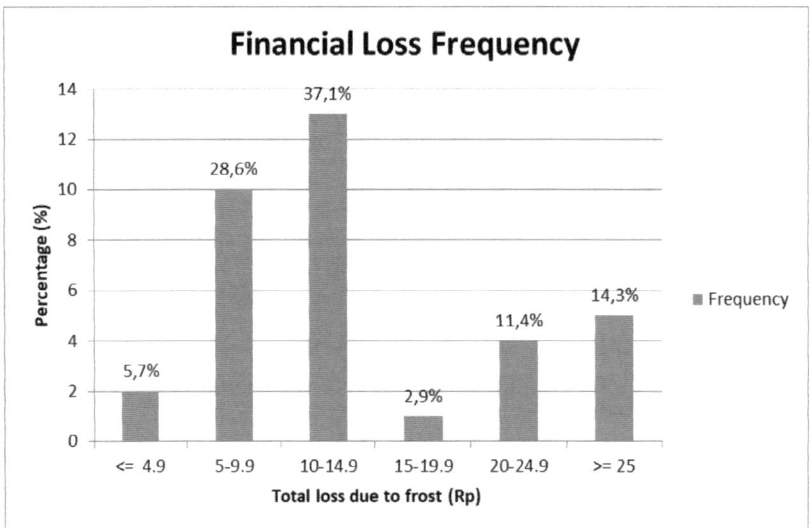

Figure 6. Financial loss due to frost.

3.2. Physical, social, and economic vulnerability and coping capacity level

Magnitude of potential damage or loss (V) of an element of risk (E) that is exposed to disaster with frequency (F) and certain quantity (M) is expressed as a degree of vulnerability [12]. Vulnerability is conceptually expressed as the degree of damage or loss of 0 (no damage) scale to 1 (total damage) occurring to the element of disaster risk due to a specific threat of frequent [13]. Vulnerability of frost strongly determined by physical, economic, and social factors, as well capacity of farmers and exposure area to frost hazard [14]. Results vulnerability level can be represented in various vulnerability level maps to reduce disaster risk and losses, among others; physical, economic and social vulnerability (figure 7). Generally physical vulnerability indicates potential damage to risk elements such as buildings, infrastructure and other physical characteristics present on the earth's surface that are beneficial to humans [15; 16]. Factors affecting physical vulnerability in this study focuses on physical aspects of agricultural land, include type of commodity (potatoes or non-potatoes), homogenity of commodity (monoculture or polyculture), the age of plants and total land area.

Physical vulnerability describes a physical condition that is vulnerable to certain disaster factors, i.e. frost. Farmland with potato farming system experience losses compared to farmland with non-potato farming, i.e. carrot, cabbage and leek. Crop damage due to sensitive frost usually occurs at temperatures of 0 °C to below -5 °C. During these temperatures, ice formation in plants from super cool water will destroy the nature of plant intercell [4]. Plants with thin leaf such as tomatoes, potatoes, peppers, and soybeans can be destroyed. Areas that have a monoculture potato commodity with young planting age (1-2) have a higher vulnerability than land that has more than two types of commodities. Agricultural land in frost prone area has a high vulnerability rate of 95.37 percent although there is 4.63 percent of land with moderate physical vulnerability. High vulnerability triggered by many agricultural lands consists of only one type of commodity (monoculture) that is potato which is very vulnerable to frost, especially young potato plants. Some lands planted with young age potatoes (1-2 months) during frost season (June until August), so the plant is not yet mature and prone to wilt when exposed to frost phenomenon. Moderate level only exists on non-potatoes farming system. Farmer chooses to plant carrots to avoid losses due to frost disaster.

Economic vulnerability is the potential financial loss suffered by humans due to disaster phenomenon [17]. Potential financial loss for frost vulnerability assessment focusing on the total financial flow within one land owner system, which is included total capital in rupiah during one harvest period (4 months), total yield in rupiah and ton during one harvest period (4 months), and total financial losses due to frost phenomenon during one year. Economic capability of individuals or communities greatly determines the extent of vulnerability to disaster threats. Amount of capital covers the whole capital of farming, such as seeding, fertilizer, plant medicine, rents of land and labor payment. Greater value of capital then increases potential loss of money if there is a frost occurrence, while the greater agricultural yield increases economic profit so as to serve as a saving economy to cope with losses in the disaster period. The results show the dominance of income from agricultural business is large enough but followed by high risk of loss due to frost almost every year. Level of economic vulnerability dominated at a moderate level of 79.18 percent. This shows losses due to frost and lost capital are large, but farmers still have the saving yield because the amount of crops during the non-frost period is much greater than the value of loss. High vulnerability to agricultural land amounted to 20.84 percent and only found on agricultural land with total losses equivalent with value of yield (100% loss). Very high level of losses generally experienced by farmers who plant young potatoes are still very young (1-2 months) in July-August. Variation of plant damage depends on plant sensitivity to cold temperatures [2]. Plants with younger age are more susceptible to damage than older plants because young plants have very soft tissue conditions that are sensitive to cold weather changes.

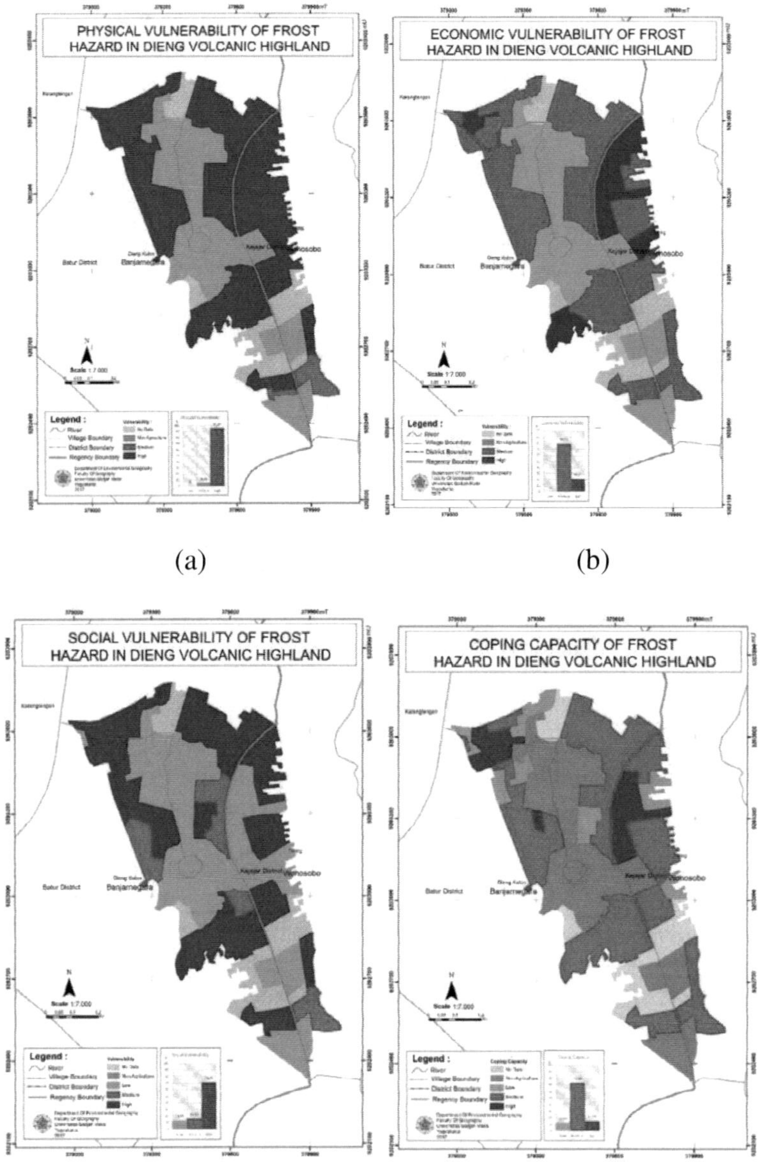

Figure 7. (a) Physical, (b) Economic, and (c) Social Vulnerability, (d) Coping Capacity.

Social vulnerability indicates the potential for loss of specific risk elements referring to human circumstances or socio-demographic condition [17]. Factors affecting social vulnerability in this research focus on social aspects of farming system such as the presence of access to loans, the ease of obtaining loans and the presence of farmer groups. Social vulnerability describes condition of social fragility in farmer community to facing the damage of frost disaster. In vulnerable social conditions, appearance of disaster will cause large losses and impact because low social capital within community. Level of social vulnerability dominated at a high vulnerability level of 70.45 percent. Main factor that affecting level vulnerability due to the absence of farmer groups who are expected to help the distribution of crops, socialize mitigation plan among farmer members and provide ease of lending capital when losses crop failure due to frost. Difficulty in obtaining capital also becomes an obstacle for farmers. Plots of land with high social vulnerability generally borrow money to banks or cooperatives with a tight lending system. High losses due to frost causes debt system can accumulate into several periods. Plots of land with moderate vulnerability of 15.92 percent can generally borrow money easily from relatives so there is no need for interest payments on the loan. Plots of land with a low vulnerability of 13.35 percent have generally been incorporated into farmer groups and have sufficient financial capability so no need to borrow money after the frost. Existence of farmer groups very helpful to direct the farmer gain preparedness effort before the frost appearance results in a loss.

Coping capacity is a combination of all potency and resources between communities or organizations that have a role in disaster risk reduction [7]. Indicators of coping capacity in Dieng communities to reduce the impact of frost disaster are assessed from the level of community preparedness and the level of community knowledge regarding frost phenomenon. Capacity assessment shows that agricultural land is dominated by farmers with moderate capacity that is 73.18 percent, while there are 12.44 percent with low capacity and 14.57 percent with high capacity. High capacity farmland only found in few farm land that already implemented frost modification using furnace with high investment input, others using sprinkle, mulch covering and shifting commodities to carrot-leek commodities so that losses can be reduced (figure 8). Farmlands with low capacity level generally do not implement preparedness efforts due to lack of understanding, awareness and financial incapacity.

(a) (b) (c)

Figure 8. Identified preparedness on agricultural land; (a) shifting commodities, (b) mulch covering, and (c) sprinkle irrigation.

Dominance of moderate level show that the preparedness efforts have been done by farmer communities are quite good. Most of community has understood the signs of weather and frost events as local wisdom through natural observations. Forms of preparedness efforts that have been identified and implemented include;

a) Shifting agricultural commodities to non potatoes, that replacing the plants are not susceptible to frost during the season, example; carrots and leek;

b) Modification of plants, that is done by intercropping to overcome a farmer who still wants to plant potatoes, by planting potatoes in conjunction with non-commodity of potato that resistant by frost on the same farm;

c) Cover the potato field with mulch or nets with a height over one meter, which is done to prevent moisture in the air contact with the ground super cooled [6];

d) Determining the appropriate planting calendar, that is ensuring planting potatoes during the season when the frost has reached adulthood so that the potatoes are more resistant to cold temperatures.

e) Modification of frost, in-depth interview showed that modification of frost was done before by destroying the forming of frost during the night, this preparedness plan have more financial or investment needed, example:

- Providing heating temperature near the ground by installing a furnace at night, in order to prevent the temperature cools below the *frost point* [2].

- Spraying potato field during the early days with the sprinkles irrigation system to interrupt process of frost formation.

3.3. Total vulnerability scenarios

Total vulnerability refers to a set of conditions and/or effects of circumstances that adversely affect disaster prevention and mitigation. The total vulnerability using Spatial Multi Criteria Evaluation (SMCE) resulted in five scenarios, namely scenario I (equal), scenario II (physical dominated), scenario III (economic dominated), scenario IV (social dominated), and scenario V (coping capacity dominated). Total scenario is strongly influenced by dominant parameters such as the level of total crop loss, type of commodity (potato or non-potato), preparedness aspect, and existence of loan. Vulnerability assessment in all scenarios shows that the study area only has moderate (0.33-0.66) up to high (0.64-1) vulnerability level (Table 2), so the potential loss due to frost can be various moderate up to very high (figure 9). Total vulnerability mapping on figure 3.7 shows that scenario I dominated by vulnerability at moderate level (56.45 percent). High vulnerability (43.55 percent) found in potatoes farmland with relatively low preparedness aspect. Scenario II shows dominant vulnerability at high level (83.05 percent) higher than other scenarios. Physical factors as dominant variable in Scenario II shows that dominant commodities which is potatoes plants that covers large areas have high vulnerability, while farmland of non-potatoes (leek and carrot) and several potato farm land with good preparedness aspect classified into moderate area (16.95 percent).

Table 2. SMCE Scenario on Vulnerability Level.

Scenario SMCE	Low (%)	Medium (%)	High (%)
Scenario I (Eq)	0	56.45	43.55
Scenario II (Ph)	0	16.95	83.05
Scenario III (Ec)	0	71.61	28.39
Scenario IV (Sc)	0	29.25	70.75
Scenario V (Cc)	0	87.57	12.43

(a) (b) (c)

Figure 9. Moderate vulnerability; (a) Non-potato farmland (carrot, cabbage, leek), and (b) 3,5 months policulture potato farmland mixed with leek with plastic covering and High vulnerability; (c) 1,5 months monoculture potato farmland.

Scenario III shows dominant vulnerability level at the medium level (71.61 percent) higher than the result in Scenario I. High vulnerability (28.39 percent) only experienced by farmland with large economic losses due to frost, which is more than 20 million rupiah and located only in few farm land with quite large areas more than 3 hectares. Scenario IV shows dominant vulnerability at high level (70.75 percent) due to agricultural land dominated by potato plots with high losses tend to have difficult condition to get access into capital loans, either from cooperation, bank or family. The existence of farmer group have relatively low contribution to minimize the impact of frost because farmer group tend to focussing on productivity socialization, have no program about frost mitigation and have lack access to help the farmers get any loans. Scenario V indicates that dominant vulnerability at medium level (87.57 percent) higher than other scenarios which is shows that increasing dominant level of capacity will reduce the number of high vulnerability farmland. High understanding of farmers related to frost and the number of plots of land that have implemented preparedness efforts causes vulnerability level in scenario V is lower than other scenario. Based on table 2, this study argues that coping capacity and physical factors of agricultural areas are the most sensitive criteria compared to other criteria because they have the most different vulnerability patterns than other scenarios (> 80 percent in percentage).

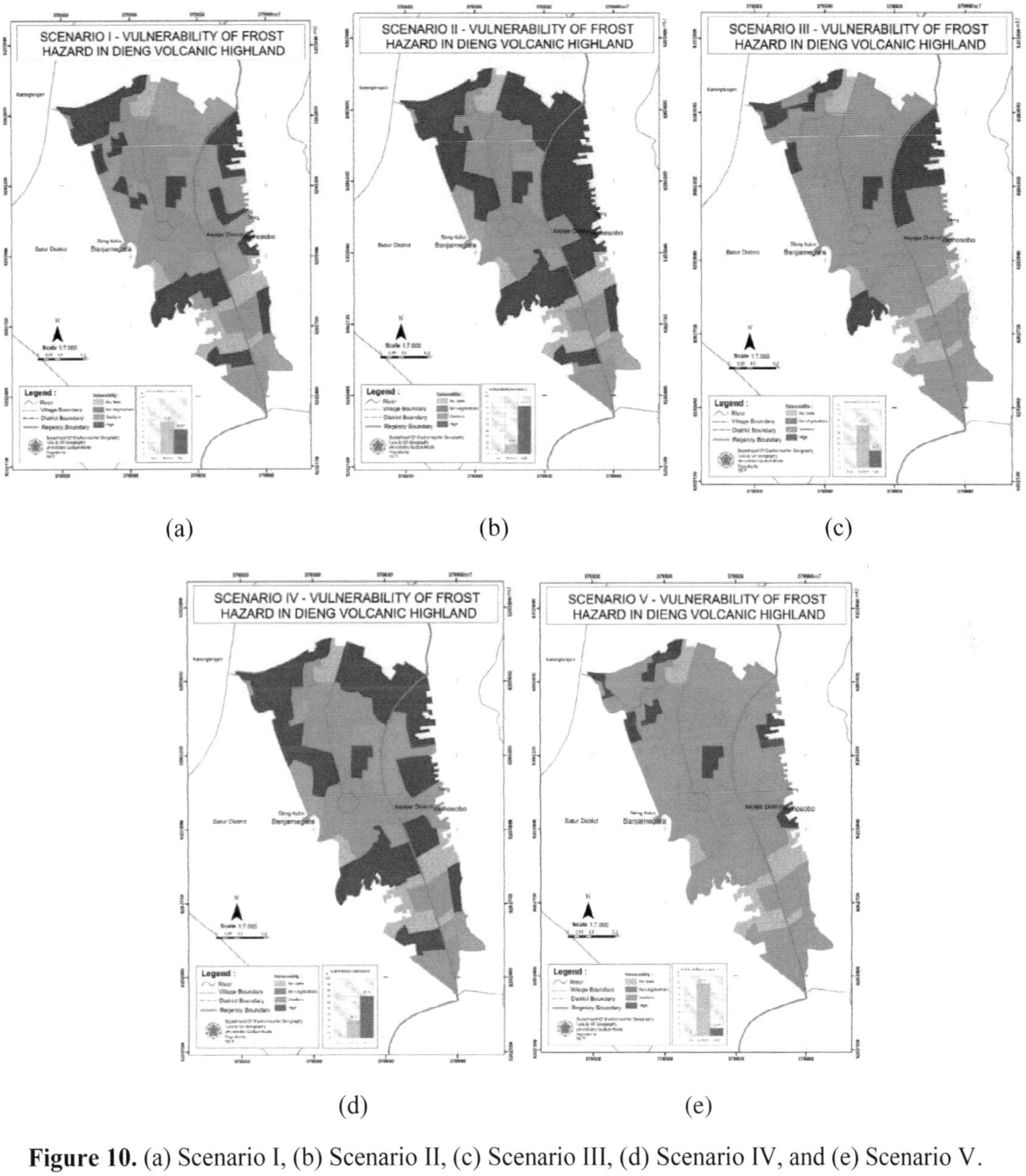

Figure 10. (a) Scenario I, (b) Scenario II, (c) Scenario III, (d) Scenario IV, and (e) Scenario V.

4. Conclusion

Impact of frost in agricultural losses mostly causes damage on potato agricultural land than any other types of commodities. Losses varies in the range of 0 million to 55 million rupiah, at most events in the range of 10 million to 15 million rupiah. Main factors determining the level of vulnerability comes from crops losses, preparedness effort, and type of commodity. Agricultural land in frost prone areas generally dominated by high level physical vulnerability (95.37 percent), high level social vulnerability (70.79 percent), moderate level economic vulnerability (79.23 percent) and moderate level coping capacity (73.18 percent). All five scenarios indicated that level vulnerability vary only on moderate level (16.95-87.57 percent) up to high level (12.43-83.05 percent). Physical factors and coping capacity of agricultural areas are the most sensitive criteria compared to other criteria determining level of vulnerability. Further appropriate preparedness need to be taken in order to reduce losses due to frost and achieve disaster resilience agriculture community in Dieng Volcanic Highland.

5. Acknowledgement

This research supported by Divisi Penelitian dan Pendidikan (PP), EGSA, as well Environmental Geography Department, Faculty of Geography, UGM, and Farmer Communities as well Village Administrations of Dieng Village, Wonosobo and Dieng Kulon Village, Banjarnegara. Thankyou for the guidance from Dr. Dyah Rahmawati Hizbaron, M.T, M.Sc and Dr. Emilya Nurjani, M.Si during this research. Thankyou for Team PKM-PE Frost Geography for contribution on frost research ideas.

6. References

[1] Tjasyono B HK 2004 *Klimatologi* (Bandung: Institut Teknologi Bandung)
[2] Synder RL and Melo-Abreu Jd 2005 *Frost Protection: Fundamentals, Practice, and Economics* (Rome: Food and Agricultural Organization of United Nation)
[3] Edy SK 2010 Usaha Tani Kentang Dengan Teknik Konservasi Teras Bangku di Dataran Tinggi Dieng Kabupaten Wonosobo *Journal Pembangunan Pedesaan* 10 115-127
[4] Lindow SE Arny DC and Upper CD 1982 Bacterial Ice Nucleation: A Factor in Frost Injury to Plants *Plant Physiology Journal* 70 1084-1089
[5] Hizbaron DR Baiquni M Sartohadi J and Rijanta R 2012 Urban Vulnerability in Bantul District, Indonesia towards Safer and Sustainable Development *Sustainability* 4 2022-2037
[6] Pradana A Mardiana A Lestari FN Sara FH Afifah S and Nurjani E 2017 Tropical Frost: Frost Hazard Assessment on Agricultural Land to Achieve Resilient Agriculture in Dieng Volcanic Highland *International Conference on Tropical Agriculture 2017* (Yogyakarta: BPP Universitas Gadjah Mada)
[7] United Nation International Strategy for Disaster Reduction (UNISDR) 2009 *UNISDR Terminology on DRR* (Geneva: UNISDR)
[8] Ahrens DC 2009 *Essentials of Meteorology: an Invitation to the Atmosphere* (USA: Broks Cole)
[9] Andrianto A 2015 *Embun ini Merusak Ratusan Hektar Tanaman Kentang* Tempo [m.tempo.co]
[10] Huwae E 2016 *Tanaman Kentang Terancam Rusak Akibat Embun Upa*s Raya Pos [rayapos.com]
[11] Ridho M 2017 *Embun Es Dieng* Liputan6 [regional.liputan6.com]
[12] Galli M and Guzzetti F 2007 Landslide Vulnerability Criteria: A Case Study from Umbria, Central Italy *Environmental Management* 40 649 – 664
[13] United Nation International Strategy for Disaster Reduction (UNISDR) 2004 *Living with Risk - A Global Review of Disaster Reduction Initiatives* United Nation [www.unisdr.org]
[14] Jiajin C Jiayi W Zhiguo M Jing L Kai Y and Lichun L 2011 Low Temperatyre and Frost Risk Division of Olive in Fujian Province Based on GIS *International Conference on Computer*

Distributed Control and Intelligent Environmental Monitoring 1585-1588

[15] Birkmann J and Wisner B 2006 *Measuring the Unmeasurable the Challenge of Vulnerabiilty* (Bonn: UNU EHS)

[16] Ebert A Kerle N and Stein A 2007 Urban social vulnerability assessment with physical proxies and spatial metrics derived from air-and spaceborne imagery and GIS data. *Natural Hazards* 48 275 – 294

[17] Sumekto DR 2011 Pengurangan Risiko Bencana Melalui Analisis Kerentanan dan Kapasitas Masyarakat dalam Menghadapi Bencana. *Seminar Nasional Pengembangan Kawasan Merapi.* (Yogyakarta: Direktorat Penelitian dan Pengabdian Masyarakat, Magister Teknik Sipil, Universitas Islam Indonesia)

Spatial modelling for tsunami evacuation route in Parangtritis Village

A Juniansah[1], B I Tyas[1], G C Tama[1], K R Febriani[1] and N M Farda[1]

[1]Department of Geographic Information Science, Faculty of Geography, Universitas Gadjah Mada, Yogyakarta, Indonesia

anwarjuniansah@gmail.com

Abstract. Tsunami is a series of huge sea waves that commonly occurs because of the oceanic plate movement or tectonic activity under the sea. As a sudden hazard, the tsunami has damaged many people over the years. Parangtritis village is one of high tsunami hazard risk area in Indonesia which needs an effective tsunami risk reduction. This study aims are modelling a tsunami susceptibility map, existing assembly points evaluation, and suggesting effective evacuation routes. The susceptibility map was created using ALOS PALSAR DEM and surface roughness coefficient. The method of tsunami modelling employed inundation model developed by Berryman (2006). The results are used to determine new assembly points based on the Sentinel 2A imagery and to determine the most effective evacuation route by using network analyst. This model can be used to create detailed scale of evacuation route, but unrepresentative for assembly point that far from road network.

1. Introduction

Tsunami is a series of huge sea waves, usually occurs because the sudden move of oceanic plate causing vertical dislocations resulting in a decrease or shift that causes the mass of water above to move in response to changes in the underlying plate [4]. Some of the factors that can cause such vertical dislocations include earthquakes, volcanic eruptions, landslides, and collisions from meteors. Indonesia as a country in the subduction zone of the plate has a high potential earthquake on the seabed until the tsunami event. Tsunami wave height that ever occurred in south coast of Indonesia ranged from 0.1 to 26.2 meters [8].

The damage caused by the tsunami is influenced by the magnitude of the earthquake that occurs. Some of the periods of the tsunami incident in 1883, due to the eruption of Krakatoa volcano, caused more than 36,000 victims and a global impact on climate change [3]. Another incident in Java Island in 2006 has killed approximately 600 people and destroyed many villages on the Coast of Java Island [6]. Given the magnitude of the loss and the high potential for the tsunami in Indonesia, appropriate disaster mitigation is needed to reduce the number of the victims. The tsunami is a secondary disaster as a result of an earthquake on the seabed. After an earthquake, there is an interval of a few minutes before the tsunami wave reaches the coast. In this interval, it is possible to provide early warning and self-evacuate to a safe area. The evacuation concept is moving people from relatively harmful locations to safety area through safe evacuation routes [9]. The success of evacuation is depend on the effectively of evacuation route. The main problem in determining the effective evacuation route is selecting the fastest route. It is not a simple case because the shortest route is not always the fastest path towards the evacuation point. Topography characteristics and land use may increase the travel time so it becomes obstacle. To solve this problem we built evacuation route model based on the network analysis. In this case road is consider as a network that can be used to determine the best route, nearest facility, service area, build the cost matrix and solve the road problems. Consequently, roads are an important variable in determining the effective routes for evacuating people and also showing the appropriate assembly point based on distance and travel time.

There are two models of evacuation method, horizontal and vertical evacuation [7]. Horizontal evacuation requires an assembly point that is far from the source of danger, while the vertical model requires the higher location only (refers to the floor in a building). Unfortunately, most of the buildings on Indonesia coast are not able to withstand the impact of tsunami waves. Therefore, the more suitable evacuation model is the horizontal model. The tsunami susceptibility model is required to determine the safe evacuation assembly point.

The main problems of making tsunami susceptibility model are the selection of representative model and the determination of wave height scenarios to illustrate the distance of tsunami waves that reach the mainland. Parangtritis Village in Bantul District was chosen as a model to determine the tsunami evacuation route in addition to its potential for a tsunami as well as having many elements at risks to the tsunami disaster (dense settlement and tourism activity). The main objective of this study is producing the most effective tsunami evacuation route recommendations based on the spatial model of the tsunami susceptibility level.

Based on the research background, then formulated three research problems, such as: a) the impact of tsunami waves determines the level of tsunami susceptibility in a coastal area, therefore the spatial model of tsunami susceptibility should be representative of the tsunami-affected landscape with the selection of certain wave height scenarios, b) the evacuation assembly points are determined based on the elevation, distance from the tsunami hazard and the vastness of the assembly point location, it is necessary to evaluate the existing evacuation points and choose the new appropriate location based on the spatial model of tsunami susceptibility, c) the main focus in determining the evacuation route is choosing the fastest route since the shortest path does not necessarily mean the fastest path to the evacuation point due to a barrier of topography and land use. Therefore, modelling is needed to determine the fastest route to reach the tsunami evacuation point.

2. Research Methods

The effective route recommendation is made with network analyst based on the evaluation of existing evacuation assembly points and then determined new evacuation points. The framework of this research is shown in figure 1. There are three data employed in this research as follows, 1) data for spatial modelling of tsunami susceptibility, 2) data to evaluate and determine the evacuation point, 3) data for the creation of the road network dataset. Details of the data and how to obtain it can be seen in the table below:

Table 1. Details of data.

No	Types of Data	Source	Other Information
1	ALOS Imagery Data	asf.alaska.edu	Resolution of 10 meters year image and recording in 2011
2	Sentinel 2A Imagery Data	earthexplorer.usgs .gov	Resolution of 10 meters year image and recording in July, 2017
3	Parangtritis Village Map	Parangtritis Geomaritime Sains Park	Map published in 2016
4	Coordinate of Tsunami Assembly Point Locations	Field survey	Data collection time is in August 2017

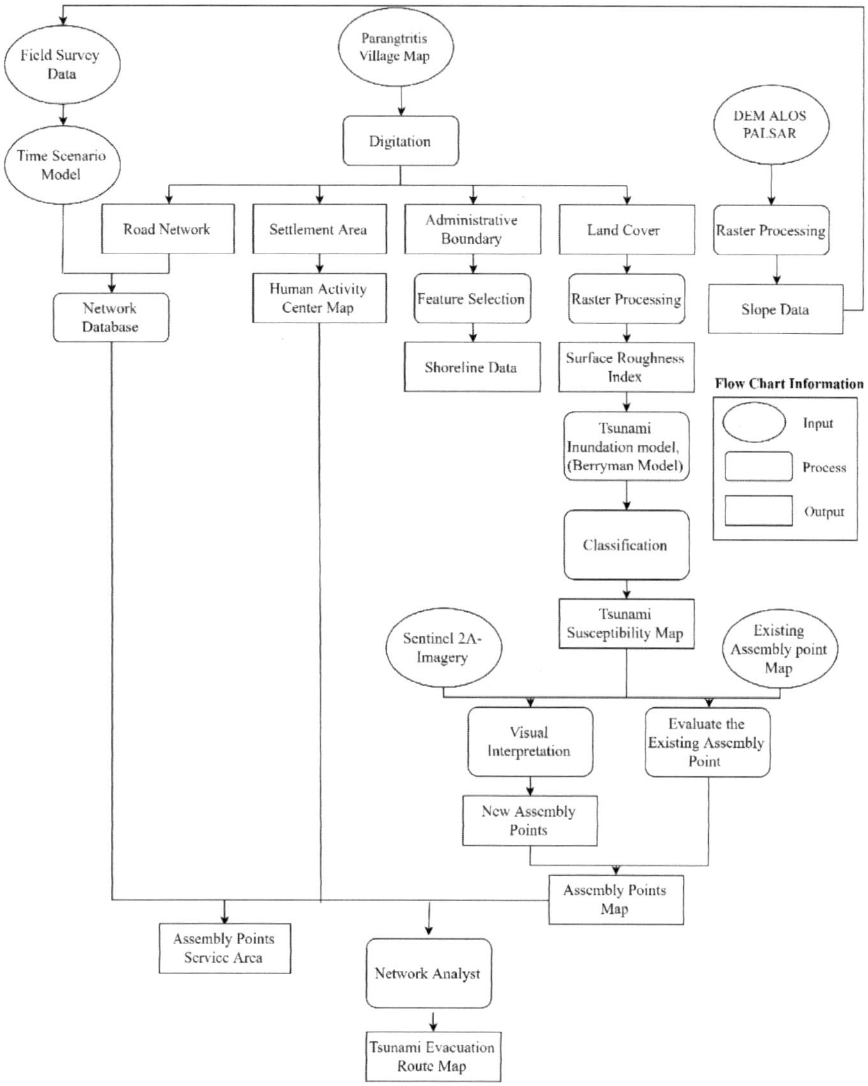

Figure 1. Framework of the research.

There are four main steps in creating network analyst model for determining the evacuation route. These steps produce three maps and it use to build the evacuation route model.

2.1. Creating spatial model for tsunami susceptibility

The tsunami spatial model is created by adopting a model developed by Berryman (2006) using coastline data, slope, and surface roughness coefficient. The tsunami wave inundation model from Berryman (2006) is a model that works by using a certain wave height scenario that starts from the coastline to the mainland then calculating the loss of wave height per one meter. The calculation considers slope, surface roughness, and elevation. The tsunami inundation model by Berryman (2006) was also adopted by BNPB (National Disaster Management Agency) to create tsunami susceptibility map through Perka BNPB No. 2 Tahun 2012.

Tsunami inundation is a maximum horizontal distance of tsunami wave that can reach the mainland [5]. Factors that strongly influence the inundation are the coastal topography, bathymetry condition, and the earthquake magnitude. Inundation is also affected by the terrain condition which is expressed by the surface roughness coefficient and has strong relationships with the inundation distance.

The classification of tsunami susceptibility level (considering the height of wave inundation) refers to the classification of tsunami hazard map by *Perka BNPB No 2 Tahun 2012*. There are three classes of susceptibility classification: high susceptibility class refers to a region that having inundation up to 3 meters, medium susceptibility class refers to an area which has inundation between 1-3 meter and low susceptibility class refers to a region that having inundation value less than 1 meter. Tsunami inundation model in this research use 20 m wave height scenario. This wave height scenario based on the highest tsunami wave statistics data that have occurred on the southern coast of Java with a magnitude of 7.8 [9].

Tsunami susceptibility modelling data are shoreline data, land cover, and slope. The shoreline data was used as the initial calculation to calculate iterations of tsunami model inundation. The coastline was obtained from administrative boundary data of Parangtritis Village Map on a scale of 1:5,000. Land cover data was used to construct a surface roughness coefficient in a tsunami model. The land cover data was obtained from the Parangtritis Village Map on a scale of 1:5,000. Conversion from land covers data to surface roughness coefficient refered to BNPB in Risiko Bencana Indonesia [1]. Table 2 shows the conversion value from land use type to surface roughness coefficient.

Table 2. conversion value from land use type to surface roughness coefficient.

Land use	Surface roughness coefficient
Water bodies	0.007
Shrub	0.040
Forest	0.070
Plantation	0.035
Bare land	0.015
Paddy field	0.025
Settlement	0.045
Mangrove	0.025
Ponds	0.010

Slope data is obtained from the geo-processing results of the DEM of ALOS PALSAR imagery. ALOS DEM data is used because it has a higher spatial resolution than other free DEM data such as SRTM or ASTER GDEM.

2.2. Mapping the evacuation point

The evacuation assembly point is an area that safe from certain disaster impacts and used as a gathering place when a disaster occurs to reduce the number of victims caused by a disaster. Evacuation assembly point data was used to build a network dataset as a destination point when composing evacuation routes. The data of the evacuation point was obtained from the map of the assembly point made by GTZ. Assembly point map then overlaid with the susceptibility of tsunami inundation model to evaluate which assembly points are really safe and does not belong to tsunami susceptibility area. In addition, the determination of the assembly point was also visually observed by sentinel 2A imagery.

2.3. Build network datasets

A network database is a set of data set needed to perform network analyst. Network datasets required some spatial and attribute data (table 3). The most important issue in making the network analyst dataset is the value of FT minutes and TF minutes. The assumption of travel time in FT minutes and

TF minutes was obtained by conducting field survey. This survey was attempting to travel by walking at slow speed on several road segments with different types of roads and terrain. A slow-track survey was used to construct the slowest evacuation movement assumptions such as elderly and children. The field survey data was arranged to create the time travel data in the network dataset (TF dan FT minutes). Also slope assumption is employed to create this network dataset.

Table 3. Network datasets data.

Data	Types of data	Information
Spatial data	Road data	Road data refers to road distribution spatial data (attribute and geometry) in Parangtritis village. There are three classifications of roads used in this research such as other roads, local and trails.
	Assembly point data	The assembly point data is derived from the existing assembly point map that has been evaluated with new tsunami susceptibility map. This modelling produces new evacuation assembly point by visual interpretation of Sentinel 2A imagery.
	Block of human centre activity	The human activity block data is obtained from settlement blocks digitization in Parangtritis Village Map. These blocks of activity are used as the starting point assumption of the community to begin self-evacuation which is modelled as the centroid point of human activity centre block.
	Slope data	Slope data is derived from ALOS Palsar DEM, showing the gradient elevation changes and steepness of a location. This data was overlaid with road network to obtain the road terrain conditions. Slope data was made using slope classification according to van Zuidam (1983).
Attribute data	Street name The road length Oneway Road types and its terrain condition FT minutes TF minutes	

The assumption used is the narrower and steeper the road, the travel velocity will be decreased. The velocity for each road segment is 1.2 m / s, obtained from the field survey data through walking simulation from the center of human activity point to the 1st and 3rd assembly point located in Mancingan Hamlet and Grogol 9. While for the velocity on the steep slope road is 1 m/s, assuming that steep slope would impede pedestrian velocity during evacuation. This velocity (v) value is a function of the travel time function (t) for each path length (s) by Equation (1):

$$t = \frac{s}{v} \tag{1}$$

IOP Conf. Series: Earth and Environmental Science **148** (2018) 012003 doi:10.1088/1755-1315/148/1/012003

2.4. *Networks Analyst*

There were two kinds of network analysis performed on this spatial modeling of evacuation route, as follows service area modelling and closest facility modelling. Service area modeling was performed on evacuation points. Service area modelling of evacuation assembly points aims to find out how far the assembly points are able to reach from the centre of human activity blocks in 50 minutes. The 50-minute assumption was taken on the average estimated time of tsunami waves on the South Coast of Java from the occurrence of earthquakes. Closest facility modelling *was* used to model the fastest route that can be travelled from the centre of human activity blocks toward the evacuation assembly points. Modelling the nearby facility will result in generating an evacuation route map.

3. Result and discussion

3.1. *Tsunami susceptibility model*

Tsunami susceptibility model was classified into three classes, high (>3 m), moderate (1-3m), and low (<1 m). The height of tsunami wave in this scenario is 20 meters based on the statistical data of maximum tsunami height that ever happened in the southern coast of Java. Based on the modelling result (Figure 2) high susceptibility class dominate the result, about 3,824,500m^2 affected by 3m of tsunami inundation also 252,113 m^2 heading to land, will affected by 3–1 m tsunami inundation.

Therefore about 34.88% of Parangtritis settlement is threatened by tsunami risk within this wave height scenario. Vertical position (elevation) really affects the modelling result beside of the land use type. The west region (Opak River and its surrounding) is classified as a high tsunami susceptibility class even though it is far away from the coastal line (figure 2).

Opak river with low elevation allows the tsunami wave to affect this region also the surface roughness coefficient of water is only 0.007 so that the wave height reduction velocity will decrease in that location. Low susceptibility class is dominant in the east region of Parangtritis village because of the hilly topographic so the tsunami wave cannot reach that location.

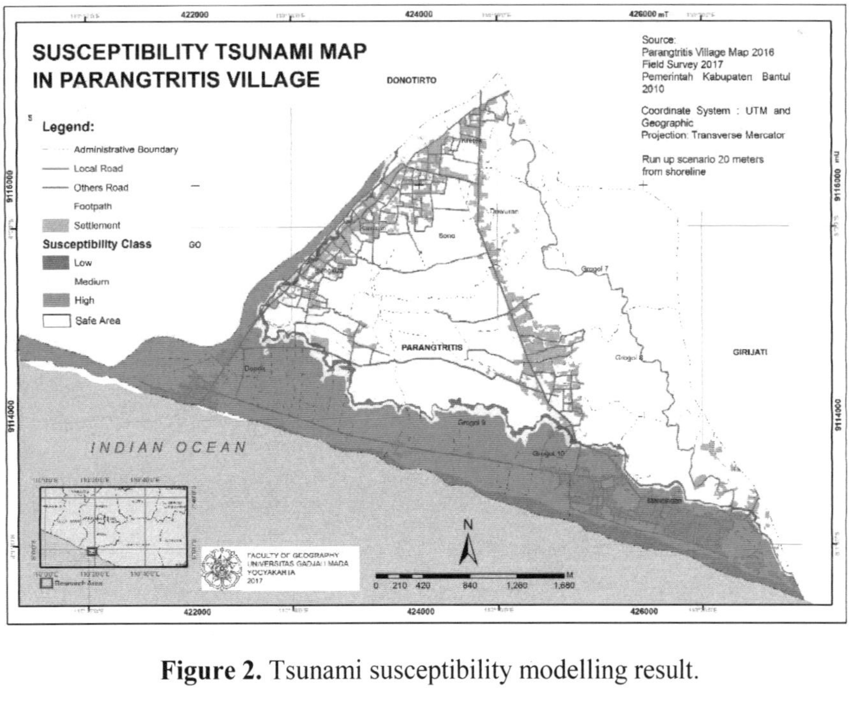

Figure 2. Tsunami susceptibility modelling result.

Area that does not belong to the high, medium and low tsunami susceptibility class is categorized as potential safe area. This area became priority in determining the evacuation assembly point. Overlay analysis of tsunami susceptibility model and land use map showed that 36.67% (of the 1,577,428.92 m^2) of the settlement area in high susceptibility of tsunami inundation, while only 8.37% of other settlement area are in medium to low susceptibility of tsunami. This model shows that land use in the form of forest, garden, and mooring can reduce the tsunami hazard level because of the surface roughness coefficient is relatively high therefore it effectively reducing the wave height.

3.2. Assembly point evaluation

Overlay analysis between susceptibility map and existing evacuation point map is used to evaluate the safety of evacuation points. There are seven evacuation assembly points (temporary assembly points and fixed evacuation point) that have been mapped by the Bantul Regency Working Team. According to this analysis, there is one evacuation assembly point in the southeast which is considered to be less suitable because it is located in high level of tsunami hazard.

Another point is then added through a visual interpretation method with a sentinel 2A imagery and DEM. Finally selected a field in Kretek hamlet, with a distance 2.812.43m from the coastline and 33m above the sea surface. Those assembly points are spreading across five hamlets such as Mancingan, Grogol 10, Duwuran, Sono, and Kretek. The result of the evaluated assembly points can be seen in figure 2. The tsunami evacuation point is located in the eastern part of the Parangtritis Village due to the presence of an escarpment with hilly relief. Another assembly point is in the northern region as it is close to the settlement as a centre of human activity and relatively safe based on distance from coastline.

3.3. Network dataset and assembly point service area

A network dataset was constructed based on the field survey data and the estimation of the slowest pedestrian speed toward the evacuation assembly point. The network dataset was used to view the service area of an evacuation assembly point and determine the fastest route toward the evacuation assembly point. There are three important attribute components in network dataset creation that is FT minutes, TF minutes, and Oneway. Slope value is also an important parameter because it's employed to assume the travel velocity on each road segment.

The centre of human activity point was obtained from the settlement, tourism activities, trade, schools and other public facilities area with the assumption that those location was occupied by a lot of people. The parameter used was a 50-minutes travel time according to DLR (2010) [5]. The tsunami occurrence in Indonesia is a local tsunami where the time between earthquake event and the intermittent of tsunami waves is less than an hour. Figure 3 shows the range of assembly point service area to the centre of human activity point created using ArcGIS network analyst tools.

The evacuation assembly point was defined as facility location meanwhile the centre of human activity point as the incident point. This model represents the area of incident point that will be able to reach the facility in 50 minutes travel time. The number of evacuation assembly points (incident) that can reach the evacuation assembly point is summarized into table 4.

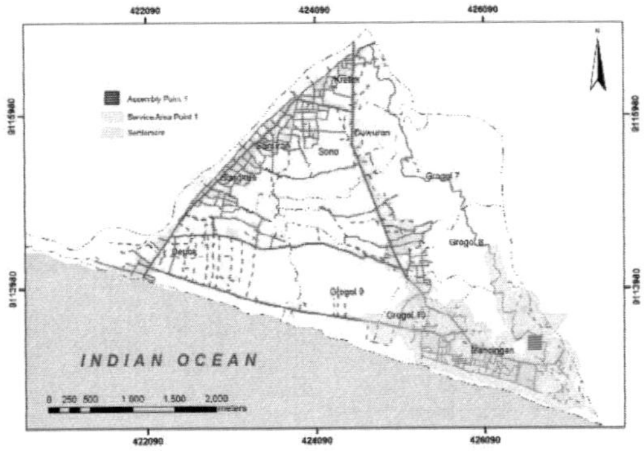
INDIAN OCEAN

(a)

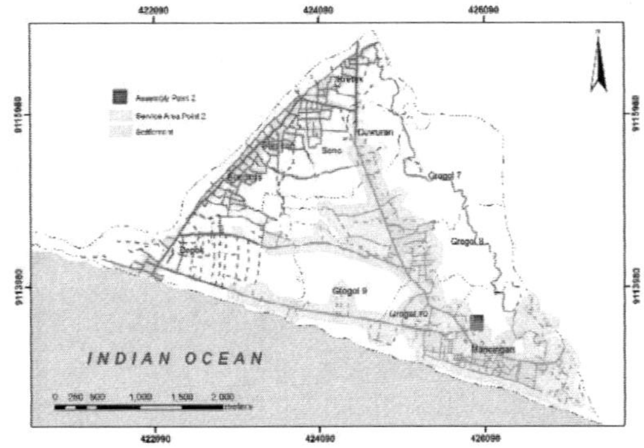
INDIAN OCEAN

(b)

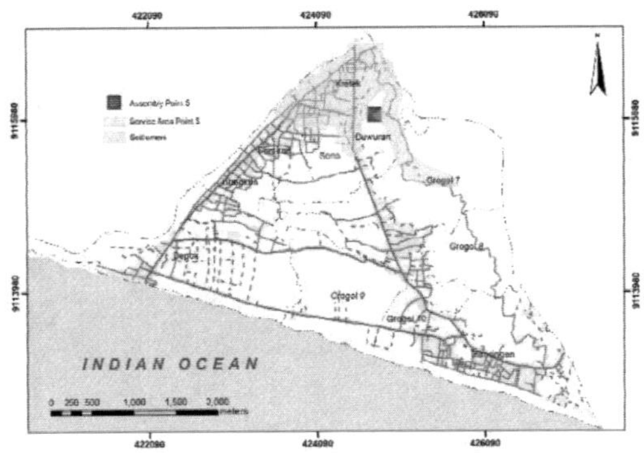
INDIAN OCEAN

(c)

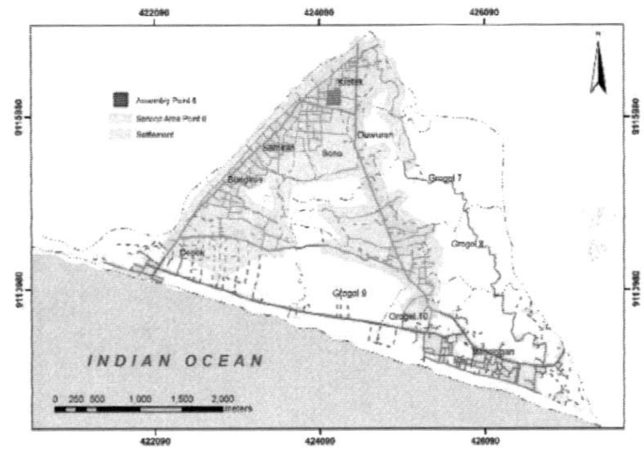
INDIAN OCEAN

(d)

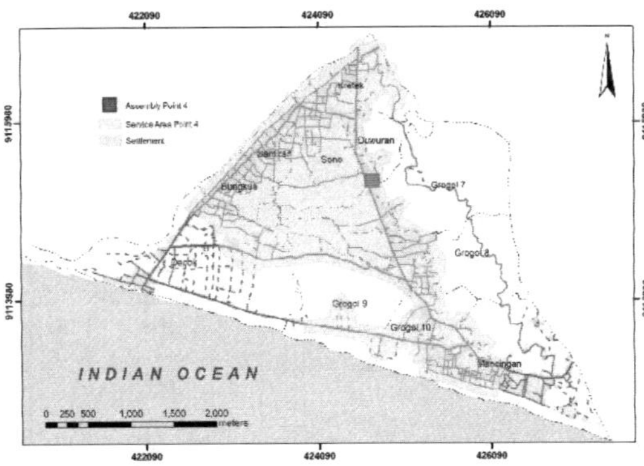
INDIAN OCEAN

(e)

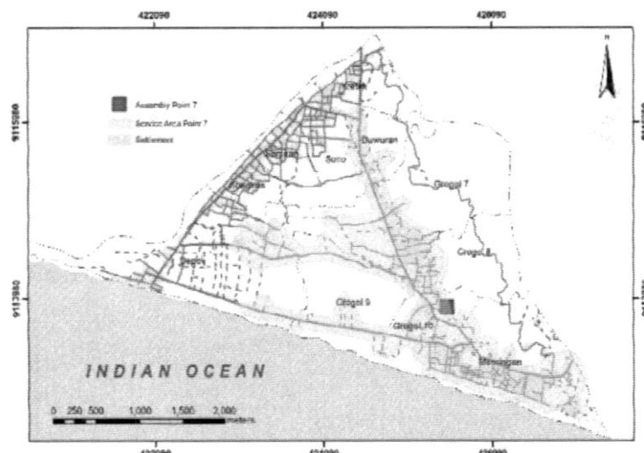
INDIAN OCEAN

(f)

(g)

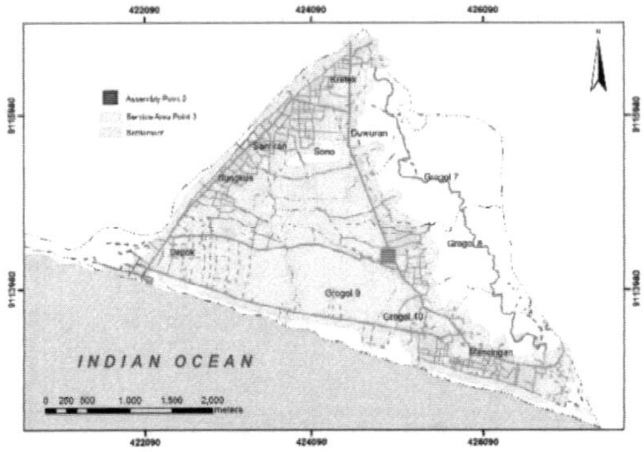

Figure 3. Service area of assembly point map (a) Assembly point 1 (hilly area) (b) Assembly point 2 (Syekh Belabelu Graveyard) (c) Assembly point 3 (Open Space area) (d) Assembly point 4 (Parangtritis Government Office) (e) Assembly point 5 (Hilly area) (f) Assembly point 6 (Field) (g) Assembly point 7 (Syeh Maulana Graveyard).

Based on table 4, it is known that the most accessible evacuation assembly point is Parangtritis Government office in Duwuran hamlet. The assembly point that located around the settlement may not necessarily serve many activity centers in the area. For example, the assembly point 1 (figure 3) only capable to serve 120 points of human activity centers while the assembly point 7 is able to serve 187 point of human activity centers although assembly point 1 is located around the denser settlement than assembly point 7.

Table 4. Range of assembly point service area.

Evacuation Assembly point	Number of center human activity reached	Center of human activity location (Hamlet)
1 (hilly area)	120	Mancingan, Grogol 10, some area of Grogol 8 and Grogol 9
2 (Syekh Belabelu Graveyard)	179	Mancingan, Grogol 7, Grogol 8, Grogol 9, Grogol 10, Dhuwuran, Sono, and Depok
3 (Open Space area)	324	Mancingan, Grogol 10, Grogol 9, Grogol 8, Grogol 7, Kretek, Dhuwuran, Sono, Samiran, Depok, and Bungkus
4 (Parangtritis Government Office)	277	Mancingan, Grogol 10, Grogol 9, Grogol 8, Grogol 7, Kretek, Dhuwuran, Sono, Samiran, Depok, and Bungkus
5 (Hilly area)	77	Kretek, Sono, Dhuwuran, Grogol 7
6 (Field)	217	Grogol 10, Grogol 9, Grogol 8, Grogol 7, Kretek, Dhuwuran, Sono, Samiran, Depok, and Bungkus
7 (Syeh Maulana Graveyard)	187	Mancingan, Grogol 10, Grogol 9, Grogol 8, Grogol 7, Dhuwuran, Sono, and Depok.

3.4. Evacuation Route

The final result of this research is the tsunami evacuation route map which is used as the recommendation of the evacuation route. Closest facility from network analyst tools produces the fastest route from the center of human activity point towards the evacuation assembly point. This model also provide direction, time, and distance information. Figure 4 shows the evacuation route map with seven assembly points which are represented by the black arrows.

The shortest route based on the model is 2 meters with <1 minute travel time and the longest route is 3.785 meters with 59 minutes travel time. One of the modeling routes shown in figure 5, the information obtained through this model is route, distance, time displayed in detail on each segment and its summary. The road's name is not identified as shown in table 5, because there are so many country lane and footpath in this road network model.

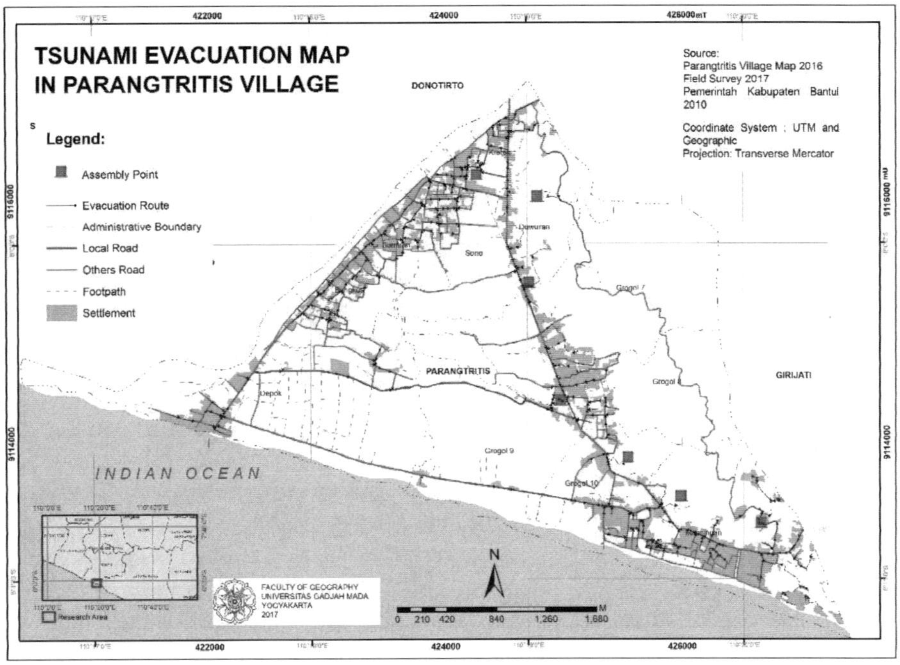

Figure 4. Evacuation route map.

Figure 5. Example of one evacuation route.

Table 5. Evacuation route detail (route example from Location 18-Location 6).

Nr.	Route Detail	Distance (m)	Time (min)
1	Start at Location 18		
2	Go northeast	55.4	< 1
3	Turn right	44	< 1
4	Continue	6.7	< 1
5	Turn left	989.1	14
6	Turn right	99.6	1
7	Turn left	3.1	< 1
8	Continue	267.2	4
9	Turn right	270.6	4
10	Turn left	80.8	1
11	Turn right	147.9	2
12	Finish at location		
	Total	1964.3	29

4. Conclusion

The tsunami evacuation route modelling with Berryman and network analyst is able to provide detailed evacuation route recommendations. However, the weakness of this model is unable to represent an evacuation route outside the road segment, whereas several locations of assembly points are not located near the road segment. So the determination of the route is only based on the distance projection between the human activity points with the assembly point on a road segment. The actual distance between the assembly point and the main road segment is not considered by this model.

5. References

[1] Amri M R Yulianti G Wiguna S Adi A W Ichwana A R Radongkir R E and Septian R T 2016 *Risiko Bencana Indonesia* (Jakarta: BNPB) p 39

[2] Berryman K 2006 *Review of Tsunami Hazard and Risk in New Zealand* (New Zealand: Institute of Geological & Nuclear Sciences) pp 52–59

[3] Bryant D E 2008 *Tsunami: The Underrated Hazard* (United Kingdom: Springer) pp 19–22

[4] Dewi R S 2010 A GIS-Based Approach to the Selection of Evacuation Shelter Buildings and Routes for Tsunami Risk Reduction; a Case Study of Cilacap Coastal Area, Indonesia. *Thesis Double Degree MSc Geo-Information for Spatial Planning and Risk Management*. Universitas Gadjah Mada and University of Twente

[5] Fernando H Braun A Galappatti R Ruwanpura J and Wirasinghe S 2008 Tsunamis: Manifestation and Aftermath, in: M. Gad-el-Hak(eds), *Large-scale Disaster; Prediction, Control and Mitigation* (Cambridge: Cambridge University Press) pp 258–292

[6] Lavigne F Gomez C Giffo M Wassmer P Hoebreck C Mardiatno D Priyono J and Paris R 2007 Field Observation of the 17 July 2006 Tsunami in Java, *Journal of Natural Hazard & Earth System Science* **7** pp 177–183

[7] National Geophysical Data Center 2009 *Tsunami Events Search - sorted by Date, Country* https://www.ngdc.noaa.gov/nndc/struts/form?t=101650&s=7&d=7

[8] Eisner R and NTHMP 2001 Plan for Evacuation, in: *Designing for Tsunami – Seven Principles for Planning and Designing for Tsunami Hazard* (NOAA) pp 45–49

[9] Webb T 2005 Recommendations for Procedures for Better Tsunami Risk Management at Regional Level, in: *Review of New Zealand's Preparedness Tsunami Hazard, Comparison to Risk and Recommendations for Treatment* (New Zealand: Institute of Geological and Nuclear Science) pp 64–79

IOP Conf. Series: Earth and Environmental Science **148** (2018) 012004 doi:10.1088/1755-1315/148/1/012004

Topographic data acquisition in tsunami-prone coastal area using Unmanned Aerial Vehicle (UAV)

M A Marfai[1], Sunarto[1], N Khakim[1], A Cahyadi[1], F S C Rosaji[2], H Fatchurohman[3] and Y A Wibowo[3]

[1]Department of Environmental Geography, Faculty of Geography, Universitas Gadjah Mada, Yogyakarta, Indonesia
[2]CV. Mitra Geotama, Yogyakarta, Indonesia
[3]Master in Planning and Management on Coastal Area and Watershed (MPPDAS), Faculty of Geography, Universitas Gadjah Mada, Yogyakarta, Indonesia

arismarfai@yahoo.com

Abstract. The southern coastal area of Java Island is one of the nine seismic gaps prone to tsunamis. The entire coastline in one of the regencies, Gunungkidul, is exposed to the subduction zone in the Indian Ocean. Also, the growing tourism industries in the regency increase its vulnerability, which places most of its areas at high risk of tsunamis. The same case applies to Kukup, i.e., one of the most well-known beaches in Gunungkidul. Structurally shaped cliffs that surround it experience intensive wave erosion process, but it has very minimum access for evacuation routes. Since tsunami modeling is a very advanced analysis, it requires an accurate topographic data. Therefore, the research aimed to generate the topographic data of Kukup Beach as the baseline in tsunami risk reduction analysis and disaster management. It used aerial photograph data, which was acquired using Unmanned Aerial Vehicle (UAV). The results showed that the aerial photographs captured by drone had accurate elevation and spatial resolution. Therefore, they are applicable for tsunami modeling and disaster management.

1. Introduction

The logical consequences of the meeting of three tectonic plates in Indonesia are, for instance, high susceptibility to earthquakes and tsunamis [8][12][19]. The meeting forms a subduction zone along the plate boundaries. With its active movement, the surrounding areas are highly vulnerable to earthquakes [6][10][18]. Moreover, when an earthquake is followed by plate deformation, it potentially develops into a tsunami [3][5].

Tsunami is one of many disasters that have enormous impact [2][7]. In Indonesia, it is associated with earthquakes that alter the submarine plates [5]. Earthquake is the leading cause of most tsunamis in Indonesia. Around 90.5% tsunami events in Indonesia were caused by earthquakes, while the remaining 8.6% and 1% were, respectively, preceded by volcanic eruptions and underwater landslides. From the tsunami in 2000 BC until the one in 2005, there were 253 events, which were the third highest occurrences after Japan (444 events) and the United States (287 events) [15]. Nevertheless, regarding death toll, the tsunamis in Indonesia are the most destructive in the world [18]. The highest casualties were during the 2004 tsunami in Nangroe Aceh Darussalam, i.e., 250.000 deaths.

The southern coastal area in Java Island is one of the most vulnerable locations to tsunami [1][9][12][14][17]. Tsunamis have struck this area for several times, i.e., in 1818, 1840, 1859, 1904, 1921, 1925, 1957, 1994 and 2006 [4][16]. Therefore, studies on hazard, vulnerability, and risk assessment as well as disaster preparedness and management for tsunamis in this area are crucial.

One of the obstacles in tsunami susceptibility and hazard analysis is the availability of high-resolution topographic data. Moreover, not every remote sensing image with high spatial resolution is available and applicable for risk analysis and mitigation planning. However, this issue can be overcome by, for instance, acquiring remote sensing images using Unmanned Aerial Vehicle (UAV).

Aside from providing the latest information at a relatively low cost, UAV produces remote sensing image with high spatial resolution and detailed Digital Elevation Model (DEM).

The coastal area in Gunungkidul Regency contributes IDR 7.1 billion per year to the local revenue, which is one of the highest incomes outside of tax revenue. The tourism sector also has a strategic role in improving the economic condition of the local people and increasing by the year [20]. On the other hand, the coastal area directly faces the open sea where the subduction zone of tectonic plates exists. Moreover, the subduction zone in the south of the Special Region of Yogyakarta is located in the seismic gaps (figure 1) in which earthquakes and tsunamis with large magnitude potentially develop [13].

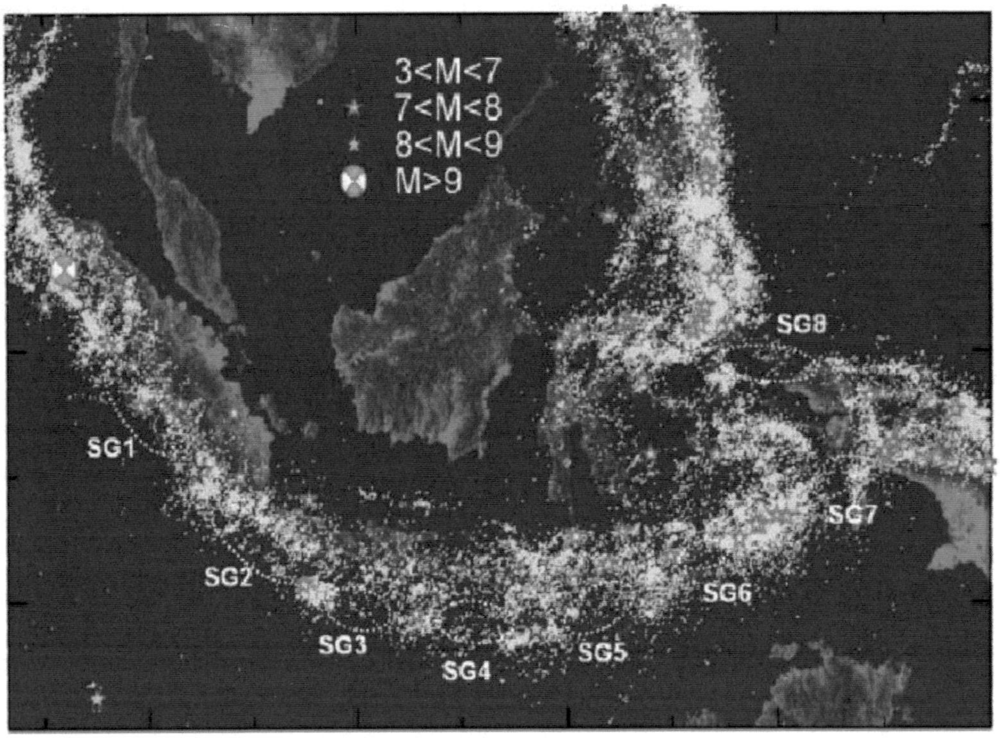

Figure 1. The location of seismic gaps (SG) in Indonesia [1].

2. Method

The aerial photography began with flight route planning and pre-mark or marker installation for the Ground Control Point (GCP). The research used two flight routes that recorded at an altitude of 300 m above the ground. There were 10 points of pre-marks distributed in the field (figure 2), which represented the topographic configuration in the study area. The coordinates were measured by geodetic GPS with Real Time Kinematic (RTK) method.

The research used a fixed-wing unmanned aerial vehicle with a Bixler airframe. The UAV had a semi automatic control, i.e., an autopilot component that directs it toward the predefined route while in the air. Therefore, the pilot controlled it during take-off and landing only. Meanwhile, the co-pilot monitored its movement and position from the Ground Control Station (GCS). The sensor was a pocket digital camera Canon Powershot A2500 with 16Mpix resolution. It was set to take photos automatically every two seconds at a shutter speed of 1/2000 to avoid blurred images.

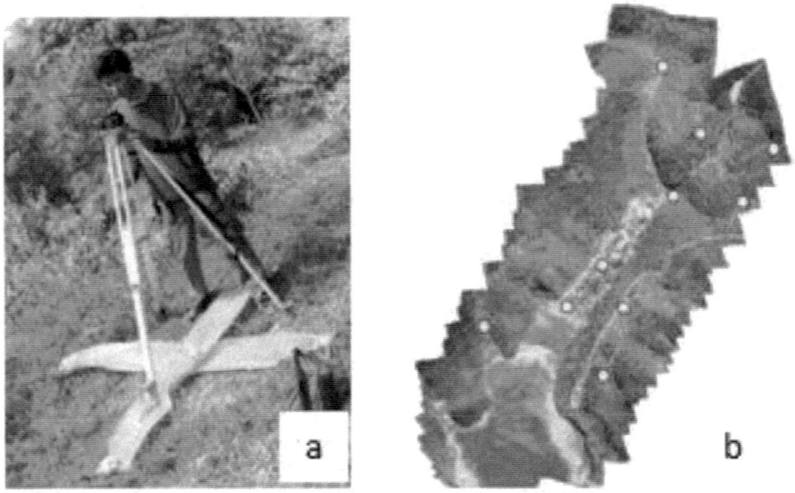

Figure 2. (a) Pre-mark Installation and Ground Control Point (GCP) Measurement Using Geodetic GPS and (b) The Distribution of GCP on the Field.

The aerial photographs were sorted to exclude widely distorted (oblique) images in further analysis. Each of the photos from this recovery was then adjusted to have equal brightness and contrast. Afterward, mosaic compilation and orthorectification were processed in Agisoft Stereoscan software. Orthorectification aims to convert the projection of aerial photos, i.e., from central to orthogonal projection, to minimalize geometric distortion on them.

2.1. UAV-assisted photography for acquiring data in Kukup coastal area

UAV-assisted photography of Kukup Beach covered an area of approximately 41 ha and took one working day in the field. The type of UAV used in this research was the Bixler (figure 3a). The aerial photography started with creating a flight plan and designing the position of GCP (Figures 3b and 3c). The plan included determining the altitude of the object capture, which was 150 m above the ground. Therefore, the resolution of the aerial photos was expected to reach <10 cm. The overlaid parameters, namely, end lap and side lap, were adjusted to 80% and 50%. Then, the GCP distribution was designed as minimal as possible to see whether three ground control points could already produce elevation data (Digital Surface Model-DSM).

Figure 3. (a) Bixler, the UAV Used for Photography, (b) Preparation, UAV component Setting, and Ground Control Station, (c) Pre-mark Installation and Ground Control Point Measurement by Team Member During Preparation.

The personnel involved in the process of data acquisition was five people, i.e. one pilot, one co-pilot, and three crews tasked with GCP coordinate measurement using geodetic GPS. The aerial photography took one day, starting from the placement of pre-marks/markers for GCP. They were made of 30x150 cm^2 tarpaulins and placed crossing each other in an opened space, so they were detectable during the UAV-assisted photography. When the team installed the pre-marks, both pilot and co-pilot prepared the UAV setting, calibrated its components, and re-checked the position of the planned flight route. The communication between the GCP team and aerial photography team used handy talkies (HT) for time efficiency. When all of the markers were in position, the UAV was ready to take off and start photographing.

The aerial photography used a fixed-wing aircraft with Bixler airframe (the main body of the plane) whose main material was EPO foam. Therefore, it is easily repairable after a crash or hard landing. The Bixler had a 1.4-meter wingspan. This size is categorized as small. Nevertheless, it has the advantage of easy equipment mobilization and maneuver control. Furthermore, for aerial photography in a small area, this kind of plane is relatively economical.

The aerial photography of Kukup coastal area took approximately 2 hours for preparation, takeoff, and image capture at an altitude of 150 m, and landing. The flight was conducted twice because the coastal area was very windy, requiring the rotor to generate a great pushing energy. Since the rotor was driven by battery, the team had to execute a landing once the battery power was low. It aimed to prevent UAV from losing control due to exhausted battery capacity. After the UAV finished its photography mission and landed, the aerial photos were checked and stored in a hard disk drive. The GCP measurement took a longer time. The measurement at each point spent ± 18 minutes.

2.2. Digital orthophoto generation with aerial photos

The aerial photos acquired using UAV were processed in photogrammetric software, Agisoft Photoscan. This software processes the aerial photos to generate an orthophoto mosaic (figure 4) and the altitude data to provide Digital Surface Model (DSM). The reconstruction of the position of each photo was also visible during the data processing in the software.

Figure 4. The reconstruction of multiple aerial photos (Orthophoto Mosaic) from the recording of Kukup coastal area.

3. Results and Discussion

The results of the aerial photography had a Ground Square Distance (GSD) of approximately 5 cm. This GSD is considered very detail for data input in large-scale mapping. The details of several objects captured and recognized from the aerial photos are presented in figure 5. They confirm that every single car, plant, and even a person in the photos are well detected.

Figure 5. (a) The visible pre-mark, (b) the recognizable distribution of tourists on a sandy beach, (c) the visible individual stand of cassava, (d) the size and type of cars identified from aerial photos, and (e) apparent road markings.

The aerial photo processing produced orthophotos with very little distortion due to the central projection of the camera. Therefore, these orthophotos could generate a mosaic with an orthogonal projection that supported an accurate mapping with minimal distortion. The orthophoto mosaic resulted in this research had a pixel size of ± 5 cm.

3.1. Digital Surface Model (DSM) generation with aerial photos

Another data acquired from photography using UAV was Digitial Surface Model (DSM) with a horizontal resolution of 0.4 m. The accuracy of horizontal resolution is affected by the camera resolution, flight altitude, and focus length. Because DSM provides a higher detail than the elevation data presented in the Indonesian Topographic Map scale 1: 25,000, it is more reliable for tsunami inundation modeling. The orthophoto mosaic, DSM, and 3D reconstruction are presented in figures 6, 7, and 8, respectively. The detail of the elevation acquired from the analysis was up to 30 cm.

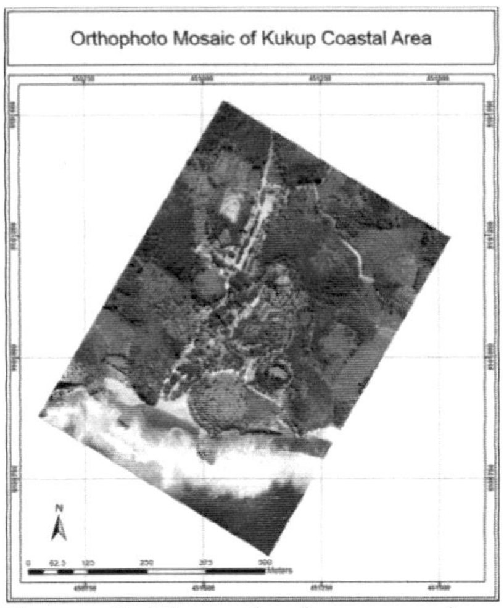

Figure 6. The orthophoto mosaic of Kukup Beach generated by UAV aerial photo processing.

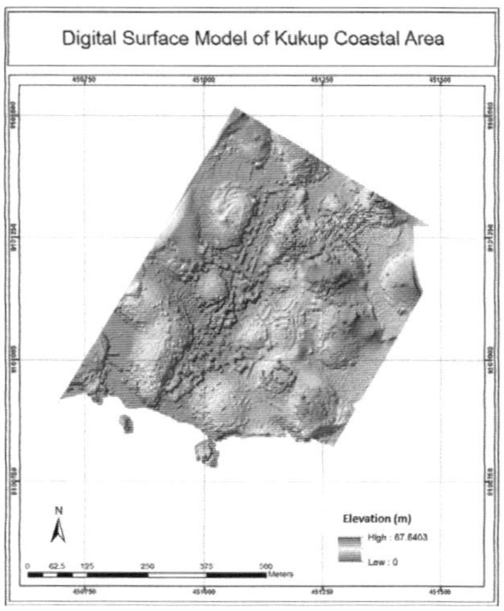

Figure 7. The Digital Surface Model (DSM) of Kukup Beach produced by UAV aerial photo processing.

Figure 8. The 3D reconstruction of Kukup Costal Area from orthophoto mosaic and DSM produced in UAV aerial processing.

3.2. The challenges of UAV-assisted topographic mapping for tsunami disaster management in Indonesia

Indonesia is vastly and greatly susceptible to tsunamis. Therefore, the large-scale procurement of supporting data requires encouragements to be able to accelerate data production using priority scale. For instance, the priority scale represents the location of the seismic gap, the subsequent potential of damage and risk, and the record of tsunami events in the past. The method used in this research is applicable for supporting the acceleration of data procurement because it offers many advantages, including the ease of operation and low operational cost. Due to the main vehicle that used for acquiring the aerial photo data, UAV is relatively cheap. The risk of flying UAV for aerial mapping is also very small. Compared to the conventional photogrammetric method, UAV method is more effective especially in small mapping area. The DSM that produced from UAV data acquisition is also accurate for generate tsunami hazard modelling. The constraints of this method are the disability to covers large mapping area. The battery endurance of the UAV is limited so it takes several flight missions to cover the whole area. Strong wind and wireless interference are also limiting the flexibility of UAV mobilization.

4. Conclusion

The topographic data acquired from an unmanned aerial vehicle has an accurate spatial resolution and elevation. Therefore, it is suitable for tsunami modelling. The results of the mapping process had a spatial resolution of 0.4 m and an altitude accuracy of 30 cm, indicating a higher detail than the topographic maps that are currently available in official publications in Indonesia. However, DSM data should be supported by bathymetry data to generate more detailed model and simulation. Furthermore, due to the excellent spatial resolution, the resultant topographic data is best suited for analyses that focus on at-risk element identification, economic valuation of potential damages, spatial planning, and disaster risk reduction and management associated with tsunamis.

5. Acknowledgement

This research is part of the Excellence Research Grant for Higher Education (Penelitian Unggulan Perguruan Tinggi-PUPT), titled "Pemetaan Risiko Bencana Kepesisiran Menggunakan Teknologi *Unmanned Aerial Vehicle* (UAV) dan GIS pada Wilayah Kepesisiran Kabupaten Gunungkidul". Authors gratefully acknowledge the research funding and support from the Indonesian Ministry of Research and Higher Education. Authors would also like to thank all parties involved in the research preparation, implementation, and evaluation.

6. References

[1] Adyan Ö 2008 Seismic and Tsunami Hazard Potentials in Indonesia with a special emphasis on Sumatra Island. *Journal of the School of Marine Science and Technology, Tokai University,* 6(3), 19-38
[2] Alberico I Di Fiore V Wavarone R Petrosino P Piemontese L Tarallo D and Marsella E 2015 The Tsunami Vulnerability Assessment of Urban Environments through Freely Available Datasets: The Case Study of Napoli City (Southern Italy). *Journal of Marine Science and Engineering* 3 981-1005
[3] Bryant E 2008 *Tsunami: The Underrated Hazard (Second Edition)* Chichester Praxis Publishing
[4] Cahyadi A Afianita I Gamayanti P and Fauziyah S 2012 Evaluasi Tata Ruang Pesisir Sadeng Gunungkidul: Perspektif Pengurangan Risiko Bencana. Paper presented at the 3rd National Seminar on Sustainable Culture, Architecture and Nature, Architecture Study Program, Faculty of Engineering, Universitas Atma Jaya, Yogyakarta, 15 May 2012
[5] Dewi RS and Dulbahri 2009 Bencana Tsunami Parangtritis. In M.A. Marfai and D. Mardiatno (Eds), Penaksiran Multirisiko Bencana di Wilayah Kepesisiran Parangtritis (pp. 65-88). Yogyakarta: Pusat Studi Bencana (PSBA) Universitas Gadjah Mada
[6] Druskha K Sprintall J Gille S T and Pranowo W S 2008 Observations of the 2004 and 2006 Indian Ocean Tsunamis from a Pressure Gauge Array in Indonesia. *Journal of Geophysical Research*, 113, 1-8
[7] Engel M and Brückner H. 2011 The Identification of Paleo-tsunami Deposits: a Major Challenge in Coastal Sedimentary Research. *Coastline Reports* 17 65-80
[8] Hartoko A Helmi M Sukarno M and Hariyadi 2016 Spatial Tsunami Wave Modelling for the South Java Coastal Area, Indonesia. International Journal of Geomate 11(25) 2455-2460
[9] Lavigne F Gomez C Gifo M Wassmer P Hoebreck C Mardiatno D and Paris R 2007 Field observations of the 17 July 2006 Tsunami in Java. *Natural Hazards and Earth System Sciences* 7(1) 177-183
[10] Liu Z Y C and Harris RA 2013 Discovery of Possible Mega-thrust Earthquake
[11] Along the Seram Trough from Records of 1629 Tsunami in Eastern Indonesian Region. *Natural Hazards*, 72(3), 1311-1328. http://doi.org/10.1007 /s11069-013-0597-y
[12] Marfai M A King L Sartohadi J Sudrajat S Budiani S R and Yulianto F 2008 The impact of tidal flooding on a coastal community in Semarang, Indonesia. *Environmentalist* 28 237-248
[13] Mori J Mooney WD Afnimar Kurniawan S Anaya AI and Widiyantoro 2007 *Seismological Research Letters* 78 (2) 201-207
[14] Okamoto T and Takenaka H 2009 Waveform inversion for slip distribution of the 2006 Java tsunami earthquake by using 2.5D finite-difference Green's function. *Earth Planets Space*, 61 17-20
[15] Subandono D and Budiman 2008 Hidup Akrab dengan Gempa dan Tsunami Bogor: Penerbit Buku Ilmiah Populer
[16] Sunarto, Marfai MA, & Mardiatno D 2010. *Multirisk assessment of Parangtritis Coastal Area. Yogyakarta*: Gadjah Mada University Press
[17] Sutikno 2009 Indonesia Negeri 1001 Bencana. Paper presented at Seminar Sistem Informasi Kebencanaan sebagai Sebuah Kearifan di Negeri 1001 Bencana, Environmental Geography Student Association (EGSA), Faculty of Geography, Universitas Gadjah Mada, Yogyakarta, 3-5 December 2009
[18] Triatmadja R 2010 Tsunami: Kejadian, Penjalaran, Daya Rusak dan Mitigasinya. Yogyakarta: Gadjah Mada University Press
[19] Verstappen H 2000 Outline of the Geomorphology of Indonesia. Enschede: ITC
[20] http://www.republika.co.id/berita/nasional/daerah/16/09/11/odcely382-pendapatan-pariwisata-gunung-kidul-melonjak-hampir-6-kali-lipat acessed: February 2018

IOP Conf. Series: Earth and Environmental Science **148** (2018) 012005 doi:10.1088/1755-1315/148/1/012005

"Smong" as local wisdom for disaster risk reduction

A Suciani[1], Z R Islami[1], S Zainal[2], Sofiyan[3] and Bukhari[4]

[1]Geography Education Department, Faculty of Teacher Training and Education, Samudra University, Meurandeh Street, Langsa, Indonesia
[2]Primary School Teacher Education, Faculty of Teacher Training and Education, Syiah Kuala University, Darussalam Street, Banda Aceh, Indonesia
[3]Mathematics Education Department, Faculty of Teacher Training and Education, Samudra University, Meurandeh Street, Langsa, Indonesia
[4]History Education Department, Faculty of Teacher Training and Education, Samudra University, Meurandeh Street, Langsa, Indonesia

ayu_suciani@yahoo.com

Abstract. The province of Aceh is located in the northern tip of Sumatera Island, Indonesia, highly vulnerable to the disasters, the so-called earthquakes and Tsunamis. This is due to the geological location of Aceh, which is located where the Indo-Australian and Eurasian plates meet. Many people learned this just after the devastating earthquake and tsunami on December 26, 2004 that killed thousands of people and also caused countless material losses. Before 2004, people in Aceh did not even notice what a tsunami was. Yet, after the earthquake in 2004 which had a magnitude of 9.2, Aceh continues to experience earthquake with magnitudes of 5-6, just as it did in Pidie Jaya on December 2016. Due to these conditions, the people of Aceh need to be informed of the real and serious threats that these disasters can cause in order to reduce the impact of these potential tragedies. Local wisdom could be an early warning for preventing risk disaster. Local wisdom could be easy to understand, adapt, and use by the society. The purpose of this paper is to publish "*Smong*" as local wisdom to reduce the risk of potential earthquake and tsunami disasters. The word is referred to Tsunami was adopted from Devayan Language. It is part of the Simeulue indigenous culture, transmitted through songs, short poems, lullabies, and stories. It is fascinating to note that the earthquake and tsunami catastrophe of 2004 resulted in only seven casualties in Simeulue, which has approximately 86.735 inhabitants. *Smong* is a key word understood by the entire population of Simeulue that describes the occurrence of giant waves after a major earthquake. During the terrible event that plagued Aceh on December 26, 2004, there was a massive evacuation of the entire Simeulue beach area within a few minutes after the earthquake. Therefore, "*Smong*" is an appropriate term to be used in order to reduce the impact of disasters, viz. earthquakes and tsunamis in high risk areas.

1. Introduction

According to the 1945 constitution no.24 year 2007, disaster was an event or series of events that threatened and disrupted the livelihood of the community. It was caused by both natural and/or non-natural factors, which resulted in the occurrence of human casualties, environmental damage, property loss, and psychological impact. Disasters can be caused by natural events or by man-made factors. Disaster-generating factors are: natural hazards and man-made hazards. According to the United Nations International Strategy for Disaster Reduction (UN-ISDR) these hazards can be grouped into geological hazards, hydro meteorological hazards, biological hazards, technological hazards and environmental degradation.

The geological location of Aceh is in the subduction area where the Asian and Australian plates meet. It is also at the end of the big fault/transform fault. This fault divides Sumatra island from Aceh all the way to the Sunda Strait. The active fault zone found in the Aceh region is the central region of Aceh Besar, Pidie, Pidie Jaya, Central Aceh, Gayo Lues, Southeast Aceh, West Aceh, Nagan Raya, Aceh Barat Daya and South Aceh. This can cause Aceh to suffer a repeated geological disaster. The most common geological disasters in Aceh are earthquakes and tsunamis that the causes described in the following figure 1.

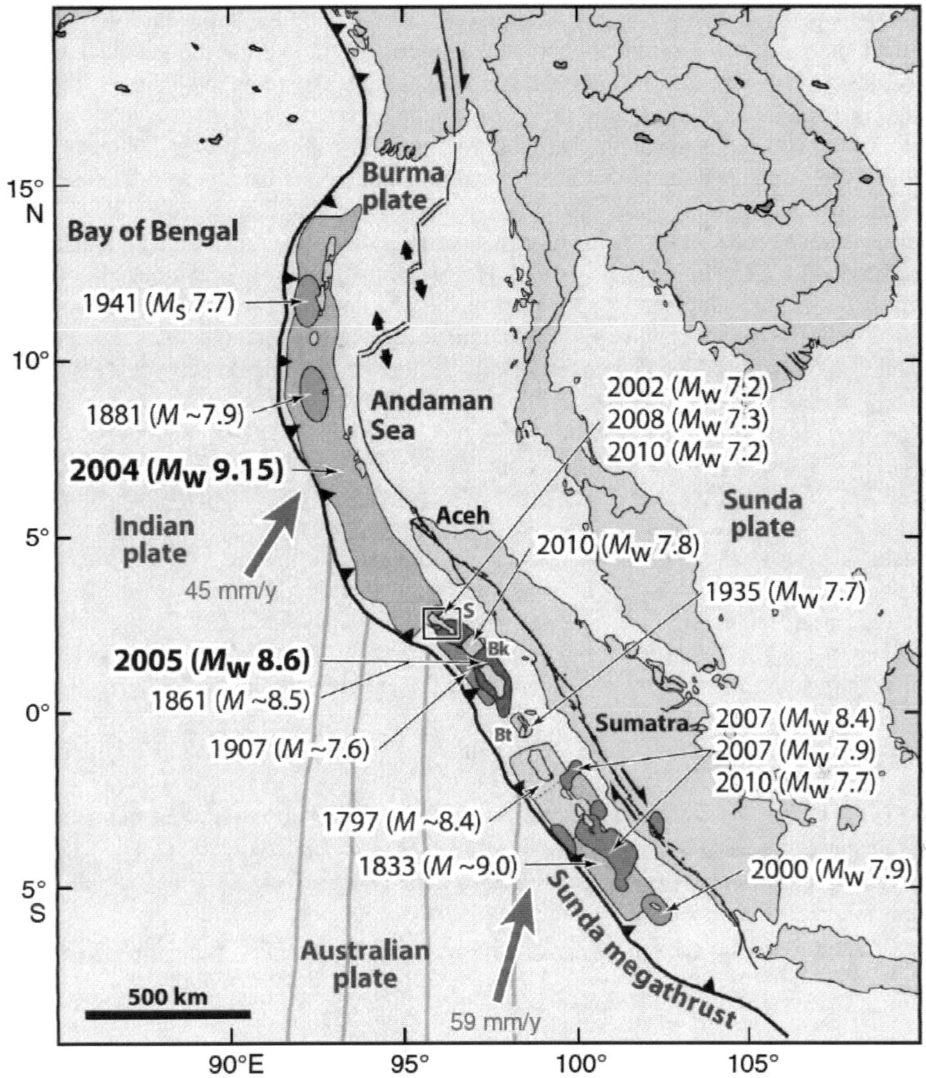

Figure 1. The active tectonics of Sumatra Island show the major sources of earthquakes in the Subduction zone. Many large earthquakes occur in subduction zones or collisions this plate. The colored ellipse shows the source of the quake and the number next to it shows the year and magnitude of the magnitude of the earthquake that occurred including the 26 December 2004 earthquake (Magnitude 9.15)[13].

Tsunami-prone areas are located along the coast of the Aceh, which face potential damage from the Indian Ocean in the west (Aceh Jaya, West Aceh, Nagan Raya, Aceh Barat Daya, Aceh Selatan, Aceh Singkil and Simeulue). Based on geological disaster records, tsunamis have occurred in 1797, 1891, 1907 and 2004. The December 26, 2004 magnitude 9.15 Sumatra, Indonesia earthquake (3.316

N, 95.854 E, depth 30 km) generated a tsunami that was observed worldwide and caused tremendous devastation and deaths throughout the Indian Ocean region. The tsunami tragedy was caused by tectonic earthquake that occurred around 160 km west of Aceh. The UN estimates that 186,983 people were killed in the quake and tsunami. The United States Geological Survey states, the death toll reached 283.100 inhabitants. Victims in Indonesia, especially Aceh, reached 167.000 inhabitants, 130.000 were killed and 37.000 are missing.

Disaster is unavoidable by human. After several disasters occured, the sustainability of society to survive from the hazard of future disasters. This modern era, protecting indigenous knowledge of one region is the main factor for the sustainability of society in reducing disaster risks. Indigenous knowledge is dynamic and continually influence by the creativity, experinces and practice. According to Baumwoll, indigenous knowledge originated within the community that contrasts from scientific knowledge which unrelated to the local culture. Indigenous knowledge could be easy to understand, adapt, and use by the society.

Early warning is an crucial factor in disaster risk reduction. Indigenous knowledge could be an early warning that is easy to adapt in society. However, the value of indigenous knowledge is getting known globally but it almost can not be seen in the practice. The International Decade for National Disaster Reduction (IDNDR) from 1990-2000 agreed for using indigenous knowledge to be developed for disaster risk reduction. This development was held and applied in disaster risk reduction policy, so that the risk of death, injuries, demage and loss can be minimized.

The effectiveness of indigenous knowledge practice for disaster risk reduction shown from the 26 November 1999 tsunamis in Baie Martelli (Pentecost Island, Vanuatu, Pasific Island) caused only five fatalities from threatened population of about 300 persons, most of whom fles inland and upslope before the waves struck. Some people who experienced it say that their Indiginous Knowledge involved in their survival which known as *kastom. Kastom* defined as cultural product that become folk knowledge was about natural hazard that told by story telling.

The recent understanding of Indigenous Knowledge for Disaster Risk Reduction emerges from a series of events organized and publications brought out during 2007-08 by UNISDR, Kyoto University, European Commission, DRH Asia, SEEDS and other Asian role players. Some of the main events in this regard are:

- Transferable Indigenous Knowledge meeting in Delhi in February 2007: initial discussion and case study experience sharing
- Transferable Indigenous Knowledge meeting in Delhi in February 2008: Indigenous Knowledge action agenda discussion
- Indigenous Knowledge Good practices document: November 2007 to June 2008: Provide key emphasis areas
- Transferable Indigenous Knowledge for DRH and Beijing workshop in February 2008: Discussion on Indigenous Knowledge usage, practice and documentation
- Indigenous Knowledge workshop in July 2008 in Kyoto University: Policy discussion on thematic Indigenous Knowledge sectors
- SAARC Disaster Management Centre study on Indigenous Knowledge for Disaster Risk Reduction in South Asia
- Third Ministerial meeting in December 2008: Indigenous Knowledge Side event to discuss the policy note

Simeulue Island and the other small islands around it are situated to in the Southwest Aceh Province; the island has been devastated from earthquakes and tsunamis. Simeulue is located on the arc of the subduction zone in western Sumatra and was so close to the source of the earthquake. Similarly, the islands of Nias, Siberut, Mentawai and Enggano are located close to the subduction zone. Tectonic earthquake shocks occur frequently due to Eurasian and Indo-Australian plate collisions. This caused several tsunami tidal waves to occur in 2004. There were seven deaths in Simeulue. Without diminishing the meaning of life, seven souls are small in comparison to the hundreds of thousands of souls that were lost in other areas. The following figure 2 describe the location of Simeulue Island.

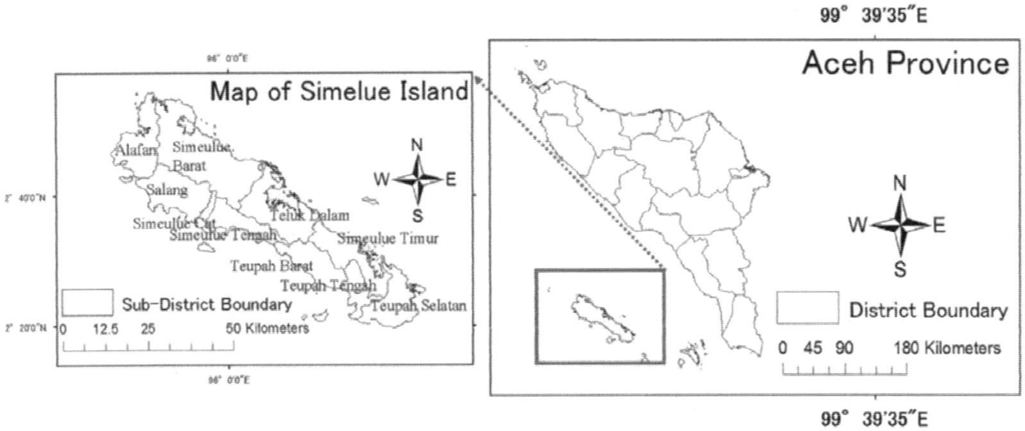

Figure 2. Map of the Island of Simeulue and the Aceh Province.[14].

The potential for disasters in Aceh will not decrease significantly in the years to come. Given the geographic, geological, hydrological and demographic conditions of Aceh, a comprehensive effort is needed to overcome these threats. This applies to when the disaster first occurs, after it has occurred, and against potential disasters in the future. The geomorphological, climatological, and demographic conditions, create hazards in Aceh that include geological, hydro-meteorological, social, and health threats.

Almost two decades after the 2004 Indian Ocean Tsunami, the world has been progressing to focus on sustainable disaster risk reduction concepts. The great story of *smong* is a practical example of indigenous knowledge of disaster risk reduction. Realizing this fact, the purpose of this study was to analyze how the *smong* story as local wisdom successfully alerted the Simeuluean people of the impending danger.

2. Method

This research uses descriptive qualitative research method. The data collections were done through documentation, study, and analysis of relevant literature. Documentation study was conducted to obtain information about local values of the *smong* in regards to the disaster mitigation efforts. The literature study was conducted by researching information about *smong* from books and journals. The data analysis technique is done by organizing, analysing, and interpreting the data obtained in the form of scientific explanation so that it can be understood by the reader.

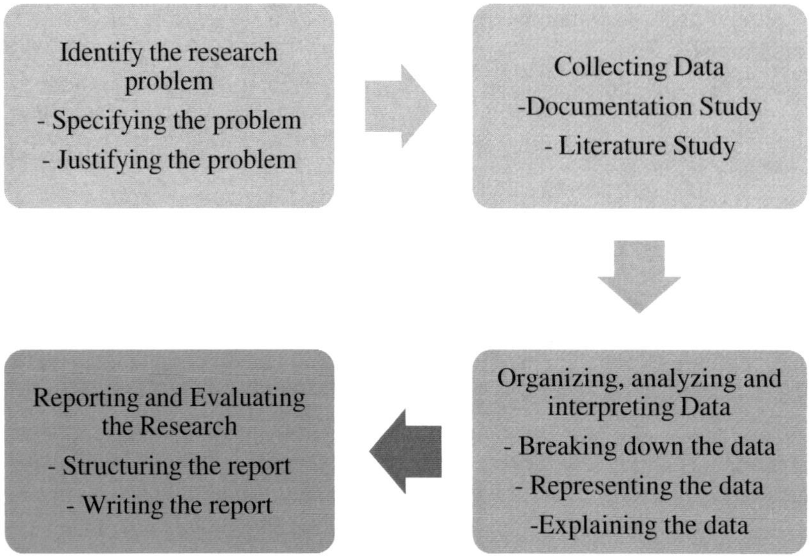

Figure 3. Data analysis technique.

3. Result and discussion

Local wisdom lately is very often discussed. Discussions about local wisdom are often associated with local communities and with varying insights. Local wisdom is local ideas that are wise, full of wisdom, good value that is embedded and followed by members of the community. Local wisdom is contained in local wisdom refers to knowledge that comes from the experience of communities and individuals that the accumulation of local knowledge.

The prevention of earthquake and tsunami disaster based on local wisdom in Simeulue community originated from earthquake and tsunami experience that occurred in 1907. Based on the Dutch record and from the story of some parents Simeulue mention the incident with the term *smong07* consists of word *smong* means wave and *07* means 1907. The story can be seen as starting from the earthquake of Magnitude 7.6 that hit off Aceh on Friday, January 4, 1907. The quake developed a giant wave that reached the land, and more than 50% of (some estimates go up to 70%) the Simeuluean people were killed. The disaster happened on Friday when most of the people were gathering in mosques for Friday prayers.

The worst impact of the 1907 tsunami was its pain that resonated in the hearts of Simeuluean people. The survivors tried to deal with the disaster by recounting the story of Smong. When the earthquake of Mw of 9.2 occurred on December 26, 2004, the Smong story successfully alerted the Simeuluean people of the impending danger, and they ran away from the coastal areas to higher places. Three people were reportedly killed, showing the success of the *Smong* story.

Indigenous story of the earthquake and tsunami that hit Simeulue in 1907 passed from generation to generation. *Smong's* storyline is sung in poetic form that contains natural phenomena that can be a tsunami. The verse that is often sung to take a child's sleep is a poem of advice to devote to both parents, and a piece of the story about the departure of parents because of the tantalizing wave of *Smong* sea. In addition, an invitation to run to a higher place if there is a strong earthquake. So it becomes the habit of Simeuleu society to run to the mountain if it feels earth shaking earthquake.

Table 1. An example of *Smong* lyrics to tell of tsunami risk from *Nandong*.

Devayan	English
Enggel mon sao curito	Please listen to this story
Inang maso semonan	One day in the past
Manoknop sao fano	A village was sinking
Uwi lah da sesewan	That's what has been told
Unen ne alek linon	Starting with earthquakes
Fesang bakat ne mali	Followed by a giant wave
Manoknop sao hampong	The whole country was sinking
Tibo-tibo mawi	Immediately
Anga linon ne mali	If the strong earthquake is
Uwek suruik sahuli	Followed by the lowering of seawater
Maheya mihawali	Please find in a hurry
Fano me singa tenggi	A higher place
Ede smong kahanne	This is called Smong
Turiang da nenekta	A story of our ancestors.
Miredem teher ere	Please always remember
Pesan dan nafi da	This message and instruction
Smong dumek-dumek mo	Smong is your bath
Linon uwak-uwakmo	Earthquake is your swing bed
Elaik keudang-kedangmo	Thunderstorm is your music
Kilek suluh-suluhmo	Thunderlight is your lamp

The tragedy on 1907, becomes a valuable learning and experience for their generation to be more careful and vigilant. Through their grandparents and their parents, they continue to tell the heartbreaking story of a century ago. This became a valuable stock for the Simeulue community. When the tsunami in 2004, people already know what steps to take when the tsunami comes.

The ancestral message of the poem mentions that, if there is a strong earthquake, followed by low tide, do not go to the beach to pick up fish that have sprung up by the beach, because soon Smong will come. If that happens, run to the mountain to save yourself. Bring the children, the elderly, and the women running away from the beach. Shout, Smong ... Smong ... Smong .... It continues to this day.

Smong poem was able to minimize casualties in Simeulue. This hereditary poetry, which in the form of advice is a local wisdom that should be maintained and preserved considering Aceh is an area prone to earthquake and tsunami disaster. However, this Smong Poem is not enough to protect the people from the disaster. To that end, the Regional Disaster Management Agency must cooperate with relevant agencies to provide a good and accessible evacuation route.

4. Conclusion

Smong is a key word understood by the entire population of Simeulue that describe the occurrence of giant waves after a major earthquake. It is part of the Simeulue indigenous culture, transmitted through songs, short poems, lullabies, and stories. During the terrible event that plagued Aceh on December 26, 2004, there was a massive evacuation of the entire Simeulue beach area within a few minutes after the earthquake. Therefore, "*Smong*" is an appropriate term that used in order to reduce the impact of disaster, viz. earthquakes and tsunamis in high risk areas.

Stories or poems like *smong* is always repeated told will always be remembered because the story or poem can be easily accepted by the public. In disaster risk areas other than earthquakes and tsunamis, stories or poems can be made based on true stories that occur in the area and make them as local wisdom. Learn from the *smong* story, can be used as a reference knowledge about disaster and policy. So that can be one of help to reduce disaster risk in disaster prone area.

5. Acknowledgement
We would like to express our sincerest thanks to Samudra University who supported us and contributed during this study.

6. References
[1] Baumwoll J 2008 The value of indigenous knowledge for disaster risk reduction: A Unique Assessment Tool for Reducing Community Vulnerability to Natural Disasters, Master Thesis, Vienna: Webster University.

[2] Dekens J 2007 Local Knowledge for Disaster Preparedness: A Literature Review. International Centre for Integrated Mountain Development (ICIMOD). Kathmandu, Nepal.

[3] Kanamori H Rivera L Lee WHK 2010 Historical seismograms for unravelling a mysterious earthquake: The 1907 Sumatra Earthquake. Geophysical Journal International 2010; 183(1): 358–374.

[4] McAdoo BG Dengler L Prasetya G Titov V 2006 Smong: How an Oral History Saved Thousands on Indonesia's Simeulue Island during the December 2004 and March 2005 Tsunamis. Earthquake Spectra 2006; 22 (S3): S661–69.

[5] Mercer J I Kelman L Taranis Suchet-Pearson S (n.d.) Framework for Integrating Indigenous and Scientific Knowledge for Disaster Risk Reduction. Disasters 201034 (1): 214–39.

[6] Mercer J D Dominey-Howes I Kelman Lloyd K 2007 The Potential for Combining Indigenous and Western Knowledge in Reducing Vulnerability to Environmental Hazards in Small Island Developing States. *Environmental Hazards 2007*; 7 (4): 245–56.

[7] Meyers K (n.d.) Simeulue, Nias and Siberut: Indigenous Practices and Lessons Learned. In Indigenous Practices and Lessons Learned for Disaster Risk Reduction. UN/ISDR and Kyoto University, (Bangkok, forthcoming).

[8] Shaw R Uy N Baumwoll J 2008 Indigenous Knowledge for Disaster Risk Reduction: Good Practices and Lessons Learned from Experiences in the Asia-Pacific Region. ISDR; 2008.

[9] Swafwina 2014 Recognizing Indigenous Knowledge for Disaster Management: Smong, Early Warning System from Simeulue Island, Aceh. Procedia Environmental Sciences 20 (2014) 573 – 582.

[10] Walshe R A Nunn P D 2012 Integration of Indigenous Knowledge and Disaster Risk Reduction: A Case Study from BaieMartelli, Pentecost Island, Vanuatu. International Journal of Disaster Risk Sciences 2012; 3 (4): 185–194.

[11] Warren D M Rajasekaran B 1993 Putting local knowledge to good use. International Agricultural Development 13(4): 8-10.

[12] Yogaswara H Yulianto E 2005 Smong: *Local Knowledge and Strategies on Tsunami Preparedness in Simeulue Island, Nanggroe Aceh Darussalam*. UNESCO and LIPI, Jakarta.

[13] McNeill L C Dugan B Petronotis K E and the Expedition 362 Scientists *Proceedings of the International Ocean Discovery Program*, Volume 362 publications.iodp.org

[14] Rahman A Sakurai A Munadi K 2017 The Analysis of the Development of the Smong Story on 1907 and 2004 Indian Ocean Tsunamis in Strengthening the Simeulue Island Community's Resilence. International Journal of Disaster Risk Reduction.

IOP Conf. Series: Earth and Environmental Science **148** (2018) 012006 doi:10.1088/1755-1315/148/1/012006

Cloud structure evolution of heavy rain events from the East-West Pacific Ocean: a combined global observation analysis

A B Sekaranom[1,2], E Nurjani[1] and I Pujiastuti[1]

[1]Department of Environmental Geography, Faculty of Geography, Universitas Gadjah Mada, Indonesia.
[2]Institute of Space-Earth Environmental Sciences, Nagoya University. Japan.

andungbayu@geo.ugm.ac.id

Abstract. Heavy rain events are often associated with flood hazards as one of the most devastating events across the globe. It is therefore essential to identify the evolution of heavy rainfall cloud structures, primarily from global satellite observation, as a tool to provide better disaster early warning systems. To identify the mechanism of heavy rainfall systems and its relationship with cloud development, especially over The Pacific Ocean, we aim to study the westward evolution of the convective systems over this area. Several datasets from Tropical Rainfall Measuring Mission (TRMM), CloudSat GEOPROF product, and ECMWF-reanalysis (ERA) interim were utilized to characterize the evolution. Geolocation and orbital time-lag analysis of the three different datasets for more than 8 years (2006-2014) could provide information related to the evolution of cloud structures associated with heavy rain events. In the first step, a heavy rainfall database was generated from TRMM. The CloudSat coordinate and time position were then matched with TRMM coordinate and time position. All of the processes were programatically conducted in fortran programming language. The result shows a transition between East and West Pacific ocean for TMI data.

1. Introduction

Indonesia is one of the countries where a large number of flood hazards occur each year [1]. The hazards are known to be not only generated by land surface characteristics but also due to the influence of atmospheric circulation. The area surrounding Indonesia, which is also known as The Maritime Continent [2], is an area adjacent to two main warm pools over the global tropics, namely the Pacific Ocean and the Indian Ocean warm pools. Although the warm pools become the source region in term of moisture, high precipitation amounts are found over islands surrounding the warm pools as generated by the influence of the land-sea breeze that transport moisture from the ocean to the land [3]. More interestingly, heavy rainfall over this region does not require a significant cloud top height to reach the "extreme" rain rate, but rather occur at a lower top height due to the humid environment at the lower troposphere [4].

One of the region that has a close relationship with rainfall in Indonesia is The Tropical Pacific Ocean, which has a significant role in transporting energy and moisture [5] in the East-West direction. The interaction occurs in the form of an atmospheric circulation which is known as Walker Circulation [6]. In a more specific, the circulation is called as The Southern Oscillation [7]. A significant rainfall increase over Indonesia occurs during The Southern Oscillation cold phase, where Sea Surface Temperature (SST) over East Pacific become colder than its normally observed. This condition triggers stronger westward circulation at the lower troposphere, mainly from The East Pacific to The West Pacific and reversed during the warm phase [8]. As a result, in the cold phase, increasing precipitation occurs mainly at the peak of the rainy season, particularly around December to February [9,10].

Although the atmospheric dynamics over the Pacific Ocean, including Indonesia, has been well understood, including its long-term trends [11], the differences related to the extreme precipitation development are still unresolved. Based on previous researchers, it seems that higher SST over the Pacific warm pool produces more rain over West Pacific (including Indonesia), but less from the East to the Central Pacific. It is therefore essential to identify its dynamics, especially in terms of extreme precipitating cloud structures and environment stability from the East to the West. This paper aims to characterize the difference of extreme precipitating cloud developments from The East to The West Tropical Pacific Ocean, and further explains the differences in terms of environmental stability.

2. Research Methods

The area domains, data, and methods of this research are described in this section.

2.1. Spatial and temporal domain

The tropical area of 15°N-15°S over the Pacific Ocean was selected as a domain in this analysis, which is divided into three regions. The regions consist of Tropical East Pacific Ocean/EastPac (90°W-130°W), Central Pacific Ocean/CentralPac (130°W-180°W), and West Pacific Ocean/WestPac (90°E-180°E). The EastPac Region covers the area surrounding the West part of North and South America. The CentralPac region covers a wide area in the middle of Pacific, while the WestPac is including Western part of Pacific Ocean and The Maritime Continent. The separation of the three regions ensures the transition between the East-West circulations over The Pacific Ocean. The time domain was selected to match the availability of the three datasets from 2006 to 2014.

2.2. Data

Three datasets were used in this research, contains precipitation data from Tropical Rainfall Measuring Mission/TRMM [12], radar reflectivity from CloudSat [13], and Environmental profile from ECMWF Reanalysis (ERA) interim [14]. The population of instantaneous rain-rate from TRMM TMI 2A12 was selected as the time base of the analysis. The TMI surface flag was selected to mask out the land and coast surfaces. 94 GHz radar reflectivity data from CloudSat GEOPROF was selected to characterize precipitating clouds over the area of interest. The ERA-interim data were further used to characterize the environmental profile of heavy precipitation among the three regions. The air temperature, specific humidity, relative humidity, and vertical velocity profiles were obtained to calculate Convective Available Potential Energy (CAPE). Detailed information about the datasets used in this analysis is shown in table 1.

Table 1. Datasets used in this research.

Platforms	Instruments	Description	Resolution	Target parameters
TRMM	TMI	Passive Microwave Imager	4.4 km (high-resolution channels) and 8.8 km (low-resolution channels)	Near-surface rain rate (TMI 2A12), TMI surface flag (TMI 2A12)
CloudSat	CPR	94-GHz radar	1.4 km × 1.7 km	Radar reflectivity (2B GEOPROF), CloudClass (2B-CLDCLASS)
ERA-Interim	-	ECMWF Reanalysis Interim	Gridded 0.75° × 0.75°	Air temperature, specific humidity, relative humidity, and vertical velocity profiles

2.3. Methods

The TRMM contains active and passive sensors with the primary function to estimate precipitation over earth surface. The data are available from 1998-2014 with a fine spatial resolution and 46 days revisiting time. This means that TRMM contains long-term precipitation data with extensive records.

However, due to a board range of data recorded by TRMM, in this research the surface precipitation data from TRMM passive microwave imager (TMI) were utilized. The TMI dataset is selected in this research since passive microwave (PMW) data are widely used by many satellites, in contrast to the narrow active radar sensor. TMI data contains instantaneous rain-rate data which is estimated from 9 Passive Microwave Channels with 4.4 km^2 resolution. Gridded precipitation data with 0.25°×0.25° resolution was generated by using this data to produce easier match-up with the CloudSat and ERA-interim data. In this research, extreme rain events are defined as the uppermost 10% rain-rate from the TMI gridded surface rain-rate population from 2006-2014. The extreme rain database was generated using this threshold, contains information about the instantaneous rain-rate, location, and time-stamp of each data.

The CloudSat contains Cloud Profiling Radar (CPR) operated in 94 GHz and stating operated from 2006. The CPR sensor produces a finer resolution than TMI (1.7 km^2), but only contain a single profile over its track. CloudSat also has a sun-synchronous orbit, measuring cloud profiles at around 2 pm local time, which is distinctly different from TMI. To match the CloudSat profiles with extreme rain identified by TMI, the corresponding CloudSat reflectivity profile are determined within the TRMM gridded 0.25°×0.25° degree resolution. The TRMM-CloudSat time differences in observing the earth surface can be beneficial to explain the cloud structure evolution from time to time [15]. A cloud fraction composite was generated to explain the development of convective clouds over time and arranged to the relative time compared to the maximum surface precipitation identified by TRMM. The highest CloudSat reflectivity within the extreme precipitating grid is selected as the coordinate center and binned within each ±1.5 hours difference. The composite cloud profiles are generated by combining data in each spatial and temporal bin. Figure 1 explains detailed illustration how composite CloudSat profile could be generated using the TRMM-CloudSat time differences. The procedure is similar to the ERA-interim data, except that the ERA-interim data records have 6 hours interval (00h, 06h, 12h, and 18h UTC) for each 0.75°×0.75° degree grid.

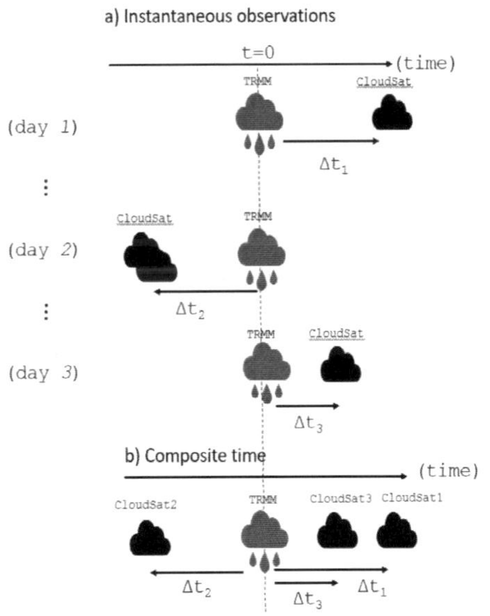

Figure 1. Construction of composite CloudSat cloud profiles using time difference with identified extreme rainfall data from TRMM TMI.

The composite cloud fractions were calculated to represent the cloud dimensions (width, height, and shape) accompanying extreme rain events using the CloudSat reflectivity data. In this analysis, the cloud profiles were combined in each horizontal distance bin and vertical height bin within 6 hours before and after the heavy rainfall events detected by the TMI. To represent the difference among the three regions in terms of environmental stability, Convective Available Potential Energy was calculated by utilizing the ERA-interim environmental profile. The following equation was utilized;

$$CAPE = \int_{zf}^{zn} g\left(\frac{Tvp - Tve}{Tve}\right) dz \quad (1)$$

where zf is the height of the level of free convection and zn is the height of the equilibrium level (neutral buoyancy). Tvp is the virtual temperature of the specific parcel, where Tve is the virtual temperature of the environment, and g is the acceleration due to gravity.

3. Result and Discussion

The results of this research are explained in terms of the number of deep convective events, convective cloud structures, and its corresponding environmental profiles.

3.1. CloudSat deep convective events

Since heavy rainfall events are often associated with deep convective clouds, it is important to observe the occurrence of deep convective events over the global area, particularly over The Tropical Pacific Ocean. In this analysis, the CloudSat cloud-class product was combined into 0.25°×0.25° degree grid from each orbit from 2006-2014. The result of this analysis is shown in Figure 2. Based on the figure, an enormous amount of deep convective clouds has been identified across the global tropics, reaching up to 2000 observed events in total.

Although the deep convective clouds were accumulated in the tropics, the distribution of deep convective clouds itself varies among the area, especially between 15°N-15°S. From the figure, a substantial number of deep convective clouds found over Indonesia and South America. The existence of deep convective clouds over Indonesia is related to the existence of two warm pools, consisting The Pacific Ocean Warm Pool and The Indian Ocean Warm Pool. Over the Pacific Ocean, most of the deep convective clouds found at this warm pool close to Indonesia, but less from the East and Central Pacific.

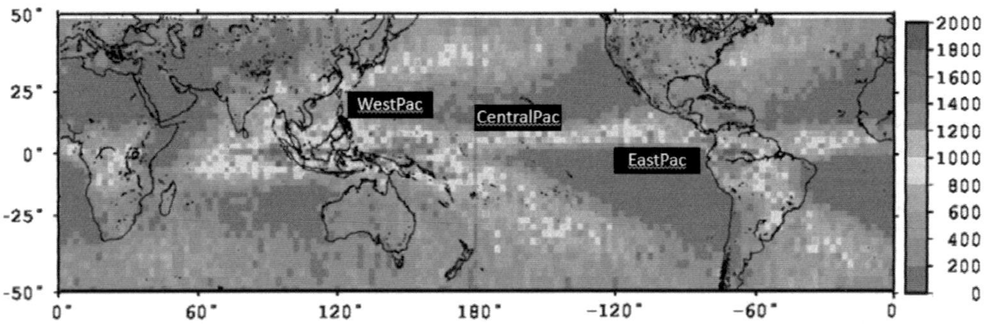

Figure 2. Number of global deep convective events obtained from CloudSat.

3.2. CloudSat convective cloud structure

Several example snapshots of extreme precipitating cloud obtained from geolocated CloudSat data is shown in figure 3. The figure shows the raw reflectivity data from CloudSat profiles, which is a range from -40 dBZ to 20 dBZ. It can be observed from the figure that heavy precipitating clouds occur with considerable variation in width, from 10 to 200 km, and could reach higher than 15 km in height.

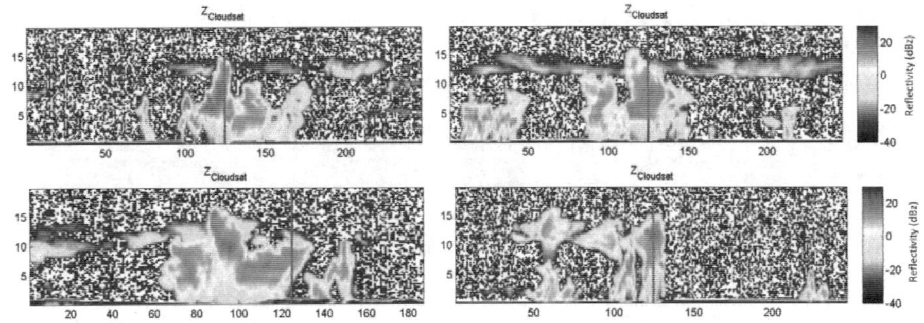

Figure 3. Example snapshot of extreme precipitating cloud from CloudSat profile data. The horizontal annotation represents distance, and the vertical annotation represents height.

To produce composite cloud fractions, highest CloudSat reflectivity and cloud-mask data corresponding to heavy precipitation grid from TMI rain events were acquired (figure 4a and 4b). The process to obtain the highest CloudSat reflectivity is intended to ensure that the convective center is positioned in the middle. Using the cloud-mask data, the cloud exterior boundaries were delineated, and each consecutive cloud cluster was labeled as a single cloud (figure 4c). Clouds that are not located at the center coordinate were masked out (figure 4d). Cloud fractions are then determined by the sum of cloud occurrence from each profile then normalized to the 0-1 range by dividing it by the maximum occurrence.

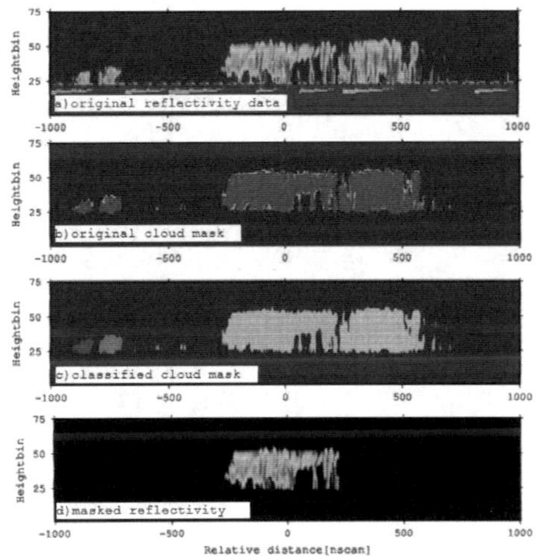

Figure 4. Cloudsat reflectivity data processing to filter out irrelevant cloud surrounding heavy rainfall and produce composite cloud fraction.

The result of composite cloud fraction is shown in figure 5. The figure represents the dimensional difference between Tropical West Pacific Ocean, Central Pacific Ocean, and East Pacific Ocean. The plot represents the development of heavy precipitating cloud from 3 hours before (top) to 3 hours after (bottom) the heavy precipitation occurs within ± 1.5 hours bin. The figure shows that heavy rainfall events over East Pacific Ocean are generated by shallower clouds with cloud top height about 7-8 km. Over Central Pacific Ocean, the cloud top heights reach up to 10 km, while over West Pacific Ocean, the height reaches to about 12 km. From the figure, we also could observe the cloud development process from the smallest at -3 hours bin to the highest at the +3 hours bin. However, extreme recipitation events have narrower width compared to the +3 hours bin.

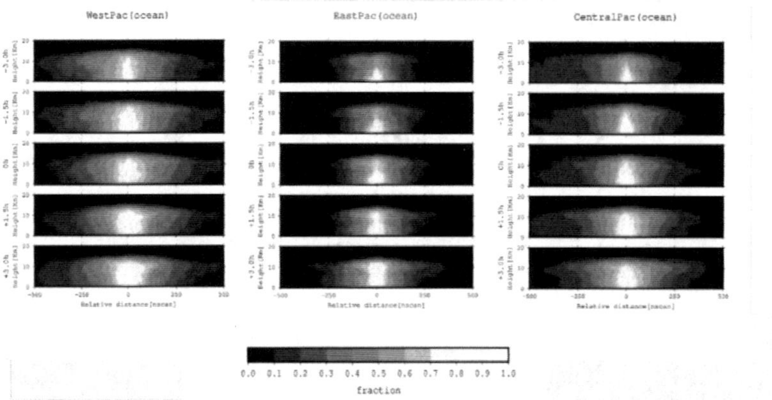

Figure 5. CloudSat convective cloud fractions from a) Tropical West Pacific Ocean, b) Tropical Central Pacific Ocean, and c) Tropical East Pacific Ocean from -3 hours to +3 hours relative to heaviest surface precipitation. The brighter color in the center represents significant cloud fraction above 70% of the total fraction.

3.3. Environmental Profile corresponding to heavy precipitation

The cloud structure differences between The East, Central, and West Pacific Ocean are further investigated using ERA-interim at the nearest grid position to the TRMM extreme precipitation grid. The initial comparison has been calculated from the vertical profile of air temperature, relative humidity, specific humidity, and vertical velocity data. The result shows a clear difference observed from the vertical velocity anomaly data (figure 6). The result shows that upward vertical velocity anomaly over The West Pacific Ocean reaches higher altitudes than the East and Central Pacific anomalies. This result could explain why the clouds growth larger over The West Pacific compared to the East Pacific in case of heavy rain events.

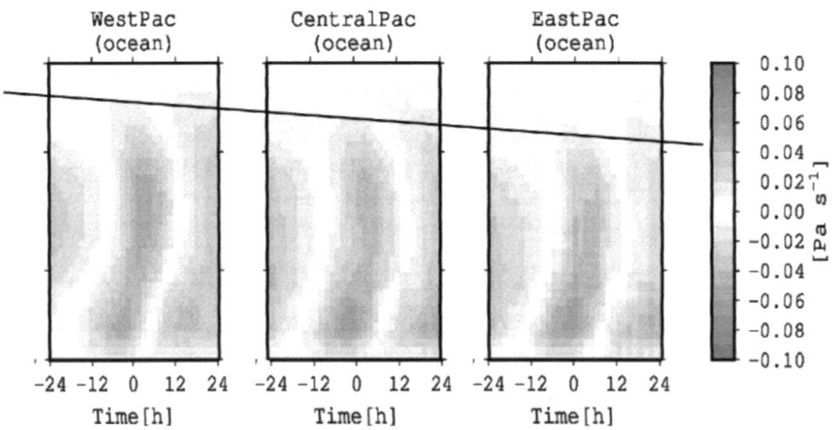

Figure 6. Vertical velocity profile anomalies from ERA-interim data corresponding to the heavy surface rain at the time center from West to East Pacific Ocean (left to right). The diagonal black line represents the difference in the height of -0.01 Pa/s anomalies.

IOP Conf. Series: Earth and Environmental Science **148** (2018) 012006 doi:10.1088/1755-1315/148/1/012006

The taller heavy precipitating clouds over West Pacific Ocean could be explained in term of CAPE. The result of CAPE differences among the West to East Pacific Ocean is shown in Figure 7. Based on this figure, large accumulation of the potential energy occurs 12 hours before to 6 hours before heavy precipitation events. The potential energy are then began to be consumed by the convective process from 6 hours before and after heavy precipitation events. Although the pattern is similar, The Western Pacific Ocean, in general, has a larger potential energy than the Central and East Pacific Ocean. The larger potential energy contributes to the stronger convective process in The West Pacific.

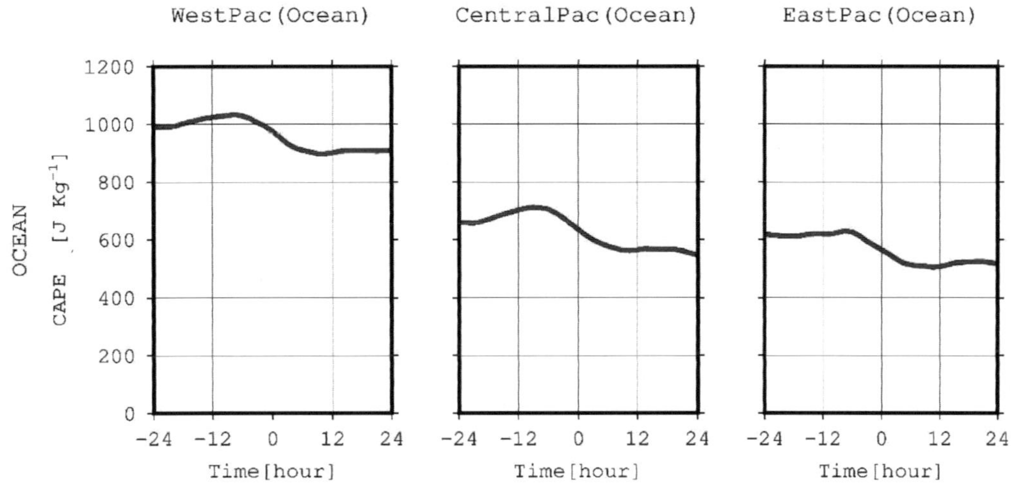

Figure 7. Temporal changes of the CAPE calculated from the ERA-interim dataset corresponding to the TMI heavy rain-rate at the time center from West to East Pacific Ocean (left to right).

4. Conclusion

In terms of hydrometeorological disaster, the interaction of atmosphere-land relationship could not be neglected. Flood hazards are known to have a strong relationship with heavy rainfall events. In the case of The West Pacific, including Indonesia, the deep convective clouds as associated with heavy rainfall are the most dominant, and possibly related to the high number of flood events in this area. The heavy precipitation over The West Pacific Ocean domain is associated with higher and taller clouds compared to the Central Pacific and West Pacific.

The higher CAPE generates the larger and taller extreme precipitating clouds over this region, which in turn, converted to a higher vertical velocity that contains larger moisture content to be transported to the upper part of the troposphere. The larger amount of moistures transported to the upper troposphere becomes the main factor in generating heavy precipitation over the West Pacific Ocean.

It is also worth to note that in this research, the only precipitation dataset is from the TRMM TMI. We select the TMI data since the PMW estimation is common and applicable not only for TRMM. However, in this case, the accuracy compared to the PR or ground observations are not conducted. A significant difference might exist when different precipitation datasets are included. Although comparison of PR-TMI data conducted by Sekaranom and Masunaga [4] showed that that PR-TMI differences over the ocean are less apparent than the land in The Maritime Continent, the differences might still exist and become a subject for the further research.

5. Acknowledgement

This research is funded by Lembaga Pengelola Dana Pendidikan (LPDP) Ministry of Finance Republic of Indonesia, with Contract PRJ-1875/LPDP/2014.

6. References

[1] Marfai M A Sekaranom A B and Ward P 2015 Community responses and adaptation strategies toward flood hazard in Jakarta, Indonesia *Natural hazards* 75 pp 1127-44

[2] Ramage C S 1968 Role of a tropical Maritime Continent in the atmospheric circulation *Mon Wea Rev* 96 pp 365-70

[3] Qian J H 2008 Why precipitation is mostly concentrated over islands in the Maritime Continent *Journal of the Atmospheric Sciences* 65 pp 1428-41

[4] Sekaranom A B and Masunaga H 2017 Comparison of TRMM-Derived Rainfall Products for General and Extreme Rains over the Maritime Continent *Journal of Applied Meteorology and Climatology* 56 pp 1867-81

[5] Alexander M A Blade I Newman M Lanzante J R Lau N C and Scott J D 2002 The atmospheric bridge: The influence of ENSO teleconnections on air-sea interaction over the global oceans *Journal of Climate* 15 pp 2205-31

[6] Julian P R and Chervin R M 1978 A study of the Southern Oscillation and Walker Circulation phenomenon *Monthly Weather Review* 106 pp 1433-14

[7] McBride J L Haylock M R and Nicholls N 2003 Relationships between the Maritime Continent heat source and the El Niño-Southern Oscillation phenomenon *Journal of Climate* 16 pp 2905-14

[8] Deser C and Wallace J M 1990 Large-scale atmospheric circulation features of warm and cold episodes in The Tropical Pacific *Journal of Climate* 3 pp 1254-81

[9] Ropelewski C F and Halpert M S 1989 Precipitation patterns associated with the high index phase of the Southern Oscillation *Journal of climate* 2 pp 268-84

[10] Chang C P Wang Z Ju J and Li T 2004 On the relationship between Western Maritime Continent monsoon rainfall and ENSO during northern winter *Journal of Climate* 17 pp 665-72

[11] Vecchi G A Soden B J Wittenberg A T Held I M Leetmaa A and Harrison M J 2006 Weakening of tropical Pacific atmospheric circulation due to anthropogenic forcing *Nature* 441 pp 73-76

[12] Simpson J Kummerow C Tao W K and Adler R F 1996 On the Tropical Rainfall Measuring Mission (TRMM) *Meteorology and Atmospheric physics* 60 pp 19-36

[13] Stephens G L Vane D G Boain R J Mace G G Sassen K Wang Z Illingworth A J O'Connor E J Rossow W B and Durden S L 2002 The CloudSat mission and the A-Train: A new dimension of space-based observations of clouds and precipitation *Bulletin of the American Meteorological Society* 83 pp 1771-90

[14] Dee D P Uppala S M Simmons A J Berrisford P Poli P Kobayashi S Andrae U Balmaseda M A Balsamo G and Bauer P 2011 The ERA-Interim reanalysis: Configuration and performance of the data assimilation system *Quarterly Journal of the royal meteorological society* 137 pp 553-97

[15] Masunaga H 2012 A satellite study of the atmospheric forcing and response to moist convection over tropical and subtropical oceans *Journal of the Atmospheric Sciences* 69 pp 150-167

IOP Conf. Series: Earth and Environmental Science **148** (2018) 012007 doi:10.1088/1755-1315/148/1/012007

Population condition analysis of Jakarta land deformation area

R F Putri[1], S Wibirama[2], Sukamdi[1] and S R Giyarsih[1]

[1]Department of Environmental Geography, Faculty of Geography, Universitas Gadjah Mada, Indonesia
[2]Department of Electrical Engineering and Information Technology, Faculty of Engineering, Universitas Gadjah Mada, Indonesia

ratihfitria.putri@ugm.ac.id

Abstract. Jakarta is located in the North West area of West Java Province which geographically positioned on 106°33'00"-107°00'00"BT and 5°48'30"-6°24'00"LS. Land subsidence has occured in several types of landuse such as trade, industrial and settlement area of the urban area of Jakarta. The land subsidence disaster is one of the consequences of building and road construction in Jakarta. This is caused by massive groundwater utilization and failure in landuse planning. This study aim to analyze the population density and settlement pattern in the urban area of Jakarta which the occurence of land subsidence has been detected. It is important to understand landuse and settlement planning processes in the area which land subsidence occured. Detection of land subsidence distribution become a necessary parameter in landuse planning. While the land subsidence area detected using Differential Synthetic Aperture Radar (DInSAR) method. The result shows the area which land subsidence occured has a very high population density and clustered and linear settlement pattern. This area is mainly used as industrial, trade, and settlement.

1. Introduction

Special capital district of Jakarta is located geographically between 106°33'00"-107°00'00"BT dan 5°48'30"-6°24'00"LS. According to the topographic characteristic of Jakarta, it shows Jakarta has a flat relief in the north and gentle slope in the south with slope between 0 and 2. The south part of Jakarta has a mean elevation of 60 above sea level, while the north part about 20 above sea level. The topographic characteristic map of Jakarta (figure 1). This sediment tends to be younger than the river sediment.

The study area has a tropical climate with average annual precipitation of 1.755 mm or about 146 mm monthly. Jakarta increased in population very fast as the city develops. According to the 2011 population census, the population density of Jakarta reached 15,400 for each square kilometer. In six districts in Jakarta, population density ranged from 2.112 to 23.312 inhabitants per square kilometer.

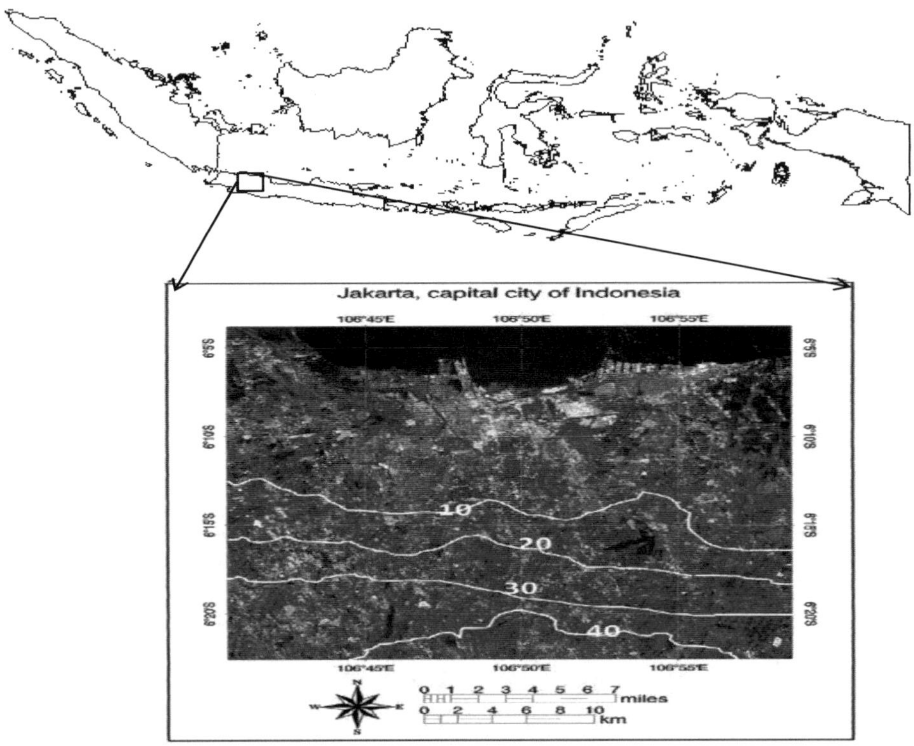

Figure 1. Intensity map of Jakarta City.

This phenomenon of land subsidence is one of consequences of urban development in Jakarta. In 1926, a Dutch surveyor estimated the land level lowering in the center of Jakarta city. The measurement using leveling method has been done since 1982 to 1999 by the scientist. The measurement of lowering land level using GPS has also been held at various point. The study of land deformation using radar image satellite data has begun since 2010 [1][2]. This study will apply DinSAR method to determine land subsidence area with TerraSAR-X data.

Jakarta is one of the biggest city in Indonesia and plays an important role in Indonesia central government and economic activities. The characteristics of Jakarta often identified as high population density, high economic activities, and center of the government. According to the statistical bureau of Indonesia, per capita income of Jakarta is about four times average national per capita income. Jakarta role as the center of governance and economy in Indonesia has become a traction for people to come and settle. On the other side, Jakarta constantly declines in physical geographic condition. The flood in Jakarta city was occure in the rainy season intensively. At the same, coastal flood occur in the north part of the city. Land subsidence has been suspected as causes of flooding in Jakarta area. The land subsidence in Jakarta caused by the increase in population density and settlement area [3][4]. Hence, the aim of this study is to analyze population density and settlement pattern in Jakarta, particularly in the area where the occurrence of land subsidence detected.

Population density is proportion between specific number of population and area. Population density has a positive correlation to carrying capacity of a certain area. Population and economic growth constantly require space expansion. On the other side, space is stagnant but the composition, both in physical, economic, and social, is dynamic as the space utilization change. The change of space utilization without considering the geobiophysic factors may cause man-made disaster and space usage efficiency may be in vain.Jakarta is one of the biggest city in Indonesia with 10,187,595 population in 2011 [5]. Jakarta, as a capital city of Indonesia, is a province which plays an important role in both economic and governance. Urban development followed by increase in population in Jakarta become a trigger to environmental problem in the last decade [6]. According to UU No. 1 -

2011, settlement is a part of settled environment which consist of more than one house provided with public infrastructure. Settlement also has other function to support activities in both urban and rural area.Settlement is a primary need in human life. It ranked third in human needs (foods, clothing, settlement, health, and education). It may be concluded that settlement is has a central role in human needs. Hence, the improvement of settlement quality will also improve living quality standard.Nowadays, settlement is not only to provide safety, but also include drinking water provision, lighting, transportation, education, etc.

Land subsidence has become an attention to both government and citizen in some big cities in Indonesia. This problem mostly caused by factors such as exploitation of groundwater exceeding its capacity and natural condition where the land become compressed [7]. Land subsidence may trigger other problems such as change in topographic gradient, land surface damage, and decrease of groundwater aquifer capacity. In Jakarta, land subsidence becomes an important factor which cause flood in January, 2003 [8]. Thus, detection and analysis of land subsidence in Jakarta become important to support flood mitigation so that Jakarta has a better development planning.This land subsidence phenomenon, and also monitoring of landuse change in Indonesia, have already been studied using ALOS PALSAR data with 10 m spatial resolution [9]. This study was using TerraSarX which provide higher spatial resolution of 1-3 meters; DLR Cluster Applied Remote Sensing, 2006. TerraSAR-X is using short wave at 3.1 cm and having highest spatial resolution between all SAR data. This research is aimed to understand the population density in the subsided land area in Jakarta, the analysis of the settlement and landuse distribution pattern in land deformation area, and create recommendation of spatial planning in the land subsidence area in Jakarta

2. Methods

This study is using a descriptive statistic as illustrating the object of study through actual sample or population data. Using this method, this study will describe characteristic of study area according to the actual condition, particularly the population density and settlement pattern. Hence, it can identify and analyze the correlation between land subsidence to settlement pattern and population density

2.1. Population density

Population density in the land subsidence area calculated according to number of population recorded in the Indonesia Bureau of Statistics. Residential area calculated according to satellite imagery interpretation in the land subsidence area.

Prediction of demographic data requires high resolution satellite imagery, for instance high resolution aerial photography or IKONOS imagery which have spatial resolution of 1 meter. This resolution means terrestrial object with size of 1 meter or bigger will be able to be shown in the imagery. If we can detect individual settlement, so we can count the numbers of resident in each house according to the population density information. The formula to calculate population density is mentioned below:

$$\text{Population Density} = \text{Number of Population/Area} \tag{1}$$

2.2. Land subsidence

Differential Interferometry Synthetic Aperture Radar (DInSAR) is a technique in satellite imagery data processing to know the lowering of land surface elevation by calculating accuracy and resolution. Benefit of this method is able to detect a wide area, whereas GPS method unable to cover research area.

The first step to do in DInSAR method is analysing local topographic condition using InSAR. Land elevation difference calculated by comparing two Synthetic Aperture Radar imagery with temporally different acquisition time. The difference between land elevation detected by comparison of temporally different images. Both radar data overlaid point to point in each location to create interferogram map which indicates lowering land surface elevation. Phase difference of the result of interferogram calculated using equation mentioned below [10] :

$$\phi_{\mathrm{Int}} = \phi_{\mathrm{disp}} + \phi_{\mathrm{topo}} + \phi_{\mathrm{atm}} + \phi_{\mathrm{noise}} + \phi_{\mathrm{flat}}, \quad (2)$$

Where $_{\mathrm{disp}}$ is lowering land surface phase, ϕ_{topo} is changing in elevation according to topography, ϕ_{atm} according to atmospheric effect, ϕ_{noise} is noise effect of the device, ϕ_{flat} is error associated phase by assumption the earth is a flat surface. Extraction process of phase difference on land subsidence, topographic, earth surface flat can be eliminated using accurate DEM data. This study usedTerraSAR-X DInSAR of the years of 2010-2015. Validation process calculated using analysis of statistic calculation result between DInSAR data and GPS data. The result of this study is to analysis of population density in the land subsidence area. The flowchart of research design (figure 2).

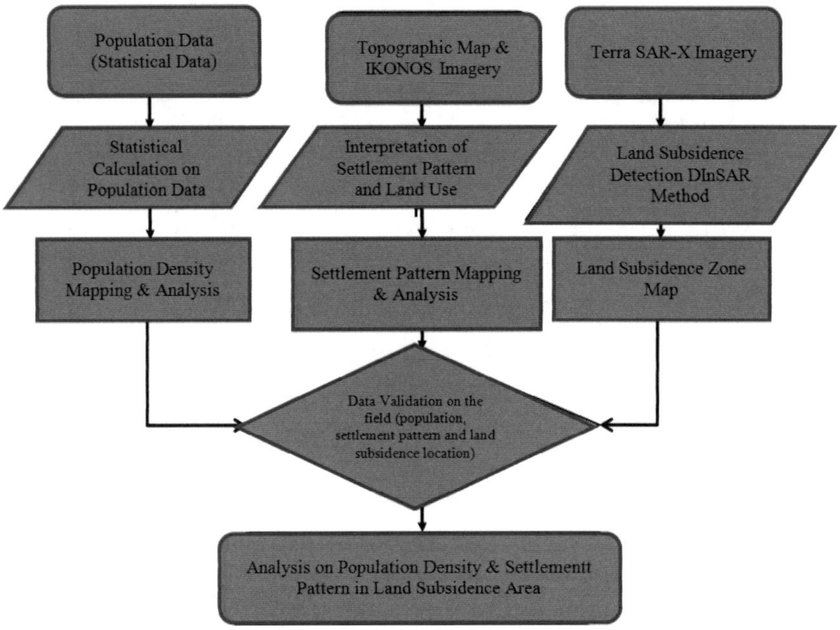

Figure 2. Research flowchart.

3. Result and Discussion

3.1. Land Subsidence Analysis

Observation on land subsidence in Jakarta focuses on three observation point, which is PantaiMutiara (P1), Cengkareng (P2), and Cakung (P3). Result of TerraSAR-X imagery data processing shows that the most severe land subsidence occurs in cengkareng (P2) with 5.6 Km^2 area and 17.5 cm/years. table 1 shows measurement result of land subsidence.

Table 1. Land Deformation Measurement Result.

Point	Location	Landuse	Land Subsidence Area (Km^2)	Land Deformation Rate (cm/year)
P1	Pantai Mutiara	Settlement, Trade and Port	1.98	10.95
P2	Cengkareng	Settlement	5.75	17.58
P3	Cakung	Industry and Settlement	0.94	9.57

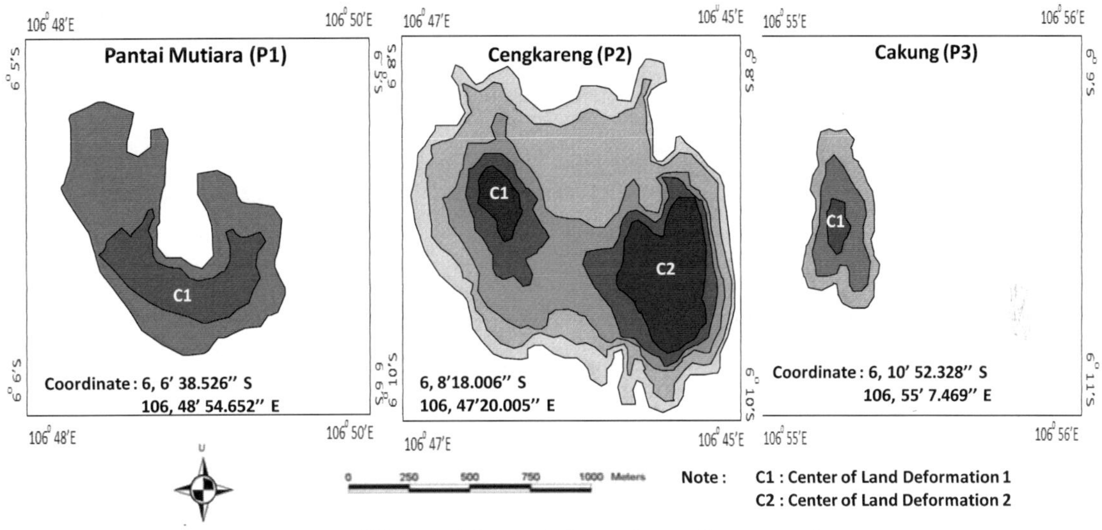

Figure 3. Land deformation pattern mapping.

The effect of land subsidence in Cengkareng (P2) is the most distinctive among the others, as shown in figure 3. Cengkareng is located in West Jakarta with an extensive area of 27.9 Km^2. The lowering land elevation surface rate in Cengkareng reaches 17.5 cm/years. This is mostly influenced by human activity in this area such as building construction and groundwater extraction. According to 2014 statistics data, the population density in Cengkareng reaches 19363 Km^2. This area is characterized with landuse of settlement (1.10 ha), Industrial area (97.25 ha), business area (328.92 ha), Agriculture (118.64 ha), Park (31.90 ha), empty land (619.12 ha) and the others (715.57 ha). Table 1 described the prediction of land subsidence rate in Cengkareng. The result is vary between 10,0 cm/years to 17.5 cm/years.

The maximum subsidence according to DInSAR calculation result is ranged from 9.5 cm/years to 17.5cm/years. Figure 3 shows the most severe subsidence occurs in Cengkareng (P2) as shown in both, GPS and DInSAR calculation. According to the government report, the land subsidence phenomenon in Jakarta mostly caused by some factor such as exploitative groundwater usage, landuse change, land consolidation and alluvial construction building.

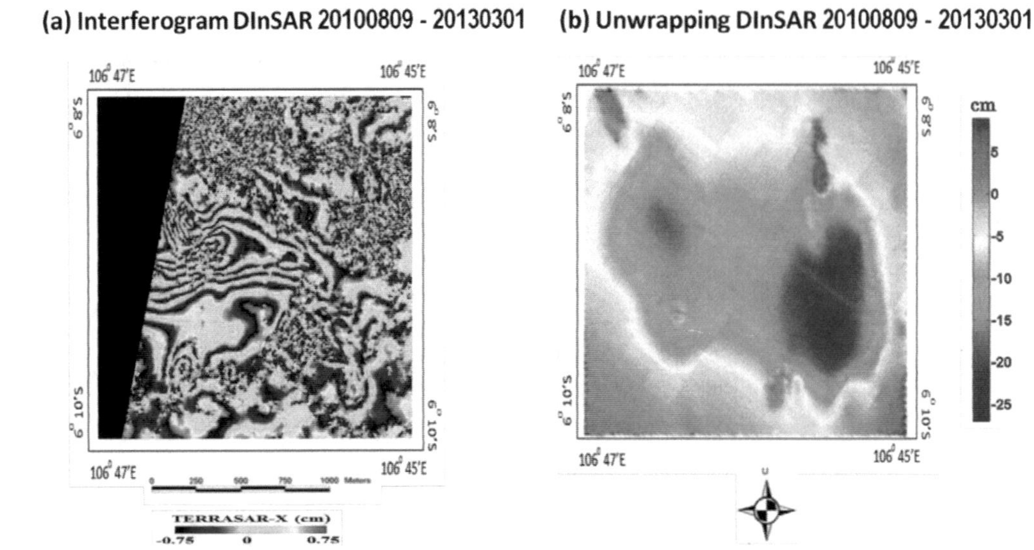

Figure 4. Land subsidence mapping.

Tabel 2. Average land deformation rate.

Pair Number	Acquisition Date: Master	Acquisition Date: Slave	Pantai Mutiara Subsidence Rate (cm/year)	Cengkareng Subsidence Rate (cm/year)	Cakung Subsidence Rate (cm/year)
1	20100809	20110613	9.5	17.5	8.8
2	20100809	20121214	8.5	11.5	9.2
3	20100809	20130301	10.5	11.7	8.3
4	20110613	20121214	9.3	10.0	7.7
5	20110613	20130301	9.9	10.5	7.5
6	20121214	20130301	10.9	11.8	9.5

DinSAR analysis results in cakung is shown in figure 4. The lowering rate of land surface elevation here is ranged 7.5-9.5 cm/years as shown in table 2. The area of subsidence in Cakung reaches 42.47 $km^{2,}$ where the landuse in cakung dominated by industrial activity. The biggest industrial area found in Pulogadung district, which established about 24.000 small and medium factories inside 6.86 km^2 area. (Jakarta Timur, 2010). According to 2010 population census, population density in cakung reaches 11.919 inhabitant/km^2. The landuse dominated by settlement (45.27%) and industrial area (24.33%).

Landuse in PantaiMutiara (P1) characterized in elite settlement, business area, harbor, and resort area tourism. Cengkareng has a landuse of settlement area with more than 30.000 households, besides business and offices area. Cakung (P3) is the most extensive industrial area in Jakarta, including an area bigger than 10.33 km^2. DInSAR calculation result for each study area, which is P1 to P3 respectively (figure 4).The phenomenon of land subsidence, as expected, caused by the forming of alluvial geologic structure. Generally, lowering of land surface elevation occur in urban area of Jakart and the estimation of land subsidence can be more accurate as temporal change. Some pairs result in noise coming from the atmosphere, but the other pairs have a distinctive result. The DInSAR result in PantaiMutiara (P1) indicate distinctive interference pattern in north west part, though the amount of the edge a bit different. The result varies between 8.5 to 10.9 cm/years. The highest rate is 10.9

cm/year which observes on December 14th, 2012 to march 1st 2013. In 2010, urban development scheme in PantaiMutiara is project funded by private institution include 30 apartment towers, 4 hotels, 5 apartment blocks and 11 hectares of waterpark. The development area is located on an offshore island in Jakarta and reclamation land begin since September of 2006. It covers areas about 0.11 km^2. Pantai Mutiara is formed by coastal area reclamation.

3.2. Population Density Analysis

According to demographic statistics of Jakarta, the population density in PantaiMutiara (P1), Cengkareng (P2), and Cakung (P3) is 8.635, 19.363, 11.919 inhabitants/km^2. Inhabitant of Jakarta increase every year. According to Indonesia Bureau of Statistic (BPS), in 2010, the population of Jakarta reaches 9.780.000. For each administrative area which is South Jakarta, East Jakarta and West Jakarta have the population of 2 million inhabitants. The requirement for population density in Central Jakarta and West Jakarta exceed 18.000 per square kilometer. The demographic composition of Jakarta dominated by productive population (15-64 years old) with proportion of 72.52% of the total population.

Theoretically, the population density of an area is proportional to land subsidence rate. The higher population density will accelerate the land subsidence. Population density is proportional to total population so that the increase in total population will directly impact the population density. An example of human activity which may induce land subsidence is groundwater overexploitation. Increase in population in an area will increase need in clean water. Indeed, the exploitation of groundwater has become a way to supply clean water.

According to 2016 statistical data Cakung divided into eight villages (urban communities). Among those urban communities, Penggilingan has the highest population density with 23.277 inhabitants per square kilometer (figure 7). While, the densest area in Penjaringan and Cengkarengsubdistrict is Penjaringan urban village with 29.868 inhabitants per square kilometer and West Cengkareng 20.217 inhabitants per square kilometers respectively. Based on the population density, those area has the highest potential for the occurrence of Land Subsidence in each subdistrict.

Table 3. Population in study area.

District	Population
Penjaringan	304,941
Cengkareng	498,130
Cakung	484,517

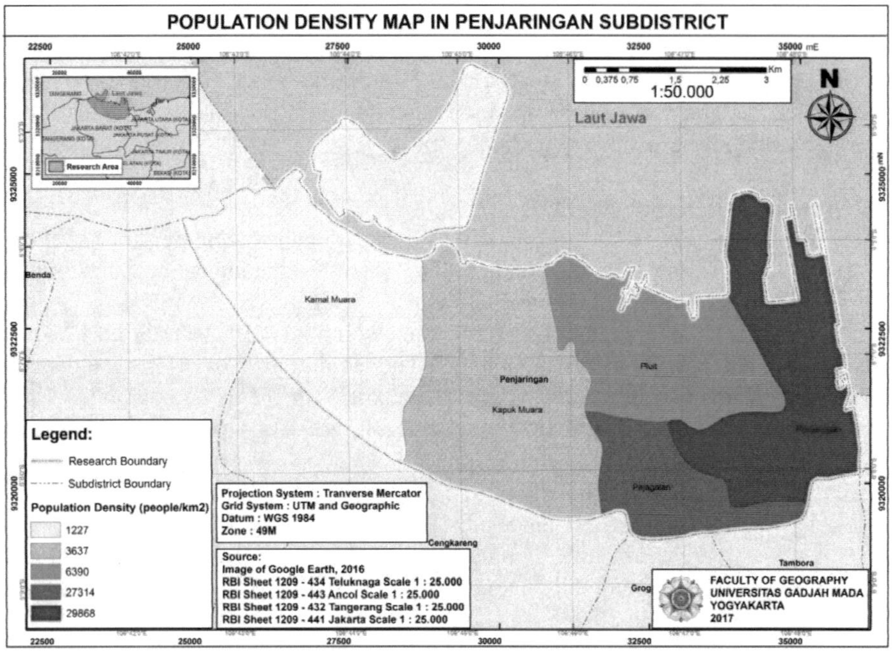

Figure 5. Population density in Penjaringan (P1).

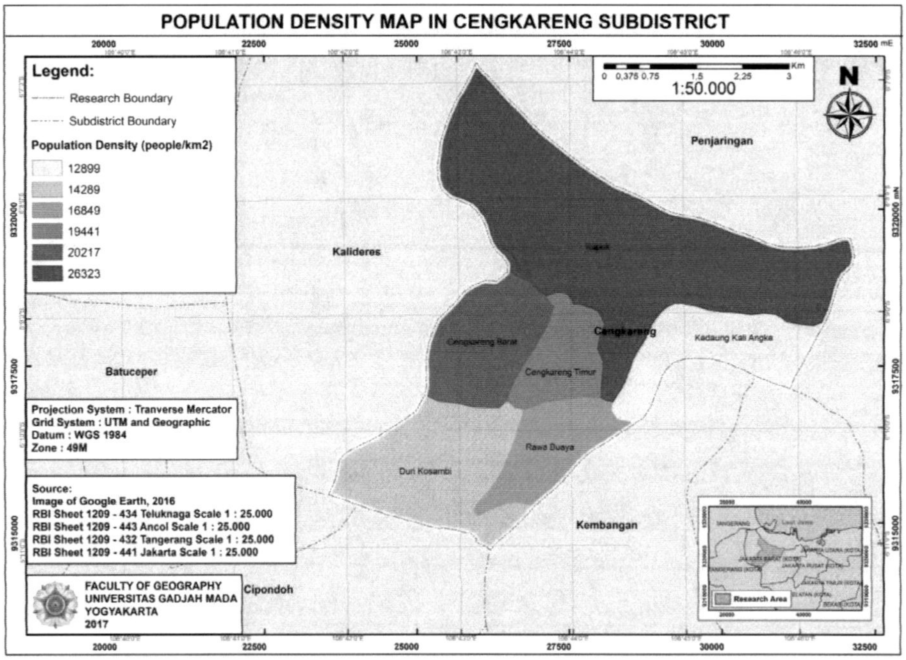

Figure 6. Population density in Cengkareng.

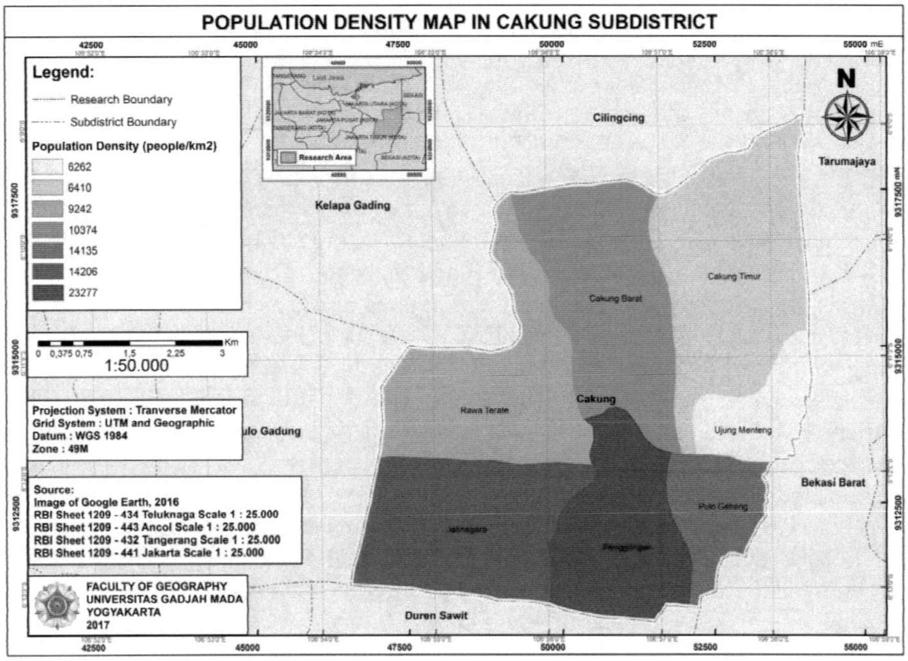

Figure 7. Population density in Cakung.

4. Conslusion

Land subsidence mapping using X-band of TerraSAR-X imagery has a higher accuracy when compared to L-band ALOS PALSAR imagery. The result of this study shows Cengkareng, Cakung, and Penjaringansubdistrict has a very high population density. Cengkareng has clustered, scattered, and linear settlement pattern cover 18.2 km^2 with the population density of 20.217 inhabitants per square kilometer. In Cakung, has the linear and clustered settlement pattern which covers 19.15 km^2 with population density of 23.277 inhabitants per square kilometer. While in Penjaringan, settlement has the clustered and linear pattern which cover 19.88 km^2 with population density of 29868 inhabitants per square. Demographic problem and environment degradation are correlated each other. Environment degradation may disrupt natural resources balance and human life in macro scale. Furthermore, imbalance between population growth and natural resources quality may destruct human civilization. Hence, we need to plan wisely to ensure the sustainability and quality of environment. This concept attempts to balance population growth and their.

5. Acknowledgement

The authors would like to thank The World Academy od Sciences UNESCO, Center for Environmental Remote Sensing Chiba University and Faculty of Geography Gadjah Mada University for the support of this paper.

6. References

[1] Putri R F Wibirama S Sukamdi and Giyarsih S R 2017 Sand dune conservation assessment in coastal area using alos palsar DInSAR technique Journal of Urban and Environmental Engineering Vol 11 (1), pp. 9-29. DOI: 10.4090/juee.2015.v11n1.9-29

[2] Putri R F Alimuddin I Sumantyo J T S and Kuze H 2013 Landslide hazard detection using alos palsar dinsar technique: Study case kayangan catchment area, Yogyakarta, Indonesia Proceeding 34th Asian Conference on Remote Sensing 2013, ACRS 2013 4, pp. 3737-3743

[3] Putri R F Wibirama S Alimuddin I Kuze H and Sumantyo J T S 2013 Monitoring and analysis of landslide hazard using dinsar technique applied to ALOS PALSAR imagery: A case study in Kayangan catchment area, Yogyakarta, Indonesia. Journal of Urban and Environmental Engineering Vol 7 (2), pp. 308-323. D0I: 10.4090/juee.2013.v7n2.308322

[4] Sri Sumantyo J T Shimada M Mathieu P P Sartohadi J and Putri R F 2011 Dinsar technique for retrieving the volume of volcanic materials erupted by Merapi volcanoInternational Geoscience and Remote Sensing Symposium (IGARSS) 6351299, IEEE International Geoscience and Remote Sensing Symposium. (Dynamics of Earth Processes and Climate Change: Geosphere Session, TU1.14: DInSAR Applications, pp. 1302-1305

[5] Dinas Kependudukan dan Pencatatan Sipil Provinsi DKI Jakarta 2011 Penyebaran Penduduk Provinsi DKI Jakarta. Available in: http://dki.kependudukancapil.go.id/index.php/jumlah-pendudk-provinsi-dki-jakarta. Accessed on 20 June 2016

[6] Anisuzzaman M D Dicky A M and Shadrina N 2013 The Role of Clay Minerals on Land Penurunan muka tanah: A Case Study in Jakarta, Indonesia Proceeding 3rd International Conference on Ecological, Environmental and Biological Sciences (ICEEBS 2013) pp. 188-191

[7] Bayuaji L Putri R F and Sri Sumantyo J T 2012 Combination of L, C and X-band SAR data for continuous monitoring of land deformation in urban area by using DInSAR technique Proceeding IEICE Technical Report SANE pp. 77-82

[8] Geospasial Badan Nasional Penanggulangan Bencana (BNPB) 2013 Peta Daerah Terdampak Banjir Available in: http://geospasial.bnpb.go.id/2013/01/18/peta-daerah-terdampak-banjir-18-01-2013-update-12-00/ Accessed on 23 May 2016

[9] Putri R F Bayuaji L Sumantyo J T S and Kuze H 2013 Terrasar-X DInSAR for land deformation detection in Jakarta Urban area, Indonesia Journal of Urban and Environmental Engineering Vol 7 (2), pp. 195-205. Doi: 10.4090/juee.2013.v7n2.195205

[10] Raucoules D Colesanti C and Carnec C 2007 Use of SAR interferometry for detecting and assessing ground subsindence.ComptesRendus Geoscience.339(5), 289–302

Vulnerability mapping in kelud volcano based on village information

D R Hisbaron[1], H Wijayanti[1], M Iffani[1], R Winastuti[1] and M Yudinugroho[1]

[1]Department of Environmental Geography Science, Faculty of Geography, Universitas Gadjah Mada, Indonesia

emmahisbaron@gmail.com

Abstract. Kelud Volcano is a basaltic andesitic stratovolcano, situated at 27 km to the east of Kediri, Indonesia. Historically, Kelud Volcano has erupted with return period of 9-75 years, had caused nearly 160,000 people living in Tulungagung, Blitar and Kediri District to be in high-risk areas. This study aims to map vulnerability towards lava flows in Kediri and Malang using detailed scale. There are four major variables, namely demography, asset, hazard, and land use variables. PGIS (Participatory Geographic Information System) is employed to collect data, while ancillary data is derived from statistics information, interpretation of high resolution satellite imagery and Unmanned Aerial Vehicles (UAVs). Data were obtained from field checks and some from high resolution satellite imagery and UAVs. The output of this research is village-based vulnerability information that becomes a valuable input for local stakeholders to improve local preparedness in areas prone to improved disaster resilience. The results indicated that the highest vulnerability to lava flood disaster in Kelud Volcano is owned by Kandangan Hamlet, Pandean Hamlet and Kacangan Hamlet, because these two hamlets are in the dominant high vulnerability position of 3 out of 4 scenarios (economic, social and equal).

1. Introduction

The risks of volcanic disaster in Indonesia are caused by two main factors; (1) the potential for large volcanic hazards and (2) the social and economic conditions of the majority of the population vulnerable to hazards. The risks of volcano disaster most often hit the poor and gendered minorities, the disabled, the elderly and the community have no access to resources and means of protection. Vulnerability as indicated in Indonesian context are hidden among hazard and risk analysis, which in turn delimits number of work in the related theme. According to applied science approach, vulnerability is a quantitative measure towards any element at risk with degree of potential loss ranging from 0 or "not vulnerable" to 1 or "very vulnerable" [3][9][10][11][7]. Once we have hazard information and vulnerability information, it is easy to measure disaster risk. Risk itself defines as degree of damage or loss which was expected if there is potential hazard with certain degree of vulnerability [5][10].

Despite high risks, communities around the volcano have developed a wide range of skills, resources and knowledge, called capacity-based communities to reduce community-based disaster risks (CBDRR). Villages in Indonesia, nowadays, are no longer at rural area per se. Administratively, an area with agriculture activity predominantly and very limited support of basic infrastructure network addressed as rural. However, it is common to have village with very much unlikely description as such. Some areas in Java, where most of it has been rapidly developed somehow indicate urban characteristics. It is "desa-kota" theory which explains characteristics as such. The village information may also address not only necessity of rural development, it is also important to support urban development.

Gunungapi Kelud is located 27 km east of Kediri, Indonesia. The volcano erupts with repeated periods of 9-75 years, and nearly 160,000 people living in Tulungagung, Blitar and Kediri are at high risk of disaster [13] [4]. Located in three administrative boundaries (figure 1), Malang, Kediri and Blitar, East Java Province, it was recorded that in the last eruption of 2014 there were 87,629 people affected and 8,452 residential units were damaged [1] [12]. Previous research in the Kelud area, has indicated that at least 29 villages along river banks have high to moderate vulnerability (figure 2).

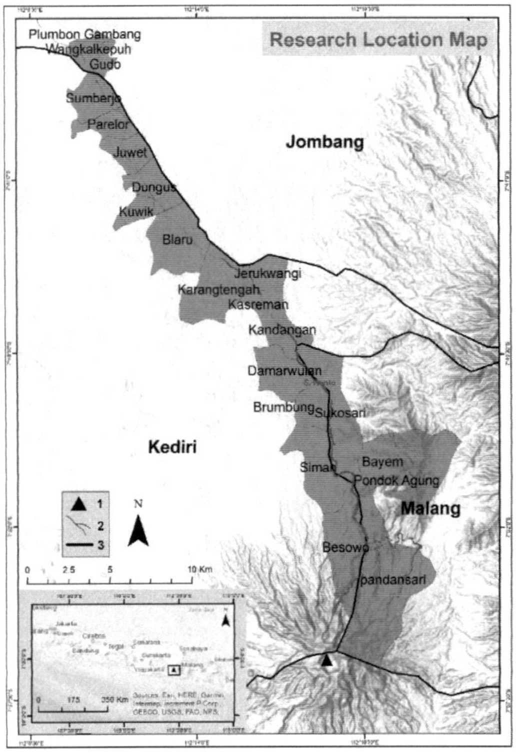

Figure 1. Research area 1- Volcano; 2- River; 3 - Regency boundary; 4 - Village boundary [7].

Taking into account towards current vulnerability distribution, this research shall also underscores the importance of adaptation strategies undertaken in local scale. The hazard and vulnerability assessments that have been carried out in the Superior Scheme of Higher Education Research entitled "Kelud Volcano Eruption 2014: Impacts, Hazards, Risks and Its Implications on Disaster Risk Reduction". The preliminary works upon the topics reveal that some part of the area surrounding Kelud had been very much subjected towards vast social and economic phenomenon. The previous research has conducted social economic vulnerability analysis towards lahar flood of Kelud Volcano in medium scale. Critically, conducting vulnerability analysis as medium scale hinders scholar to conduct physical vulnerability analysis. The physical vulnerability analysis focus to observe the element at risk of buildings at local scale. The condition of building, its location toward the river, its existing adaptive strategies are one among series of elements to be observed. Detailed vulnerability mapping, as argued within this research allows scholar to identify range of vulnerability at very micro or local scale using village information which has been established earlier.

IOP Conf. Series: Earth and Environmental Science **148** (2018) 012008 doi:10.1088/1755-1315/148/1/012008

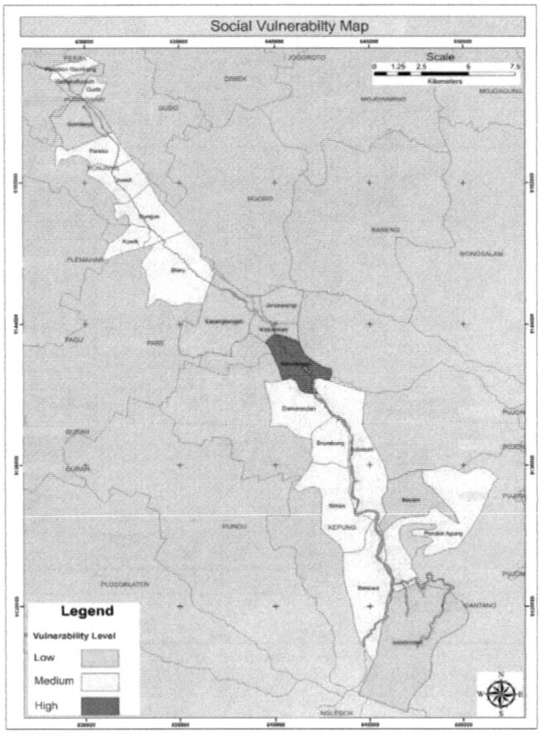

Figure 2. Social Vulnerability Scenario in Banjir Lahar at Kelud Volcano at Medium Scale [7].

The village information currently incumbent for Indonesia, taking into account "one map policy", "one village one product" and "resilience village program". There are several inline policies which establish based upon village unit. The main idea to introduce importance of village information are as follow: 1) village information either in urban and rural context should be equipped with various baseline information, including hazard, vulnerability and risk information as an integral part of development input; 2) village information is likely change local mindset to put more attention towards their environment; 3) village information generally conducted continuously, prior to support national statistic baseline data and many inter-sectoral policies in Indonesia.

2. Research Method

This study employed spatial and temporal approaches, applied participatory geographic information system or PGIS techniques to collect data. PGIS is a cartographic approach that involves local knowledge of disaster information related to adaptation strategies and disaster mitigation efforts [6]. In the context of risk analysis, there are two important aspects need to be evaluated, such as hazard and vulnerability. To come up with vulnerability, there are essential objects need to be observed entitled as element at risk. This research focused upon human being with their economic condition, buildings characteristics as well as environmental condition (figure 3). In detailed, there are three influential form of energy prior to cause hazard potentials, such as endogenous, exogenous energy and technological failure. This research as mentioned above focused upon lahar flood, which influenced both by endogenous (post earthquake and volcanic eruption), and endogenous (driven by flow of liquid and or mass into lahar flood). In return, there are several elements at risk potential to be threatened by the hazard potential as such. The elements at risks are divided into social, physical, economic or financial and environmental aspects. Herewith, the research observes level of inability from each element at risk to survive from lahar flood which entitled as vulnerability level.

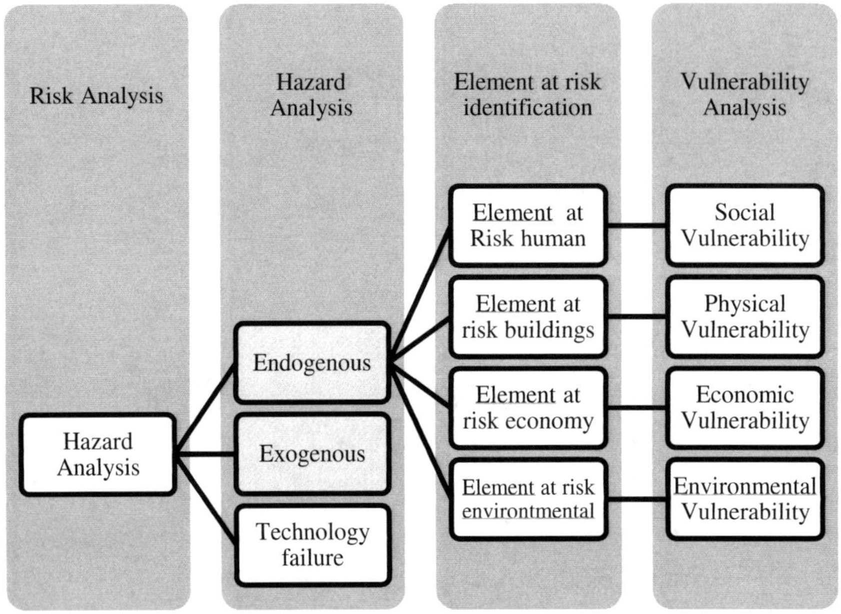

Figure 3. Research Framework.

Based upon previous studies, Ngobo River and Konto River are actually prone towards lahar flow [7]. The previous research generated social economic vulnerability using Spatial Multi Criteria Evaluation (SMCE). This research, hence, employed also SMCE to generate physical, social and economic vulnerability at local scale. SMCE assists in conducting multi criteria assessments that can facilitate the process of spatial tabular data and attribute data. The applied technique of SMCE consists of these following steps: problem definition, factor standardization, assign weighting, and generate vulnerability classification (figure 4).

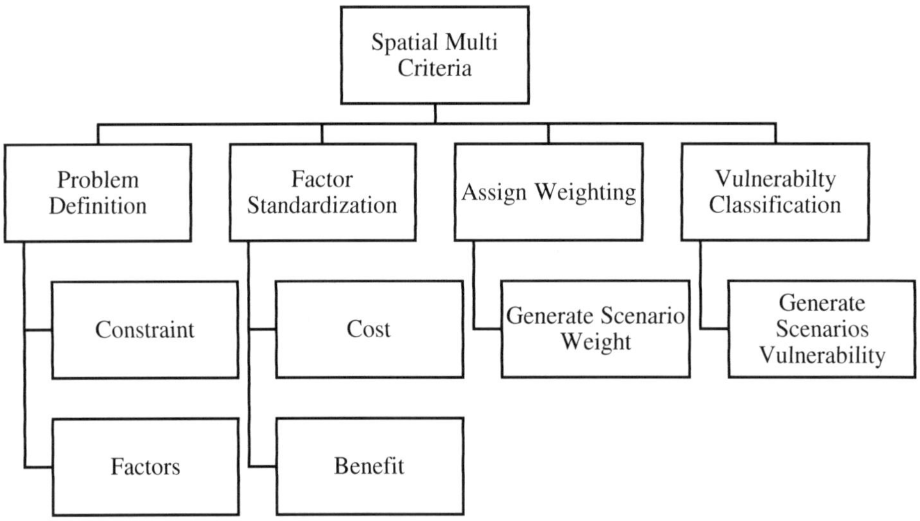

Figure 4. Aplied Technique of Spatial Multi Criteria Evaluation.

The use of the SMCE method begins with determining criteria, which can be constraint and factors (figure 4). There are social, physical and economic criteria. Each of criteria breaks down into multi factors, according to available data in the research area. Before assigning weight, criteria and factors are structured in a problem tree analysis in order to get clear objectives. The objectives aim at providing clear guidance towards setting up research assumption between each factor. The research set objective into measure total vulnerability between 0 (not vulnerable) – 1 (vulnerable or lethal). Henceforth, each factor for example total population get standardized using the logic assumption that the more population the higher its vulnerability (table 1). Standardization as such helps scholar to formulate accumulative assumption from diverse criteria, which then equips with weight. Multi criteria assessment also makes possible for scholar to generate more than a scenario using weighting scenarios. The main intention to generate more scenario is to ensure its robustness and foreseen possible overlapping areas from diverse assumption. This research employs four scenarios, such as vulnerability from social criteria, physical criteria, economic criteria, whereas each of the criteria assign dominant weighting. Meanwhile, the last scenarios are equal scenarios, which explains equal weighting between demographic, physical and social aspects.

Table 1. Factor and Research Assumptions.

No	Criteria	Factor	Research Assumptions
1	Social	Total population	The higher the population, the higher the vulnerability (+)
		Population density	The higher the population density, the higher the vulnerability (+)
		Disabled	The more disabled group, the higher the vulnerability (+)
		Infants	The more number of infants, the higher the vulnerability (+)
		Elderly	The more number of elderly, the higher the vulnerability (+)
		Education, knowledge	The higher the level of education and knowledge, the lower the vulnerability (-)
2	Physical	Number of buildings	The more numbers of buildings, the higher the vulnerability (+)
		Building density	The more of building density, the higher the vulnerability (+)
		Distance to river	The closer to the river, the higher the vulnerability (+)
		Affected area	The higher the area of impact, the higher the vulnerability (+)
3	Economic	Poor family	The higher the number of poor families, the higher the vulnerability (+)
		Agriculture area	The more agricultural land, the higher the vulnerability (+)
		Number of farmers	The higher the number of farmers, the higher the vulnerability (+)

What is the meaning of vulnerability from social aspects? This relates to an "if condition". Many scholars aim at predicting future condition, which should be close to real situation. Herewith, if the real situation in the future heavily affected by dynamic social aspects, thus the spatial distribution of vulnerability for the area will be shown in the vulnerability from social aspects scenarios. There are in total 4 scenarios generated in this research, such as vulnerability from social, physical, economic, and equal vulnerability scenarios.

3. Results and Discussion

For preliminary observation, the research collected information from satellite imagery. Two images obtained from Worldview-2 Image Year 2014 and Unmanned Aerial Photo Image Year 2017 (figure 5). These images indicate comparative vertical transects towards the research area before eruption (2014) and after eruption (2017) of Kelud Volcano. Through this comparative image, Kandangan Village was clearly seen to have pile of sediment at 2014, while the riverbeds was lessen its depth in 2017 with wider area of alluvial plain. During field observation, the area along Konto River at

Kandangan was subjected to mining activities. The initial river flow of the subsequent river was narrowed down due to shifting river body. Such morphological dynamics towards Konto River was taken into account towards increasing element at risk upon lahar flood. Before the eruption the riverbed was quite steep, whilst it getting fairly steep these days due to this shifting characteristic. This refers to increasing potential of lahar flood towards the area once an eruption occurs.

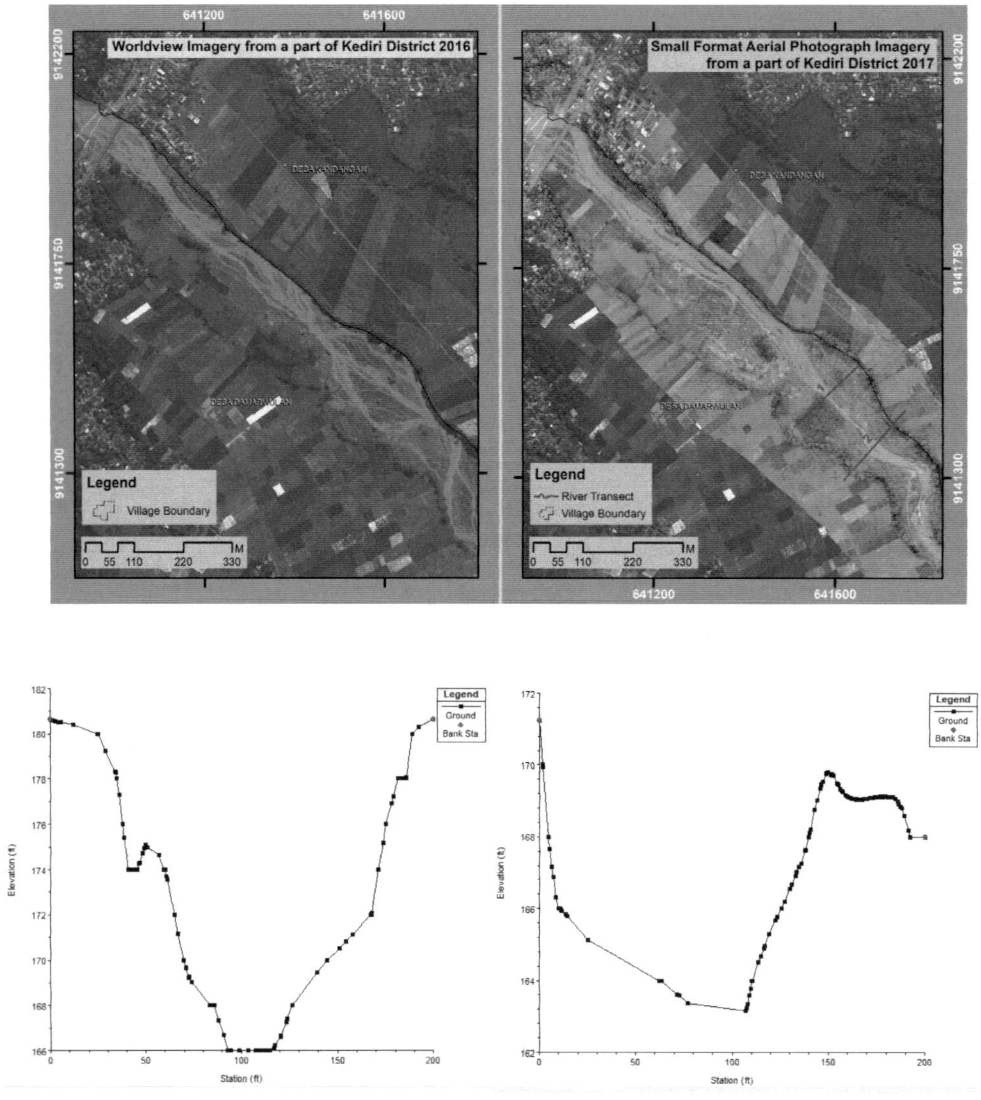

Figure 5. Worldview-2 Image 2014 (left) and Aerial Unmanned Photo 2017 (right) to depicts vertical transect towards Kandangan Village.

The possible aspects influential towards morphological setting of the river are flow velocity, structure of lava material, and river bed characteristics. The research conducted two transects areas, northern line and southern line. Northern line of transect indicated that Konto River experiencing massive accumulation of post-eruption material. It was indicated the middle riverbed, characterized by deeper and steeper topography, due to massive erosion at the riverbeds. Southern line of transect indicated that Konto River experiencing greater energy that shapes its riverbed into fairly steep and increase width compare to northern side. Along Konto River are utilized for agricultural land, and settlement area. To the south, potentials of getting more inundated land are greater compared to the northern area.

After observing research area using aerial photo, this research explores many ancillary data for the research area. Accordingly, its demographic feature was experiencing rapid population growth from 2013 to 2016. Village monograph recorded in 2013 total population in Kandangan was 46,352 people, with total density of 1,112 people/km². In 2014, total population increased into 46,586 people, and in 2015 increased into 48,333 lives. Meanwhile the latest village information recorded that total population of 2016 approached 50,010 people. This number is quite high for a village administrative, whereas population density rapidly increases up to 1,144 people/km² in 2014, 1,159 people/km² in 2015 and 1,200 people/km² in 2016.

Table 2. Demographic trend of Kandangan Village 2013-2016

Year	Total Population	Density (people/km²)	Paddyfield (ha)	Non paddyfield (ha)
2013	46,352	1,112	1,890	2,277
2014	46,586	1,144	1.848	2,319
2015	48,333	1,159	1,888	2,279
2016	50,010	1,200	1,888	2,279

In terms of land use, there are several actual activities such as agriculture area especially for paddy field, mix-cropping, fields and built up areas. The area has predominantly use for agriculture activities, and it has not much shifting after the eruption. Prior to this ancillary data, we may argue that agriculture activity has been very much resilience to support local livelihood although it had impacted heavily at the aftermath of Kelud Volcano eruption. The following figure 6 indicated that the number of buildings at the research area were severely decreasing since 2014. It indicated number of houses or buildings which had been damaged due to lahar flood of Kelud Volcano.

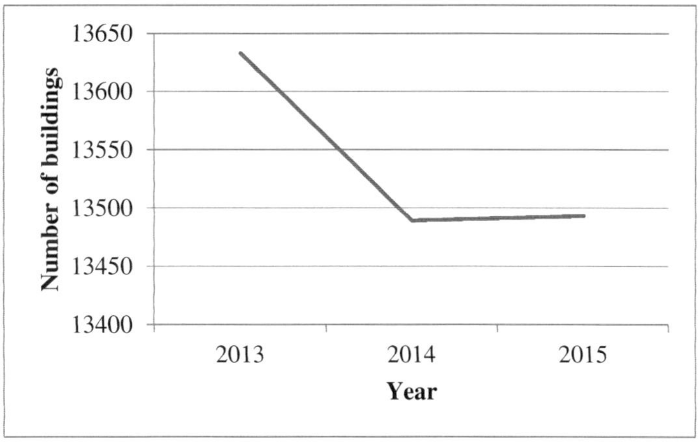

Figure 6. Building distribution at Kandangan Village, 2013 - 2015.

Evidently, this baseline information taken from village information provide brief overview of temporal dynamics within the area. Volcanic eruption did not change much the natural morphology, even the land use. However, the man-made morphology had been shifting at the aftermath of the eruption. Lahar flow, diverted river morphology, but not necessarily impacted so much to the land use. In fact, as noted earlier, lahar flow also put dynamic change towards alluvial plain along river streams.

Taking into such phenomena, the research thus tried to foreseen future condition of vulnerability. The vulnerability from physical aspects refers to condition of future vulnerability if physical criteria is dominant compared to other criteria. There are four factors supporting physical criteria, such as the distance of the hamlet to the river, the extent affected, the number of buildings and the density of the building. The assigned weights of each factors upon physical criterion are as follow, whereas, the distance to the river (0.56) and the weight of the affected area, the number of buildings, and the densities respectively are as follows, 0.26, 0.12 and 0.06. According to the physical scenario, the area to the north and to the south of the Konto River justified as highly vulnerable area. The main object evaluated in this scenario are buildings. Its location, distance and impacts were emphasized heavily within this scenario (figure 7). The red area shown on the map indicated a high vulnerable area. Additionally, the research was also looking for any reasoning that cause buildings at Kandangan Village become very much vulnerable towards lahar flood. In fact, according to field survey, there was no early warning system established in the research area. Prior to early warning system, it is important to monitor and control channel debit and material sedimentation before and during lahar flood occurrence. The remaining Sabo Dam was located at Biyoro Hamlet, the most populated neighbourhood unit in Kandangan Village. This Sabo Dam also not in good condition, although Biyoro Hamlet located directly adjacent to Konto River. Additionally, the research area also had not yet equipped with evacuation routes. The research also learned from Participatory Geographic Information System that the Pandean Hamlet and Biyoro Hamlet were the most impacted area in the aftermath of lahar flood.

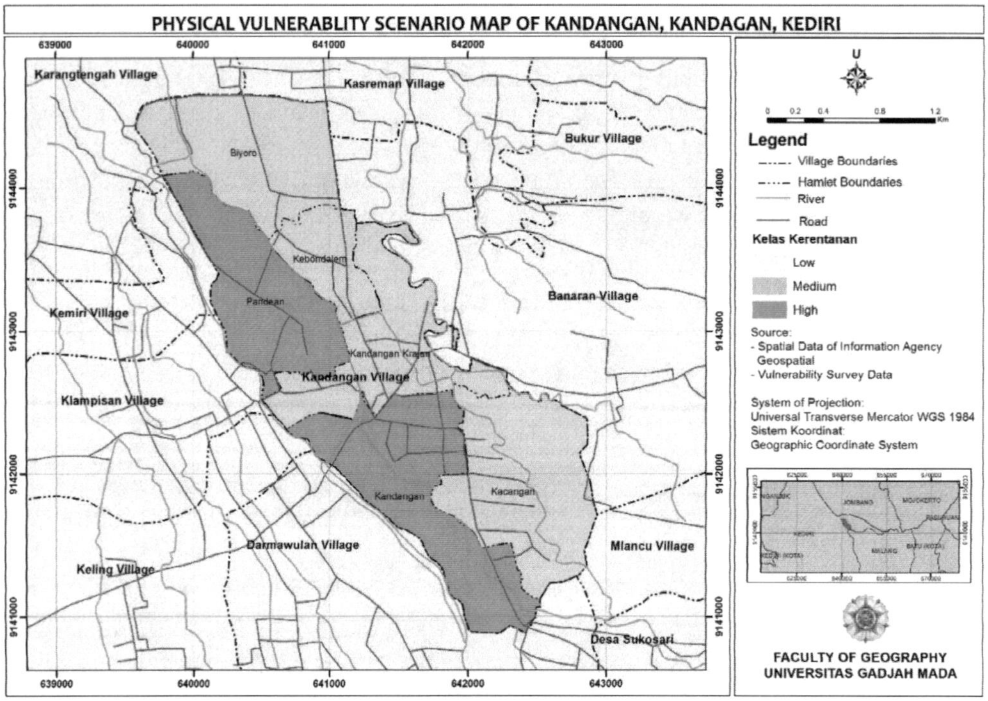

Figure 7. Physical vulnerability scenario map.

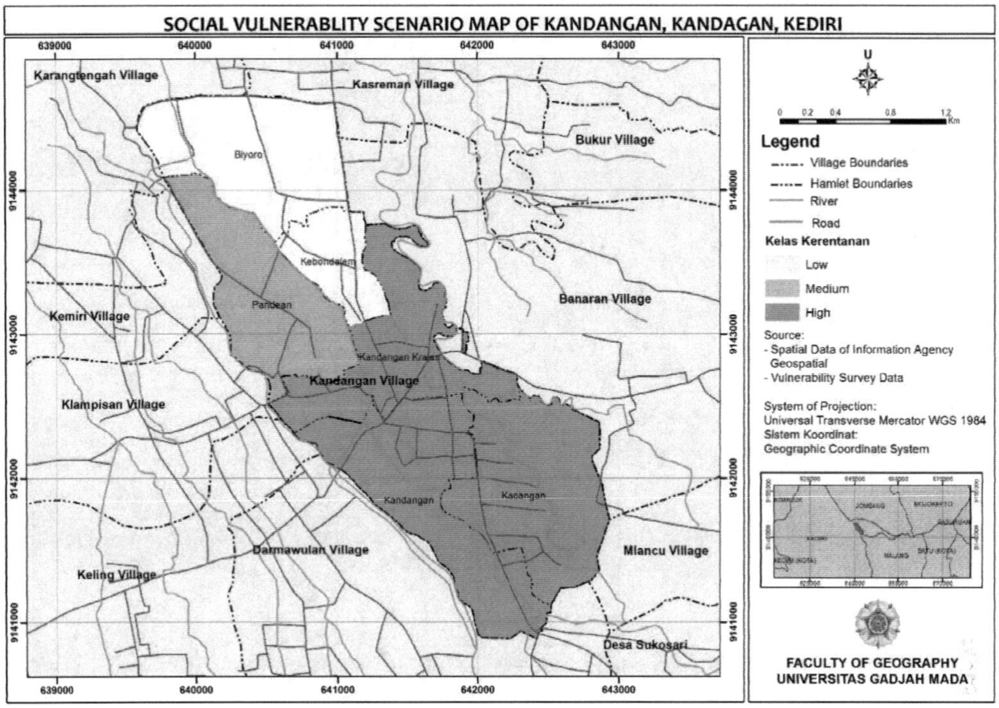

Figure 8. Social vulnerability scenario map.

The second scenario is vulnerability from social criteria. The overarching goals in determining social vulnerability is to evaluate spatial distribution of people's aggregate. Factors supporting this criterias are total population, density, disabled, elderly, and toddlers. The spatial distribution towards these selected demographic criteria off course quite different with physical criteria, therefore, the vulnerability pattern coming out of the computation also different to the previous scenarios (figure 8). Spatially, the southern part such as, Kandangan-Krajan Hamlet, Kacangan Hamlet, and Kandangan Hamlet subjected to high vulnerable area due to accumulation of demographic features and vulnerable groups. Meanwhile, the northern part of the research area is less vulnerable compared to the other area.

The third scenarios are vulnerability from economic criteria. As indicated by baseline information, the most impacted area is building, whereas each of the household livelihood fully dependant upon agriculture activities. Therefore, within this economic criterias, the factors address farmer as subjected groups together with poor families as vulnerable groups from economic perspective. Administratively, number of farmers and poor families are accumulated in two neighbourhood units at Kandangan Village as show at figure 9.

IOP Conf. Series: Earth and Environmental Science **148** (2018) 012008 doi:10.1088/1755-1315/148/1/012008

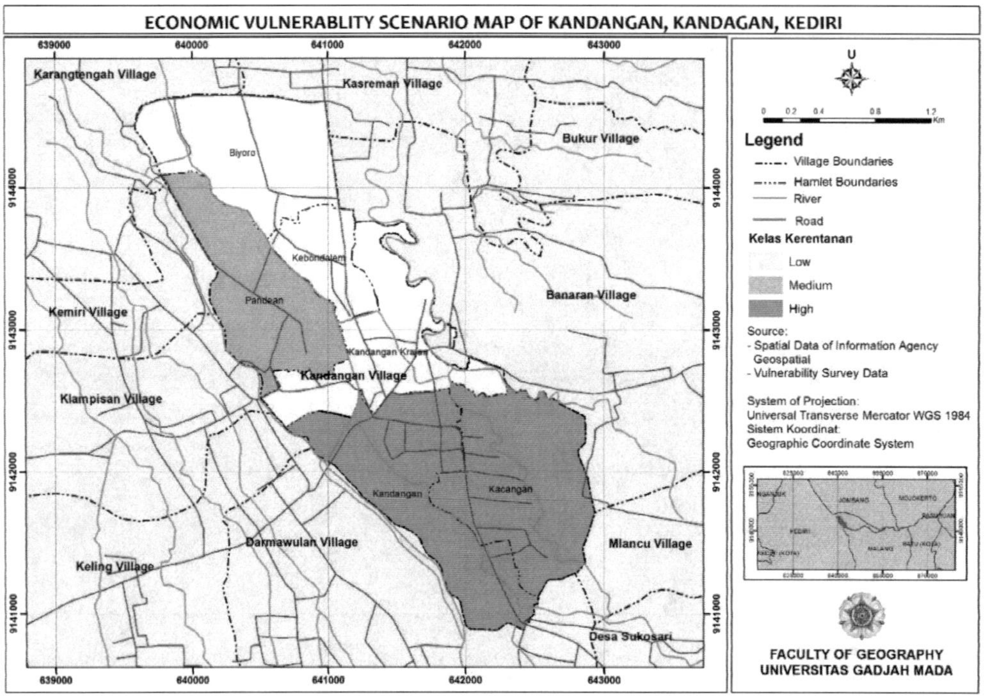

Figure 9. Economic vulnerability scenario map.

According to table 2 below, some of the villages exposed to composite types of vulnerabilities. Kandangan for example, whereas it exposed to high physical, social and economic vulnerability at once. Meanwhile, Biyoro is exposed to low physical, social and economic vulnerability. This indicated that these villages have robust vulnerability pattern.

Table 2. Recapitulation of vulnerability by village.

Type of Vulnerability		Low	Medium	High
1.	Physical	Biyoro Kebondalem Kandangan-Krajan Kacangan	-	Kandangan, Pandean
2.	Social	Biyoro Kebondalem	Pandean, Kacangan	Kandangan, Kandangan- Krajan
3.	Economic	Biyoro Kebondalem Kandangan-Krajan Pandean	-	Kandangan, Kacangan

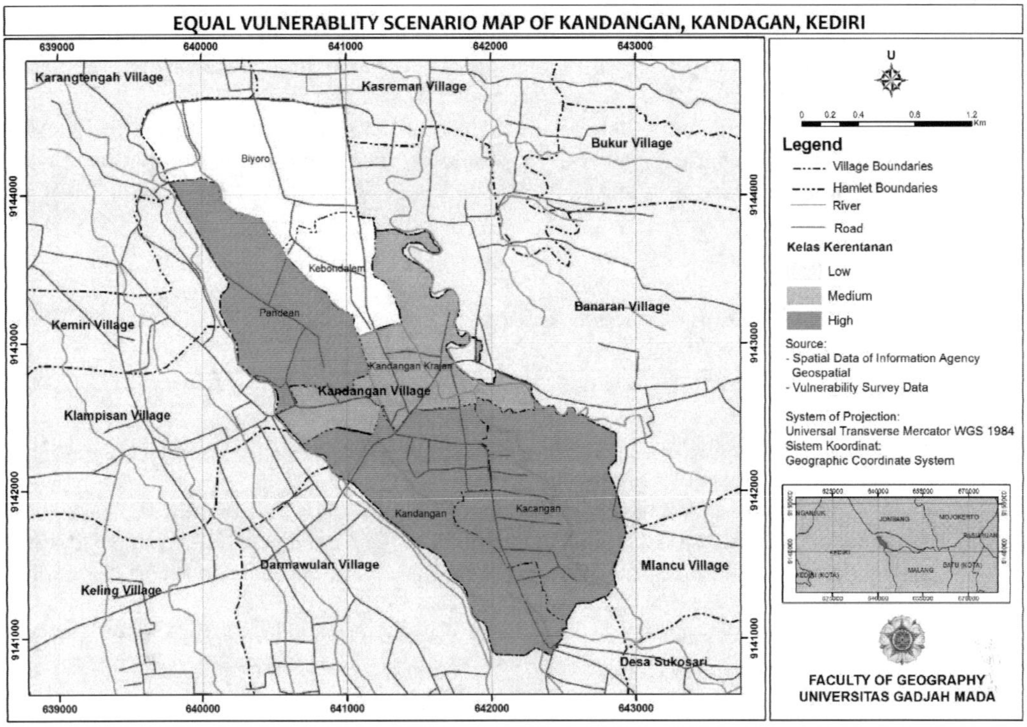

Figure 10. Equal vulnerability scenarios.

The equal scenarios steward scholar to understand fairly common spatial pattern with one or more spatial pattern shown in the previous scenarios. Administratively, Kandangan Hamlet justified as the most common vulnerable area from different scenarios (figure 10). If the research compared four scenarios at once, it is clearly seen that Kandangan Hamlet deserve to have priority action in the future if any lahar flow occurs. This highest vulnerability area, followed by Pandean Hamlet. While the rest of area considerably less vulnerable.

Derived from this result, the village information is considered essential to predict future trends of vulnerability. The availability of village information should not be limited to statistics ancillary data. It should also be equipped with spatial information, either taken directly from satellite imagery or unmanned vehicle imagery. Additionally, spatial data also possible to be taken from participatory process.

Taken back the result to the "desa-kota" theory, whereas urban and rural areas are differed from its accumulation of assets and vulnerability, it is argued that, the most populated and developed area among these rural setting are happened to be the most vulnerable. It means that this pseduo-urbanization process occurs in the area somehow put more burden on the area.

For what its worth, the vulnerability analysis itself is easily conducted using multi criteria evaluation. The advantage of applying such method is to reduce the possibility of non-exist datasets in village information. Multi-criteria make possible for scholar to select any available data using standardization to construct the logic with its main objectives. The disadvantage of applying multi criteria evaluation has been very much political. The Indonesian Government provides guidelines to evaluate vulnerability analysis which employed mainly weighting scenarios to particular element at risk observation. Therefore, such method will be very much challenging to add into the formal datasets.

4. Conclusion

1. The highest vulnerability to the lava flood disaster in Gunungapi Kelud is owned by Kandangan Hamlet, Pandean Hamlet, and Kacangan Hamlet, as these two hamlets are in the dominant high vulnerability position of 3 out of 4 scenarios (economic, social and equal).
2. The vulnerability analysis indicated that for any chance of lahar flood in the future, the Kandangan Hamlet, Pandean Hamlet and Kacangan Hamlet should be prioritized in order to minimize potential risk arisen in the area.

5. References

[1] BNPB 2013 *Indeks Risiko Bencana Indonesia* Badan National Penanggulangan Bencana. Jakarta: BNPB
[2] Cutter S L 1996 Vulnerability to Environmental Hazards *Progress in Human Geography* 529-539
[3] Cutter S L 2003 The Vulnerability of Science and the Science of Vulnerability *Annals of the Association of American Geographer* 1-12
[4] De Bélizal E Lavigne F Hadmoko D S Degeai J P Dipayana G A Mutaqin B W Marfai M A Cooquet M. Le Mauff B Robin A K Vidal C Choelik N and Aisyah N 2012 Rain-triggered lahars following the 2010 eruption of Merapi volcano, Indonesia: a major risk. *Journal of Volcanology and Geothermal Research (2013)*
[5] Kates R W 1971 Natural Hazard in Human Ecological Perspective: Hypotheses and Models *Economic Geography* l47 No.3 438-451
[6] Haase D 2009 Participatory Modelling of Vulnerability and Adaptive Capacity in Food Risk Management. *Nat Hazards* 67 77–97
[7] Hizbaron D R and Hadmoko D S 2016 Mapping Vulnerability for Kelud Volcano, Indonesia. Approach Towards Disaster Resilience. *6th International Conference on Building Resilience: Building Resilience to Address the Unexpected* pp. 52 Auckland: Massey University of New Zealand & University of Auckland New Zealand
[8] Hizbaron D R and Hasanati S 2016 *Menuju Kota Tangguh di Sungai Code Yogyakarta: Perencanaan Integratif Perkotaan dengan Pendekatan Pengelolaan DAS dan Pengurangan Risiko Bencana* Yogyakarta, Indonesia: Gadjah Mada University Press
[9] Kubat A S Eyüboglu E Ertekin Ö and Sari F 2008 Earthquakeas Fact of Istanbul: Risk-based strategy and an actionplan for the vulnerable Zeytinburnu area. In T Kidokoro, et al. Vulnerable Cities: Realities, Innovations and Strategies Tokyo: *Springer Publisher* 275-299
[10] Sarris A Loupasakis C Soupios P Trigkas V and Vallianatos F 2009 Earthquake Vulnerability and Seismic RiskAssessment of Urban Areas in High Seismic Regions: Application to Chania City, CreteIsland, Greece. *Natural Hazards*
[11] Schmidtlein M C Shafer J M Berry M and Cutter S L 2010 Modeled Earthquake Losses and Social Vulnerability in Charleston, South Carolina. *Applied Geography (InPress)* 1-13
[12] Sulaksana N Sukiyah E Sudrajat A and Syafri I 2014 The Crater Configuration of Kelud Volcano, East Java, Indonesia. *International Journal of Science and Research* 2319-7064
[13] Thouret J C Abdurachman K E and Bourdier JL 1998 Origin, Characteristics, and Behaviour of Lahars Following the 1990 Eruption of Kelud Volcano, Eastern Java (Indonesia) *Bull Volcanol* **59** 460-480
[14] Thouret J C Abdurachman K E Bourdier J L and Bronto S 1998 Origin, Characteristic and Behaviour of Lahars Following the 1990 Eruption of Kelud Volcano, Eastern Java (Indonesia) *Bulletin of Volcanology* 59 460-480

IOP Conf. Series: Earth and Environmental Science **148** (2018) 012009 doi:10.1088/1755-1315/148/1/012009

Riparian planning in Yogyakarta City

R Rachmawati[1], E Prakoso[1], M I Sadali[1] and M G Yusuf[2]

[1]Department of Development Geography, Faculty of Geography, Universitas Gadjah Mada, Yogyakarta, Indonesia
[2]Regional Development Study Program, Faculty of Geography, Universitas Gadjah Mada, Yogyakarta, Indonesia

rinirachma@ugm.ac.id

Abstract. Riparian is a potential for slums in urban areas. The city of Yogyakarta is passed by three rivers namely Code, Gajahwong, and Winongo, crossing the city. Riparian in the three rivers are potential for slum if the area is not well managed. This paper is based on the survey results of the structured interview with the people living in the riparian area in Yogyakarta City. They were 75 respondents from the three riparian. The result shows that several reasons why people prefer to remain living in the area are limited spaces and high land price in the city as well as inherited from their parents. The facts that there are still several problems related to the condition of settlement environment in the riparian, i.e., The condition of densely-populated areas, limited availability of land, and limited public spaces. Efforts that can be done to solve problems related to the riparian planning are anticipating disasters like flood and landslide, paying attention to densely-populated and unwell-planned areas, and handling garbage that has been abandoned into the river. The program expected by those living along both riversides is intended to give priorities on providing some aid for those whose houses are not in good condition, controlling buildings without a permit, and building a dike along the river. Efficiency can be made by making use of the space adequately between the one for settlement and the other one for open-green space for both aesthetic and economic purposes.

1. Introduction

The riparian is a potential for slums in urban areas. It is caused by several reasons, i.e., high population, the short distance between houses and river, people's misbehavior in discarding garbage into the river, limited well-arranged paths, and susceptibility of the area to flood threat. Three rivers cross the City of Yogyakarta, namely Code River, Gajahwong River, and Winongo River. The riparian area of Code is the most susceptible to flooding, particularly lava flood. The last lava flood took place when Mt. Merapi erupted in 2011. Such flood also once happened some years before 2011, and it caused the settlement on both sides of the river to be affected [1][2]. Since Mt. Merapi is the most active volcano, it is possible that lava flood will happen again in the future. Therefore, several recommendations need to be made to be implemented in the area [3]. They are; 1) the grand design for spatial planning to minimize the society-based disaster risk needs to be made, 2) the technical studies and analysis of disaster risk in an integrated way need to be done, 3) it is necessary to strengthen the society capacity in the spatial planning to minimize the disaster risk, 4) regulations on the implementation of space use and control over the space use in the disaster-susceptible area need to enforced. Thus, further discussion related to the efficiency of space use is of primary importance to be done.

In relation to the settlement in Code Riparian area, there are three *rusunawa* (flats) that have been built and the people who previously lived in houses on riparian are given priority to move and live in the *rusunawa* [4]. Those three *rusunawa* are located closely to one another, less than 1 kilometer far

[4]. They were built not far from the river, an area susceptible to lava flood as a result of the eruption of Mt. Merapi [1]. The three *rusunawa* were built to help solve the problem of a slum in the riparian area. In relation to the problem of the slum, it can be viewed from the aspect of facilities based on the criteria of slum area set out by the Department of Public Works in 2007 [5]. The condition of most facilities in one of the areas of riparian area belonging to the village of Tegal Panggung is categorized medium-slum [5]. Tegal Panggung is one of the villages in Yogyakarta City which is located in riparian.

On the other side, the riparian planning shows a successful result [6]. One of the areas where the neighborhood used to be considered slum shows a successful result of maintaining the settlement to be a well-managed area [6]. The same case is also found in one of the settlement area belonging to the riparian at Gajah Wong river. The area, also known as a Green Kampong, shows that the management of the area has been well implemented due to the presence of the river- structuring community and the leader of the community [6]. The management of the area shows that self-reliance and creativity of the community can result in clean and beautiful public space and neighborhood [6].

Studies over the problems and the need for the program to help solve the problems need to be done. Therefore, studies over the survey results related to the riparian planning of the three rivers are of primary importance to be done. Several issues on riparian planning that are presented in this paper are related to; 1) people's home status, 2) the condition of the settlement environment, 3) communities concerned with the riparian planning, 3) the availability of public spaces, 4) arising problems related to the riparian planning, and 5) efforts that need to be done and priorities that need to be given related to the riparian planning.

2. Method

This paper is documenting the part of results of Collecting People Aspiration Survey in Yogyakarta City in 2016, especially for the cases of riparian at Code, Gajahwong, and Winongo Rivers [7]. The results of this survey are used by the Yogyakarta municipal government as one of the inputs in preparing an annual development plan the following year. The analysis used in this research is descriptive quantitative.

In the survey of Collecting People Aspiration of Yogyakarta City in 2016, there were 1000 respondents from 14 sub-districts. There are six issues exposed in the event, i.e., education, health, people-based economy, culture-based tourism, planning for river area, and regional development. This paper focuses on the issue of the planning for river area with 75 respondents, 25 respondents from each riparian. Meanwhile, the number of respondents in each area of the riparian of the respective village can be seen in table 1. The respondents are therefore those living on the riversides of Winongo, Code, and Gajah Wong. The research location can be seen in figure 1. They are taken randomly as purposive sampling, comprising the board members of RW (Rukun Warga/administrative unit at the next-to-lowest level in city), RT (Rukun Tetangga/neighborhood association), PKK (Pendidikan Kesejahteraan Keluarga/program at village level to educate women on various aspects of family welfare), youth figures, common people, and those in charge of maintaining the environment and security. In collecting the data, a structured interview is conducted.

Table 1. Distribution of the Number of Respondents.

Name of River	Village	Sub District	Number of Respondents
Winongo	Bener	Tegalrejo	5
	Bumijo	Jetis	5
	Pringgokusuman	Gedong Tengen	5
	Ngampilan	Ngampilan	5
	Gedongkiwo	Mantrijeron	5
North Code	Terban	Gondokusuman	2
	Cokrodiningratan	Jetis	2
	Kotabaru	Gondokusuman	2
	Gowongan	Jetis	2
Middle Code	Tegalpanggung	Danurejan	3
	Purwo Kinanti	Pakualaman	2
	Keparakan	Mergangsan	3
South Code	Brontokusuman	Mergangsan	5
	Sorosutan	Umbulharjo	4
Gajahwong	Rejowinangun	Kotagede	10
	Prenggan	Kotagede	15
Total			75

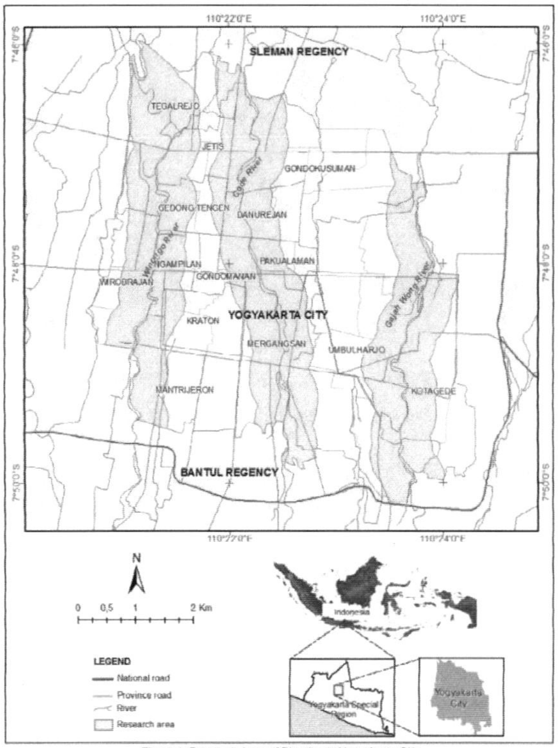

Figure 1 Research Area of Riparian in Yogyakarta City

Figure 1. Research Area of Riparian in Yogyakarta City.

3. Result

The survey results show that the respondents consist of 19 from RT board members, four from RW board members, four from PKK board members, two youth figures, 13 common people, and one from environment and security cadres respectively. The respondents are between 21 and 77 years old. Most respondents are SMA (Sekolah Menengah Atas/Senior High School) leavers (28 respondents). For occupation, most respondents are housewives (27 respondents). Furthermore, in this paper will describe characteristics of the settlement, the condition of settlement environment, communities for structuring river area, the availability of public spaces, problems related to the structuring for the river area, efforts to improve the structuring for the river area and program priority.

3.1. Characteristics of settlement

Based on the survey results, 93% of the respondents reveal that they have been living in the riparian area for more than ten years. Meanwhile, only 7% of the respondent state that they have been living in the area for less than ten years. It shows that some of the citizens of Yogyakarta city have been living around riparian for long. There are several reasons why they prefer to live there. They are, among others, limited space, high land price, and ancestor descent. It is these factors that cause them to find it difficult to move out. 70% of the respondents say that they have possessed a certificate of ownership for their land. Meanwhile, 13% of the respondents say that their land status is Magersari (living on state-owned land), 2% state that they have possessed a letter of HGB (*Hak Guna Bangunan*/ Building Use Right), and 5% of them are tenants. Figure 2 shows the percentage of the home status of the people living in the riparian area in Yogyakarta city.

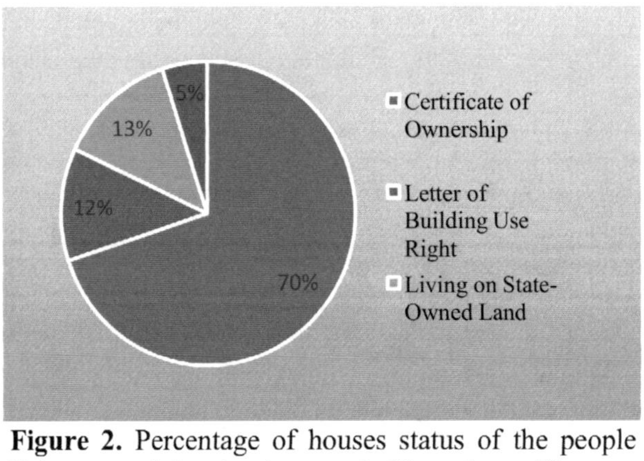

Figure 2. Percentage of houses status of the people living around the riparian area in Yogyakarta City.

3.2. Condition of settlement environment

Based on the survey results, 61% of the respondent state that their settlement area and the river have been well managed. In addition to the role of local people, the city government of Yogyakarta has played an essential role in the success of the structuring of the river and their neighborhood. One of the local people's roles in the managing of the river and its surrounding area is the establishment of community focusing on maintaining the condition of environment and river. About the problem of settlement area and river management, the main problem is the condition of densely-populated settlement area. In addition, the availability of land and public space that is limited also leads to another crucial problem.

Based on the Mayor's Decision No. 393 the Year 2014 on the stipulation for the location of an area that should not be used adequately for building houses in the city of Yogyakarta, 3,304 houses are not adequately used for living in [8]. In 2015, the City Government of Yogyakarta recollected the data of the area with dirty settlement and houses that are not adequately used for living in. Based on the result of datum collection, in the city of Yogyakarta, there is an area with slum of 264.90 hectares and 3,194 houses that are not adequately used for living in [8]. The area with slum and not adequate houses for a living in the city of Yogyakarta is associated with the riparian area; this is Winongo River in the western part, Code River in the central part, and Gajah Wong River in the eastern part of the city. Houses that are closely located to one another and paths that are too narrow are factors leading to creating a densely-populated settlement area, and that slum is inevitable. Several problems related to the structuring of the area around the river are shown in figure 3.

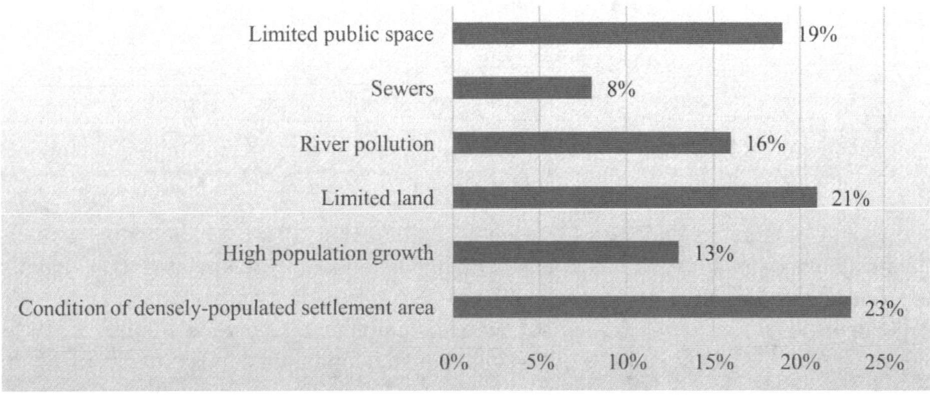

Figure 3. Percentage of problems related to the planning for the settlement area around the river.

In addition to the problems mentioned above, the people living in the area on both sides of the river also reveal other problems. They are, for example, inadequate infrastructure, too narrow paths, different path race, and sanitation problem. Some of them do not have their sanitation facilities and that they go to the river to take a bath, do the laundry, and have a bowel motion. The respondents claim that any effort that needs to be done to help solve the problems is by establishing excellent cooperation between the local government and the local people in facing the problems related to managing the settlement area, improving people's awareness, restructuring dirty settlement, and increasing the number of public spaces. Therefore, optimizing the role of the society is of primary importance in solving the problems mentioned above. The involvement of the society in every planning process is still limited, whereas, in fact, they know more about their environmental condition, problems, and situation of the area.

In relation to sewers from household, based on the survey result, 33% of the respondents throw their liquid household waste into the communal IPAL sewers, and 32% of the respondents throw their liquid household waste directly into the river. Meanwhile, 15% throw their liquid household waste into their septic tank, 12% into the sewage system, 6% into the gutter, and 2% into the well. The information mentioned above is shown in figure 4.

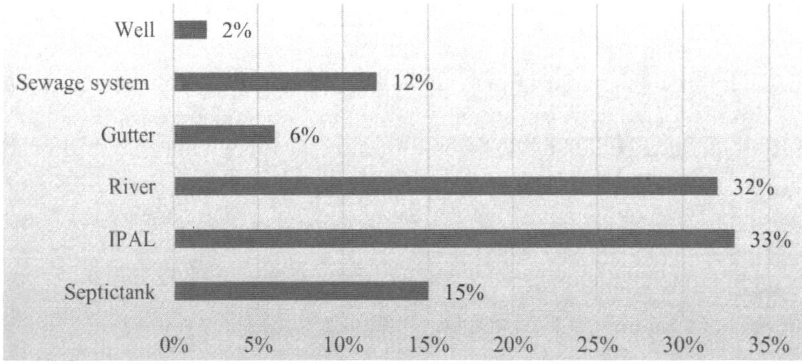

Figure 4. Percentage of where people living in riparian area of Yogyakarta City throw their liquid household waste.

Furthermore, for throwing their solid household waste, 71% of the respondents throw their solid household waste onto TPS (*Tempat Pembuangan Sampah*/garbage-throwing points). Twenty percent of the respondent state that their solid household waste is taken by a garbage picker. In addition, 8% of the respondents send their solid household waste to garbage bank, and 1% burns it. Based on the percentage presented above, it can be said that most of the community in the riparian area have been aware of the importance of throwing their solid household waste to TPS with the help of garbage pickers. This condition shows that the awareness of throwing their garbage onto the right place is high. They are aware because the impact of dumping garbage into the river will lead to siltation of the river, aesthetic damage and the possibility of water rise during the rainy season and that flood is inevitable

3.3. Communities for structuring river area

Based on the survey result, 61% of the respondents reveal that there are river communities around their area and 39% state that river communities are not found in their area. The communities found there are, among others, Lintas Winongo, FKWA (*Forum Komunikasi Winongo Asri*/Communication Forum of Winongo Asri), Berskap, Tagama (*Tanggap Bencana*/Disaster Alertness), Komunitas Dewa Bonto (Dewa Bonto Community), Pemerti Code, Komunitas Sungai Gajah Wong (Gajah Wong River Community, and Mertikali. Those communities are found in the area of those three rivers. The presence of the communities is of primary importance for both the City Government of Yogyakarta and local people since they can become a partner in developing the local potentials and solving any arising problem related to the river.

Fifty-seven percent of the respondents reveal that they are actively participating in the existing communities. Meanwhile, 42% are not. Thus, more people are participating in the communities. It shows that they are becoming aware of the importance to make the river cleaner and more beautiful. Up until now, the communities have undergone various activities. Based on the survey result as shown in Figure 5, 60% of the respondent state that they have been focusing on clearing up the river. It is the result of their awareness of how to keep the river clean, irrespective of more and more garbage thrown into the river. River communities like *Merti Kali* Code and Code *Boyong Muda* have even done a regular activity to walk along the river for clearing it up from garbage.

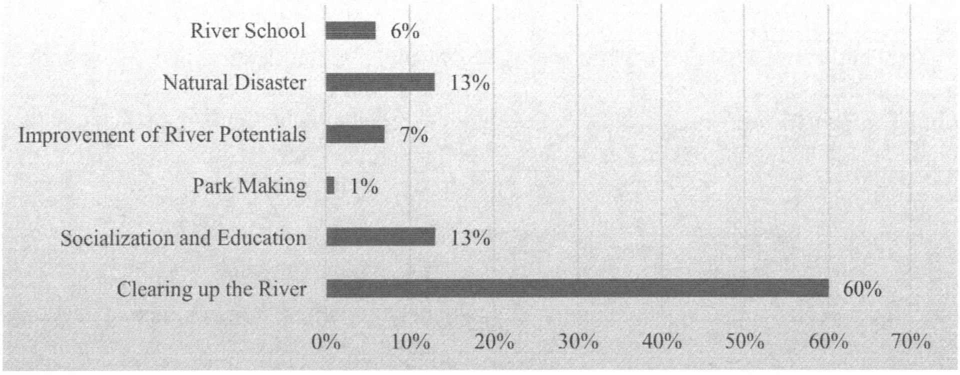

Figure 5. Percentage of Activities Done by River Communities According to the People Living in the Riparian Area in Yogyakarta City.

The presence of these communities is of primary importance since they can strengthen the empowerment and participation of society in developing their area. In relation to this matter, in the area of riversides. The participation of the people should be implemented by empowering local communities [9]. The existing communities need to be coordinated into one formal organization that is supported by the local government [9]. Supports from the government can be in the form of facilitating programs, creating a network, and providing the fund. Meanwhile, other sites like universities, private sectors, NGOs, and the like are also able to play an essential role in strengthening the communities by setting up empowering programs [9].

3.4. The availability of public spaces

The survey result shows that one of the problems related to the settlement area is limited public spaces. However, 72% of the respondents, based on the survey result, claim that there are enough public spaces in their neighborhood and the people there make use of them. Only 28% of them do not use the public spaces. According to the respondents, the public spaces available show that the city government of Yogyakarta has been successful in providing public service for the people. Meanwhile, the public space is mostly used for field, playground, and park as shown in figure 6. The public spaces need to be optimally used in such a way for the sake of the people's interest.

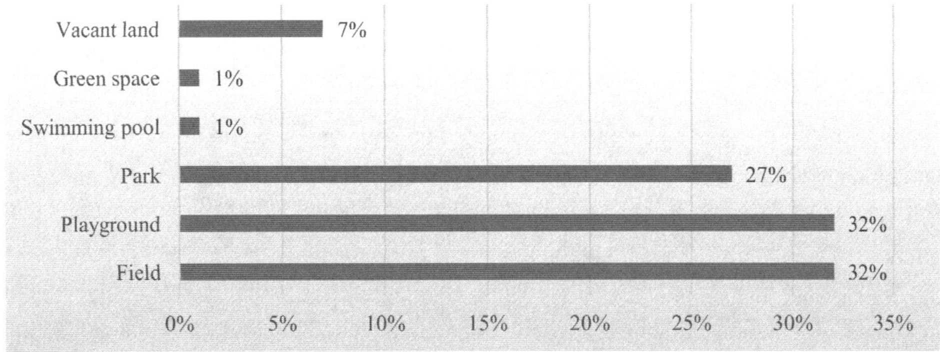

Figure 6. Percentage of public space use around the riparian area in Yogyakarta City.

3.5. Problems related to the structuring for the river area

Based on the survey results, the majority problems related to the structuring of the river area are river pollution, much garbage dumped into the river, frequent flood and landside. Even though it has been described previously that most people have adequately thrown their garbage, much garbage thrown into the river is still found, as shown in figure 7. It is possible that the garbage does not come from the people there, but from outside. Based on the survey result of the collecting people aspiration, various problems can be the guidelines for the city government of Yogyakarta in taking measures of the structuring for the river area.

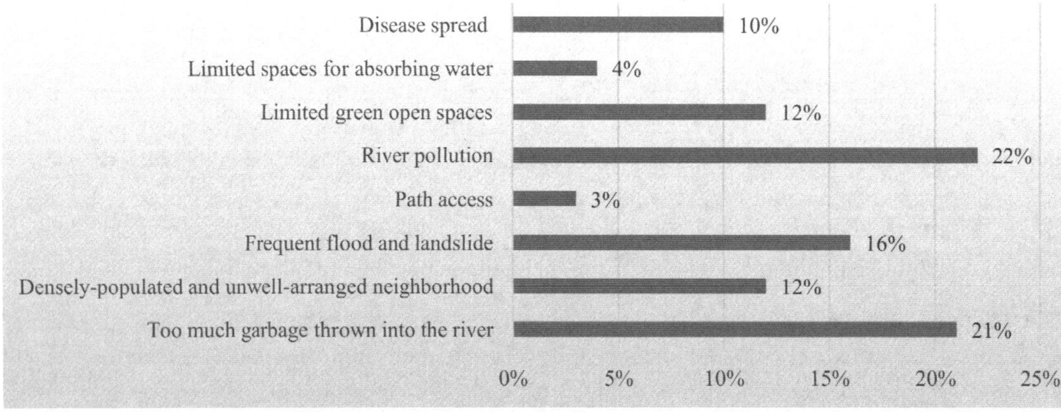

Figure 7. Percentage of Problems Related to the Structuring of Riparian Area in the City of Yogyakarta.

3.6. Efforts to improve the structuring of the river area and program priority
According to the result of the interview (shown in figure 8) there are three highest efforts to solve the arising problems related to the structuring for the river area i.e. anticipating the flood and landslide threats (21%), paying particular attention to the densely-populated and unwell-arranged neighborhood (19 %), and handling garbage dumped into the river (17 %). Other efforts to improve the structuring of the area on both sides of the rivers in the city of Yogyakarta are providing green open spaces, making more water -absorbing facilities, building flats, solving river pollution, and paying special attention to narrow and broken paths.

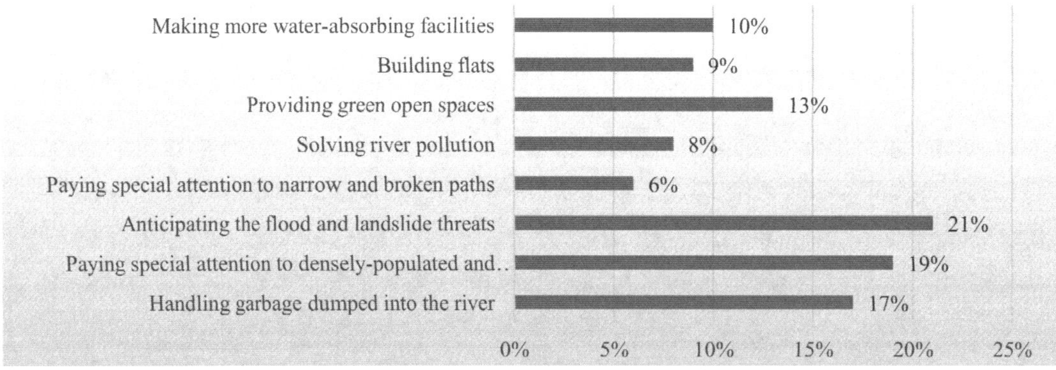

Figure 8. Percentage of efforts to improve the structuring of the river area in the City of Yogyakarta.

In relation to the program priority, the people living around the riversides are expecting the following; giving aids to those whose houses are not adequately used to live in, controlling houses and any other buildings uncompleted with a permit and constructing dikes for the river. Figure 9 shows the percentage of program priority of the structuring for the riparian area in the city of Yogyakarta.

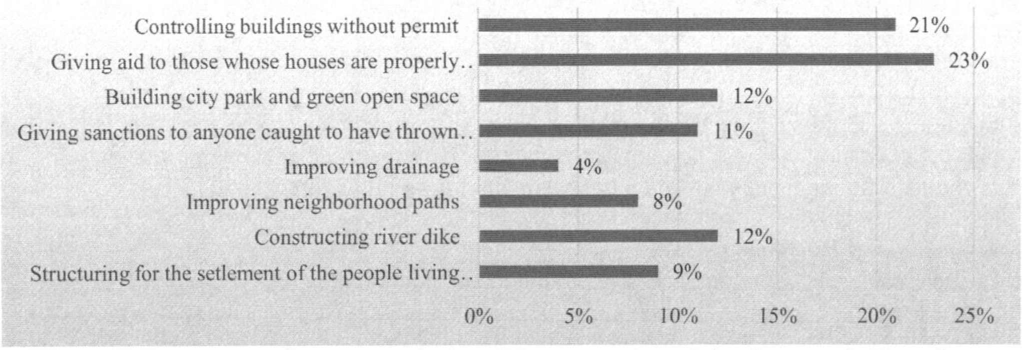

Figure 9. Percentage of program priority of the structuring of riparian area in Yogyakarta City.

In relation to these motions for program priority, it is necessary to synchronize them with the strategy related to the structuring for the river area based on the earlier study. For example, the result of research conducted in a slum area of the riversides in the sub-district of Sungai Raya, the Regency of Kubu Raya, recommended planning strategy in the structuring for the respective slum area, i.e., planning strategies for the building construction and infrastructure [10]. The handling settlements with waterfront concept can be expected to be one of the solutions that will minimize the risk of flooding. He states that this strategy is efficient, easy to apply and compatible with the environment, and can incorporate into the long-term development program for the city [11]. Therefore, this effort requires consistency of all stakeholders since the participatory approach of stakeholders is a key to the success of the efforts [11].

Evaluating implementation of the program is necessary. The implementation of the Program for Structuring and Renovating Facilities of Settlement in the Riparian Area of Code, especially the area located in the Village of Tegal Panggung, is considered being satisfying enough by the people there [5]. It is shown by the fact that 56% of the people there regarded the program for structuring and renovating the facilities in their neighborhood is an excellent success.

Meanwhile, local wisdom is a critical factor to be considered in the structuring for the river area. Local wisdom is fundamental in planning and designing site so that the space created can improve the quality of the urban environment and enhance the image of the area [12]. The aspect of local wisdom in Spatial Urban Design, which is the combination of the spatial analysis using Geographical Information System (GIS) and site planning is essential [12].

Besides the participation of the local people and local wisdom, it is also necessary to give understanding and knowledge to the local people about the existing plan products related to the space structuring in the riparian. It is referred to the research result at Krueng Aceh river [13]. The result shows that the respondents agree to the structuring for the border of Krueng Aceh river based on RTRW (*Rencana Tata Ruang Wilayah*/Regional Spatial Plan) of Banda Aceh City years 2009 – 2029 as well as the importance of improving the knowledge of the local people about the space structuring based on the plan. Nevertheless, plan product should be in the form of more detailed spatial plan. In addition, it is also necessary to integrate the risk analysis and spatial distribution of the risk included in the detailed spatial plan considering that the riparian area is an area susceptible to flood threat [1].

It is necessary to require society-based planning to make information, policies, and risk-minimizing program by integrating traditional aspects like the knowledge of the public [14]. The social capital of the river bank of Code has been empowered since 1970 with the active role of community leaders and supported by the regional institutions in achieving a better quality of environment [15]. Evaluation of the program achievement of the infrastructure development in the northern part of Code River related to structuring and rehabilitating slum area consist of water supply, path renovation, and solid waste handling [16]. The evaluation was involved the local society. The evaluation is focused on the problems before the program was set up, the benefits and impacts of the program, and responses to the program as well. The results show that the program of supplying water and renovating path could be accomplished. Meanwhile, the program of handling solid waste could not be continued.

Considering that those living on the riparian are mostly from low-income people, it is necessary to increase their economic condition to support the better settlement neighborhood. The better neighborhood in floodplain area through improving based economic activities [17].

4. Conclusion and Recommendation

The research results show that the problems found in the riparian area are a high density of population and limited land space. However, the people living in the area have done many things in structuring their neighbourhood. It is shown by their awareness and the presence of several communities in structuring the rivers and their surrounding area. Several steps they have taken are, for example, anticipating flood and landslide threats, giving particular attention to the densely-populated and unwell-arranged settlement, and handling garbage dumped into the river. They do hope to live in the area and that they do not mind being involved in structuring their neighborhood. Therefore, they propose a program to the city government so that it provides some aids for those whose houses are not proper to live in, controls any building which is not allowed (no permit), takes measures in structuring their neighbourhood, and builds river dikes and open green spaces.

Some recommendations resulting from this research and others that have been done previously are related to efficiency, meaning that it can be done by building *rusunawa* intended to those who previously resided in the area of riversides. Nevertheless, in implementing the recommendation to build a *rusunawa*, it is necessary to enforce the existing rules. They are, for example; the people should be from the local area (proven by their identity card), they should live in the area no longer than six years, they should be encouraged to have a place of their own to live in after living in the *rusunawa* for a specified period, and they should have income with the amount in line with the stated requirements [18].

Efficiency can be achieved through the balanced use of land space between the one for settlement and the one for green open space in the form of the park for both environmental beauty and economic purpose. In addition, efficiency can also be achieved by structuring the area with the concept of ecotourism [1] and community innovation. Integrating the economic values that are still possible to push but still based on the aspects of environmental preservation, empowerment of the local people, and sustainability is the key to the success of the program. One of the city government's steps through Bappeda (*Badan Perencanaan Pembangunan Daerah*/Regional Development Planning Board) is setting up a plan for developing green open space. An idea to build Code River View is one of the plans [19]. This idea is one of the efforts to make the riparian area at Code River well-arranged and that the economic values can grow through various events of ecotourism and innovation. The same case has been proven successful in the area of both sides of the river in other cities, like the one in the city of Malang. The concept of Kampung 3D and colorful kampung is, in fact, able to increase the economic condition of the local people there and improve the popularity of the kampung in particular and the city in general. Therefore, efficiency in land space use along with added value needs to be realized in its implementation in the context of achieving urban space structuring as one of the viewpoints in managing urban resources for better condition.

Research to the examined public perception of the disasters risk on the socio-cultural and regional environment needs to be done [20]. Some recommendation to the people who live in the high disaster risk area is raising awareness of hazards and the risk of disasters, including education and training, encouraging voluntary work and improving public access to vital information resources [20].

It is essential in the riparian area of Yogyakarta City related to improving public access to vital information resources. It is also will help the community in the area to get some informal education through a training program that can encourage them in increasing valuable social-economic condition.

A requirement to community resilience and to prepare them to cope with disasters is the crucial issues for the future research [20].

The same theme research is also needed to be done in the riparian area of Yogyakarta City. This theme is also an essential thing for the risk area of flood and lahars [21]. In relation to the arrangement of riparian area in Yogyakarta City, several risk reduction programs through adequate land-use planning and early-warning systems supported by education, capacity, and confidence-building efforts are needed, for handling flood and lava flood areas.

In relation to the programs that have been in progress, it is necessary to evaluate them so that the programs stipulated for the next steps are in line with the real needs for better implementation. Inputs and feedbacks from local society are of primary importance for the continuation of the programs and events. They constitute an effort to improve the participation of the local people in development, and that local wisdom can indirectly be accommodated. Thus, by doing so, the development that is implemented matches the real needs of the local people so that efficiency can be achieved from both sides, programs in line with the real needs and the arising problems. In relation to the survey of collecting people aspiration from the local people, it is right as one of the methods and efforts in accommodating aspiration from the local people in development, especially in the structuring for river area.

5. Acknowledgements

The authors would like to thank Bappeda of Yogyakarta City which has given opportunities in handling the Project of Survey of Collecting People Aspiration in the City of Yogyakarta. This paper is a part of the survey results that are rewritten and intended to disseminate the survey results with more analysis by referring to several references and previous research resultant.

6. References

[1] Rachmawati R and Budiarti C V 2016 Use of Space and The Need for Planning in the Disaster Prone Area of Code River, Yogyakarta, Indonesia Indonesian Journal of Geography (IJG)

[2] Lavigne Franck 1999 Lahar hazard micro-zonation and risk assessment in Yogyakarta City, Indonesia. GeoJournal. 49 173–183

[3] Rachmawati R 2014 Pengembangan Hunian Vertikal di Kota Yogyakarta: Kajian Aspek Lokasi, Ruang dan Perilaku (Yogyakarta: Fakultas Geografi Universitas Gadjah Mada)

[4] Rachmawati R, Budiarti C V, Febrita D and Sulistyani E 2017 Inclusive Development through Providing Vertical Housing for Low Income Family in Yogyakarta Urban Areas. Forum Geografi. 31 (2)

[5] Yulianti B P, Baiquni M, Ritohardoyo S, Yunus H S and Setiawan, B 2015 Evaluasi Penataan Kawasan Permukiman Kumuh (Studi Kasus: Program Peremajaan Kawasan Tegalpanggung di Kota Yogyakarta. Jurnal MGI (Majalah Geografi Indonesia). 29 7-18

[6] Rachmawati R 2015 Community Participation in Managing The Space and Settlement Environment on The Riparian Area: A Case Study in Yogyakarta, Indonesia. KKU International Journal of Humanities and Social Science. 5

[7] Bappeda Kota Yogyakarta 2016 Laporan Akhir Hasil Penjaringan Aspirasi Masyarakat Kota Yogyakarta (Yogyakarta: Government of Yogyakarta City)

[8] Bappeda Kota Yogyakarta 2015 Laporan Akhir Pendataan Perumahan dan Permukiman Kumuh Kota Yogyakarta (Yogyakarta: Government of Yogyakarta City)

[9] Widodo B, Lupiyanto R and Wijaya D 2010 Pengelolaan Kawasan Sungai Code Berbasis Masyarakat. Jurnal Sains dan Teknologi Lingkungan. 2 7-20

[10] Putro Jawas D 2011 Penataan Kawasan Kumuh Pinggiran Sungai Di Kecamatan Sungai Raya. Jurnal Teknik Sipil Untan. 11 19-34

[11] Irwansyah M et al. 2016 Int. Conf. on Alternatives for Settlement Area Management of Meureudu River watersheds towards Flood Mitigation (Case Study of Meureudu City

Center, Pidie Jaya–Aceh) p 63-71
[12] Wardhani D K Yudono A and Priambada, C K 2010 Spatial UrbanDesign pada Area Sempadan Sungai (Penerapan GIS dalam Urban Design). Jurnal Local Wisdom. 2 36-46
[13] Moulwian I Irwansyah M and Azmeri 2015 Kajian Penataan Sempadan Sungai Krueng Aceh Ditinjau Dari Rencana Tata Ruang Wilayah Kota Banda Aceh. Jurnal Teknik Sipil Pascasarjana Universitas Syiah Kuala. 4 95-104
[14] Cronin S J Gaylord D R Charley D Alloway B V Wallez S and Esau J W 2004 Participatory Methods of Incorporating Scientific with Traditional Knowledge For Volcanic Hazard Management On Ambae Island, Vanuatu. Bulletin Volcanology. 66 652–668
[15] Salmah and Sjarifah 2012 Social Capital: Strength and Defence in Riverbank. Jurnal Kesehatan Masyarakat Nasional 7 31-36.
[16] Rahayu T Sudaryono and Baiquni 2003 Evaluasi Program Penataan dan Rehabilitasi Permukiman Kumuh Studi Kasus Kawasan Bantaran Sungai Code Bagian Utara, Yogyakarta. Jurnal Manusia dan Lingkungan. 10 53-62
[17] Poedjioetami E 2008 Penataan Ulang Kawasan Bantaran Sungai dengan Menghadirkan Sentra Ekonomi dan Rekreasi Kota. Jurnal Rekayasa Perencanaan. 4
[18] Rachmawati R Sukoco E and Budiharto T 2014 Proc. Penataan Ruang dan Kapasitas Masyarakat untuk Pengurangan Risiko Bencana: Kajian di Kawasan Bantaran Sungai Code Kota Yogyakarta (Riau: Universitas Islam Riau)
[19] Bappeda Kota Yogyakarta 2014 Laporan Akhir Penataan Kawasan Pasar Terban (Yogyakarta: Government of Yogyakarta City)
[20] Alshehri S A Rezgui Y and Li H 2013 Public Perception of The Risk of Disasters in a Developing Economy: The Case of Saudi Arabia. Nat Hazards. 65 1813
[21] Kunzler M Huggel C and Ramırez J M 2012 Risk Analysis for Foods and Lahars: Case Study in the Cordillera Central of Colombia. Nat Hazards. 64 767–796

IOP Conf. Series: Earth and Environmental Science **148** (2018) 012010 doi:10.1088/1755-1315/148/1/012010

Monitoring the transformation of Yogyakarta's urban form using remote sensing and Geographic Information System

B Rozano[1] and W Yan[2]

[1]Regional Development Planning Agency of Sleman Regency, Jl Parasamya No. 1, Sleman 55511, Yogyakarta, Indonesia
[2]Environmental Design and Governance, Graduate School of Media and Governance, Keio University, 5322 Endo Fujisawa City Kanagawa Prefecture Japan, 252-0882

boby.rozano@gmail.com

Abstract. Urban form transformation can be seen as the results of urbanization spatially, where the land changed into an urbanized one. Monitoring its changes, however, require many human and financial resources. Accordingly, this research aims to identify urban form transformation using GIS/remote sensing and its spatial implications to the peri-urban area. In order to analyze the land cover changes, this research uses multispectral images from 1990-2016 for built-up extraction using New Built Up Index (NBUI) analysis and population data 1996-2015 combined with primary data from the respondents and key informants. Based on the analysis, it is seen that the compacted Yogyakarta Urban Area scattered predominantly to the Northern part of its periphery with the increase of urban area from 21.19% in 1990 to 50.91% in 2017. While this urbanization is an on-going process, the population of urban core showed a de-concentration phenomenon in 2015, spreading to its periphery causing some negative implications to the peri-urban area.

1. Introduction

Urban form affects habitat, ecosystem, endangered species, and water quality as reported on Our Built and Natural Environment by the U.S. Environmental Protection Agency in 2001, through land use changes and ecosystem segregation, thus, its transformation may also impact the functionality of people's living environment. To some extent, urban form transformation can be seen as the results of urbanization spatially, where the land changed into an urbanized one. The rapid one causes land transformation from agricultural, rural, and natural landscapes into urban areas [1]. Furthermore, it may lead to regional (urban – peri-urban – rural) problems due to its dynamic land use changes.

As one of the indicators on improving urban economic, urbanization is a global phenomenon especially in developing countries and it may lead to urban sprawl resulted in the decrease of vegetation area and deteriorating environmental quality. Just like an organism, the urban area may grow as its inhabitant grows with its certain structure. However, it could affect not only the urban itself, but also the peri-urban area, a transitional area between urban and rural, mixed areas beyond urban influence but with rural morphology in which many conflicts may arise, such as land use, economic, and social activities so that affecting environmental quality. Thus, one of the challenges is how to identify problems as implications of urban form transformation.

Urban expansion due to rapid urbanization also influences peri-urban area such as development problems from the insufficient infrastructure to provide its growing inhabitant to the air quality due to the increase in road traffic. This rapid urbanization also occurred in Yogyakarta, Indonesia, in which its urbanization level increased from 57.6% in 2000 to 66.4% in 2010, already passed the national level at 53% in 2010 [2].

Yogyakarta is a capital city of Yogyakarta Special Region inhabited by 412.331 people in 2016 known as the center of education and its classical Javanese art and culture. With the area of only 46 km^2, it is a compact city where its infrastructure and facilities fulfilled its citizen's needs and made this city as the most livable city in Indonesia [3]. The urban structure affecting the urban form is planned to fulfill the needs of its citizen.

However, the rapid growth of population of Yogyakarta urban area may influence the environmental quality both within or surrounding its periphery and at the same time results in problems such as land use conflict between agriculture activities and vertical housing, urban activities oriented in rural environment, to the social conflicts that objected to the urban development surrounding their residential areas.

In order to comprehend the urbanization processes spatially, monitoring on its spatial aspects need to be assessed so that data required for further spatial analysis can be conducted. Population data to explain the urbanization processes is important, however, it cannot describe the location aspects such as where is the direction of the urbanization processes. Monitoring of its changes also require many resources due to the vast area and occurred rapidly, thus, it needs to be identified and analyzed. Further, the implication of this urban form changes to other areas especially the peri-urban area also important to be assessed. The local community or the stakeholders may get the implication of this transformation. Shortly, based on the problems stated previously, this research has the main objective: "To identify urban form transformation of Yogyakarta Urban Area using GIS/remote sensing and assess the implications of urban form transformation to the peri-urban area.

Figure 1. Problems in peri-urban area due to urbanization (land use conflicts, social conflicts, urban-oriented land usage).

The urban form can be defined as the spatial organization of permanent elements within a metropolitan region including the spatial pattern of land uses and their densities and, as the spatial pattern of human activities in general, it can be categorized into density, diversity, and urban spatial structure pattern [4]. It can be differentiated into three archetypal forms (based on population and employment density):

1. The concentric city (urban economic theory). Central Business District is the main focal: the location with maximum employment density, the maximum number of trip ends, and the maximum rent.
2. The radial city. While the central business district is the main focus but sectors of intense land use stretch out along major lines of transport from CBD, leaving areas of sparse development between them.
3. The multinucleated city. This form is a more complex, hierarchical system of transport infrastructure where not all routes are oriented toward the CBD – the higher overall level of connectivity in the city.

The Urban form can also be viewed from various geographical scales and classified into such levels as metropolitan area, city, and neighborhood [5]. In metropolitan scale, urban form can be differentiated into several dimensions i.e. metropolitan size, density, unequal distribution, centrality, continuity, and other measures such as spatial autocorrelation.

Monitoring urban form transformation could be a burdensome work in planning and managing the urban area since it involves vast areas and needs a long-time period of observation. However, the advancement of technology in earth observation can be utilized accordingly. Multi-temporal satellite images can be useful in studying urban growth and planning [6], such as the Landsat images provided by the USGS.

There are many methods to analyze remote sensing data started from NDVI (Normalized Difference Vegetation Index), NDBI (Normalized Difference Built-up Index), to NBUI (New Built-up Index). The NDVI firstly developed by Tucker in 1977 to detect healthiness of vegetation by utilization of Near Infra Red/NIR Band and Red Band of satellite images. While the NDBI was proposed to detect the impermeable surface by utilization of Near Infrared/NIR and Shortwave Infrared/SWIR Band [7]. The least, NBUI, purposely to detect impervious surface material, green vegetation, bare soil, and water body (urban land use) [8]. Table 1 shows the three examples of methods used in satellite image analysis comparison.

Table 1. Comparison of satellite images analysis methods

Indicator on Landsat images analysis	Developed by	Usage
NDVI (Normalized Difference Vegetation Index)	Tucker, 1977	Vegetation healthiness detection
NDBI (Normalized Difference Built-up Index)	Zha, et.al., 2003	Urban impermeable surface detection
NBUI (New Built-Up Index)	Sinha, et.al., 2016	Impervious surface material, green vegetation, bare soil, and water body (urban land use) detection

This research adopted the NBUI analysis to monitor Yogyakarta urban form transformation from 1990 to 2017 utilizing Landsat 5 TM and Landsat 8 OLI/TIRS Level1. The NBUI use almost all wavelengths of Landsat images to represent major urban land use. The equation [8] is as follows:

$$NBUI = \frac{SWIR - NIR}{10 * \sqrt{SWIR + TIR}} - \left(\frac{(NIR - R) * (1 + l)}{NIR - R + 1} + \frac{G - SWIR}{G + SWIR} \right) \quad (1)$$

where, SWIR = Shortwave Infrared Band; R = Red Band;
NIR = Near Infrared Band; G = Green Band;
TIR = Thermal Infrared Band; l = 0-1 (l = 0 high, l=1 low density vegetation).

2. Research Site

This research focuses on Yogyakarta Urban Area, Special Region of Yogyakarta, Indonesia centered at Latitude 7°48'14.54"S and Longitude 110°21'51.78"E. According to Special Region of Yogyakarta Regional Spatial Plan 2009-2029/RTRW DIY 2009-2029, the Yogyakarta Urban Area consists of three municipalities, the Yogyakarta City as the city center, the Sleman and Bantul Regency as the peri-urban area, with the total of 71 villages while the relative position on its regional perspective can be seen in figure 2.

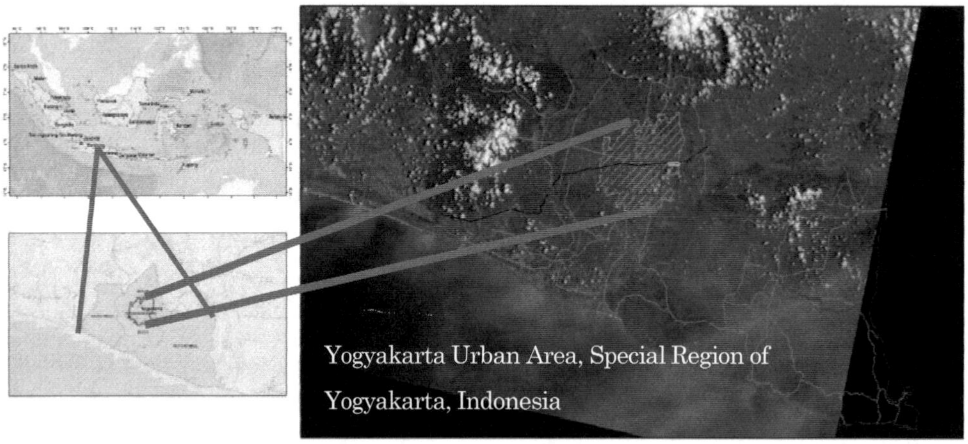

Figure 2. The research sites.

3. The Method

Based on the literature review on the previous section, urban form transformation is influenced by the growth of urban core and its interaction with surrounding city or villages represented by the road network. The urban growth leads to the expansion of its spatial interface into the peri-urban area and may overlap the existing village nearby. This peri-urbanization, not only result in some implications to the community inhabit this area, but also influence the overall environmental quality. Thus, the increase of the population will develop urban/built up density and the peri-urban will get negative implications due to unplanned/uncontrolled urban form transformation. Monitoring urban form transformation needs to be conducted so that the urban expansion can be anticipated in the future and possible impacts may be lessened. Diagrammatically, the conceptual framework of urban form transformation can be seen in figure 3.

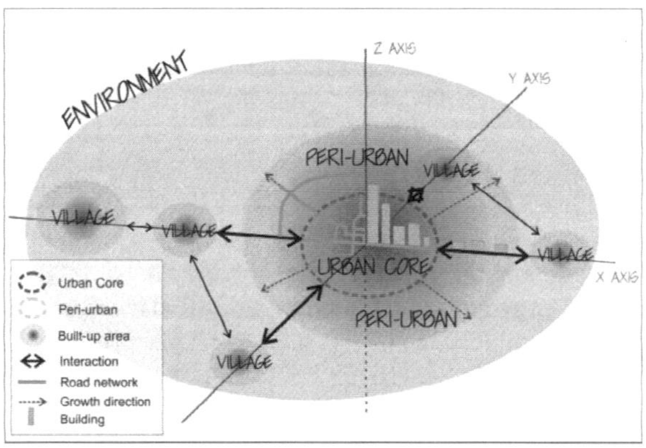

Figure 3. Urban form transformation conceptual framework.

In order to identify urban form transformation of Yogyakarta Urban Area and assess its implications to the peri-urban area, this research uses an integrated analysis system consists of several quantitative analyses. This research used multi-temporal Landsat 5 TM and Landsat 8 OLI/TIRS from 1990-2017 as the input for urban extraction using NBUI analysis to identify the urban density changes coupled with population data to describe the population density changes from 1996-2015 and the urban area size changes. Finally, the results of this research disseminated to the society through the utilization of 5D world map system [9].

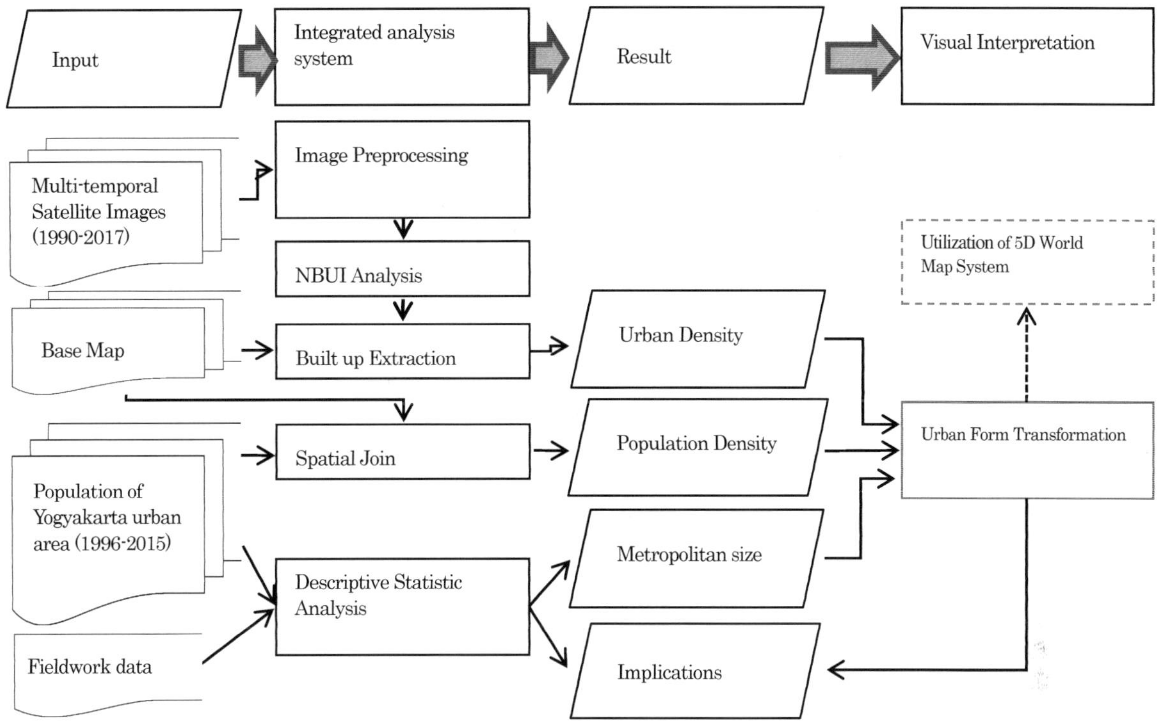

Figure 4. Research method.

Table 2. The research data.

Nr.	Data	Data Coverage	Period	Samples	Data Source
1	Satellite Images (Landsat TM 5 & Landsat 8 OLI/TIRS C1 Level1)	Yogyakarta urban area (Path 120, row 65)	1990, 1995, 2000, 2007, 2011, 2017	-	USGS - https://earthexplorer.usgs.gov/
2	Population	Yogyakarta urban area	1996-2015	-	Central Agency on Statistics, Indonesia
3	Base Map	Yogyakarta urban area	Till present	-	Yogyakarta Special Region, Local Government Office
4	Urban built up-development	Yogyakarta urban area	2016/ 2017	-	Field study/ Observation
5	Local community perceptions	Peri-urban area (Sleman Regency)	2017	120	Questionnaire (Proportionate sampling)
6	Land cover ground truth	Yogyakarta urban area	2017	113	Field study

Satellite images data (Landsat TM5 & Landsat 8 OLI/TIRS C1 Level 1, path 120, row 65) of the year 1990, 1995, 2000, 2007, 2011, and 2017 in this research were obtained for the built-up area and transformation detection and processed using Arc GIS 10.2. All of the data was pre-processed so that all images similar and able to be considered taken at the same environmental conditions, and by the same sensors and its digital numbers (DNs) was changed into radiance and/or surface reflectance for a quantitative analysis of multiple images on different acquisition time [8]. The radiometric calibration of Landsat 5 in this research followed the equations suggested by Chander and Markham [10], while the conversion of Landsat 8 into the TOA Radiance/Reflectance used the equation provided by the USGS. The acquired data also was geo-referenced to the WGS 1984 UTM Zone 49 S projection system so that it harmonizes with other map sources available especially the governmental data. More detailed data used in this research can be seen in table 2.

4. Results and Discussions

Urban development form in developing countries can be divided into the controlled residential-commercial area; unplanned peri-urban area; and satellite town [11]. It includes a number of physical and non-physical aspects including size, shape, scale, density, land uses, building types, urban block layout, and distribution of green space [12]. The growth of an urban influenced its form in the long term, and it may transform from one form to another.

The term of urban sprawl is also used, including contiguous suburban growth, linear patterns of strip development, and leapfrog or scattered development and sprawling forms can be reckoned to lie along a continuum from relatively compact to completely discrete developments [13]. A typology based on continuous dimensions can be used to describe urban forms: settlement density (high and low) and physical configuration (ranging from contiguous and compact to scattered and dis-contiguous).

Conventionally, in understanding the relationship between urban areas and their hinterlands, or people's activities in such area has been explained based on a simple urban-rural dichotomy by some social scientists studying on urbanization, assuming the peri-urban area as a short-term transitional area that had little interest or importance [14]. The peri-urban area is the term used to describe the transitional areas, mixed areas beyond urban influence but with rural morphology. It can be defined as areas around or outside the city center that are ecologically and socio-economically integrated into their core city [15]. It also describes the interface between urban, rural, and natural areas with relatively rapid growth, dynamic and mixed physical and socio-economic attributes [16].

Urban growth may expand into its periphery and transform its physical and non-physical condition into a more urbanized one. Its transformation which took place outside the urban cores refers to peri-urbanization processes [17]. It generally refers to the process of urban growth in contiguous transitional areas between city and countryside. The changes in peri-urban area are mostly caused by intensified pressure towards urban development [18]. The urban growth also explains the urbanization process due to its economic activities. Urbanization economies can be explained as factors stimulating many activities to be located in big cities where population concentrated and efficient urban infrastructure provided, thus strategic spatial plan on decentralization of urban core is needed [21]. The urbanization process also occurs in Yogyakarta due to its student and tourism city predicate as the strong pull factors [22]. However, the concept on urbanization based on population activities often neglecting the spatial aspects on the analysis. Low population in-migration many times interpreted also as low urbanization, conversely the land use change occurred rapidly especially in the peri-urban area and transforming the urban form. By utilizing built-up maps, the spatial aspects of urbanization can be analyzed.

As described earlier, by using remote sensing analysis, built up area maps can be extracted from Landsat 7/8 satellite images. However, compared to the real land coverage, the results may produce some errors compared and lower the accuracy. Thus, in order to assess the accuracy of maps generated, the ground truthing processes were conducted by comparing the image analysis result with actual land cover in the field with the total sample of 113 locations. It is found out that the accuracy for built-up is 87.5 % while in overall, the accuracy is 83.19 %.

Tabel 3. Accuracy assessment on NBUI analysis map,

	Ground truth:		Total
NBUI map:	Built-up	Non-built-up	
Built-up	49	12	61
Non Built-up	7	45	52
Total	56	57	113
Accuracy (%)	87.5	78.95	83.19

In 1990, the first period of monitoring, Yogyakarta showed a relatively compact city with the built-up area occupied 23.19% of the area while the non-built-up area 76.81% of the observed area. The core, the Yogyakarta city, showed a densely urbanized area while the peri-urban area dominated by non-built up area. In 1995, the urbanized area started to increase significantly, compared to previous year of observation, especially in the peri-urban area. The built-up area recorded at 33.51% more scattered in the peri-urban area while the non-built up area 66.49%. The built-up area in 2000 reached 37.72% and the non-built-up area 62.28%. There is an increase compared to previous year and its form almost the same, compacted in the city center and scattered in the peri-urban area. There is no increase recorded in the built-up area in 2007 compared to previous period. The built-up area almost remains the same at 37.21% and the non-built-up area at 62.79%. So, does the urban form, compacted in the core and scattered in the peri-urban area. Even though there is no increase in 2007, however, there is a significant increase in 2011 where the built-up area recorded at 47.65 % and the non-built-up area 52.3 %. While it is more scattered in the peri-urban area, the urban area denser in the Northern part of the urban. Figure 5 shows the detail on the increase of urbanized area. The increasing trend of the urbanized area continues in 2017, where the built-up area occupying 62.19% of the area, while the Non-built-up area continues to decrease to 37.81 % of the area. The peri-urban area now is dominated by a more urbanized land usage such as settlement, shops, hotels, restaurants, and education institutions especially in the Northern part of the urban area.

A well-planned city will have a good structure and land use pattern arrangement, of which the urban form is generated so that it will efficiently support its citizen. In the long term, the development of infrastructure will also improve the city condition and increase the pull factor of people to migrate by the creation of new opportunities. This urbanization phenomenon, however, explained by many researchers with the focus only on population, few of them focus on describing the urbanization spatially. This will explain the location factor, where the urbanization took place and which direction its trend of growth. Monitoring the urban form transformation is the first step to understand, not only describing the urbanization spatially, but also define its growth trend and pattern.

Based on its spatial pattern, the Yogyakarta Urban area in 1990 is more compacted in the core or the city of Yogyakarta. The urban density is higher in the center but is lower in the periphery. However, in 1995 the peri-urban started to be more urbanized where the built-up area started to increase significantly, from 22.88 % in 1990 to 33.40 % in 1995. One factor that influences the increase is the development of Yogyakarta Ring Road (the Northern and Western part) in 1994-1996. The development of new road improves people's accessibility and lead to both direct effects in the form of reduced journey time, reduced costs, and improve reliability and indirect effects, changes that are stimulated by the direct effects: changes in investment, transport service supply and demand, competition, etc [19].

The increasing trend continues in 2000 but remains steady in 2007. Nonetheless, there is a significant increase of built-up area in 2011, from 37.21 % in 2007 to 47.65 % in 2011, and in 2017 the peri-urban become more urbanized, the built-up area becomes 62.19 % in this year. Figure 5 and figure 6 illustrate the pattern and the changes of urban form in Yogyakarta Urban Area.

The images extracted from the previous analysis show the form of Yogyakarta urban area transformed from a relatively compact city in the early year of observation to be more scattered in the periphery in the last year of observation, even though the center is still compacted. The Yogyakarta city remains as an urban center while its size enlarges to the periphery.

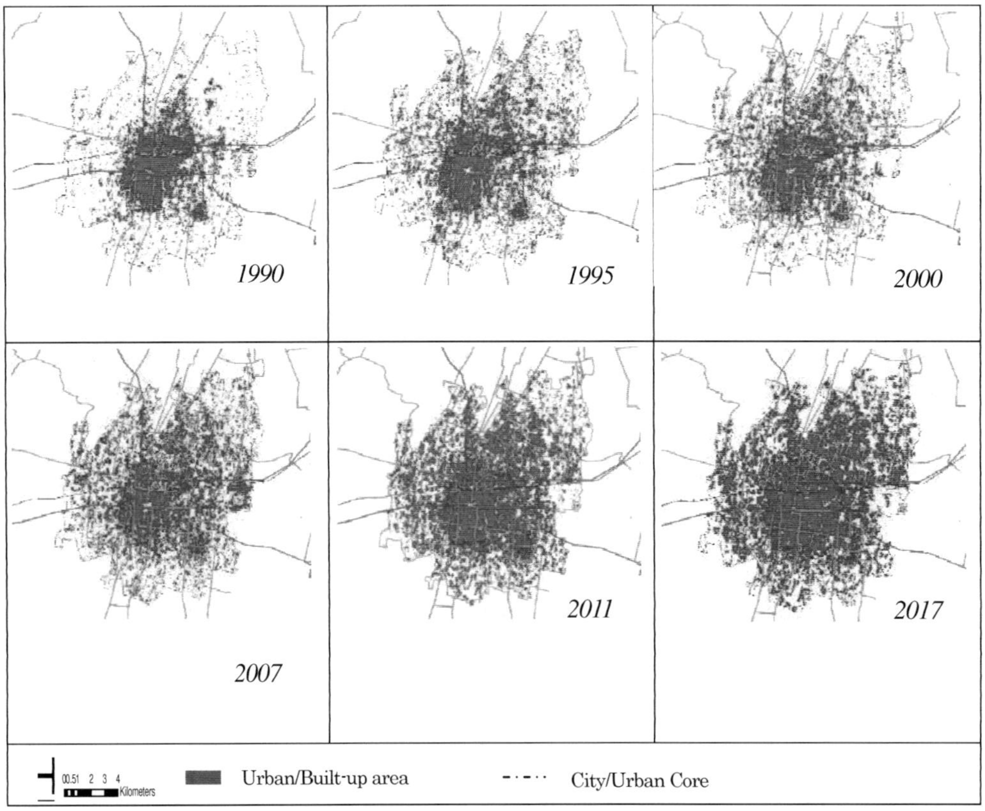

Figure 5. Yogyakarta urban form transformation 1990-2017.

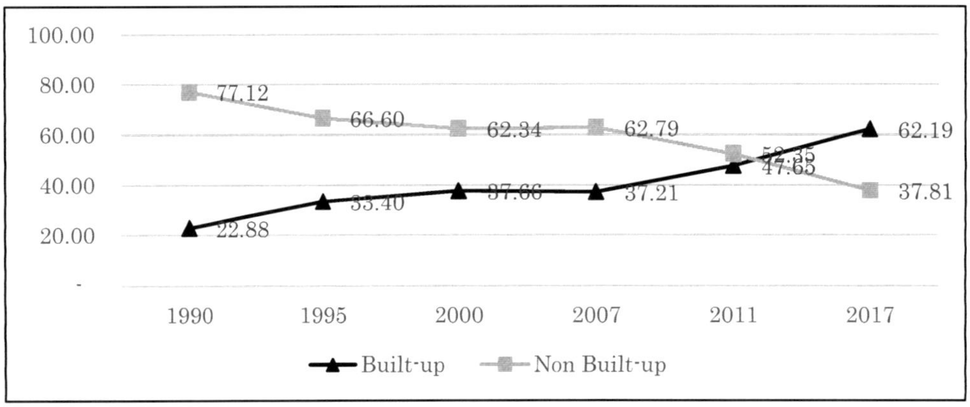

Figure 6. Yogyakarta Urban Area changes 1990-2017

Using population data as other indicator in monitoring urban form transformation, Yogyakarta Urban Area was concentrated in its core and classified into five categories: very low, low, medium, high, to very high. The area with higher population density located in the urban center varies from medium to very high density. In contrast, the peri-urban area is dominated by very low and low-density population. Using 1996 data as based on observation and to normalize following year data, population growth trend can be monitored spatially. It started to sprawl to the Northern and to the Eastern Part of the area in 2000 and continue in 2005. The expansion of population in 2010 undertook to the Eastern and some Western part of the area. However, the density trend in 2015 started to expand to the Southern part of the area.

By observing figure 7, it can be seen that the density at some part of the area both in the urban and peri-urban area increased during 1996-2005. Nonetheless, while the peri-urban density continued to grow until 2015, the density at some part (eastern) of urban core started to decrease. The density de-concentrate in the urban core and may indicated as the occurrence of counter-urbanization, migration from the city center to the peri-urban.

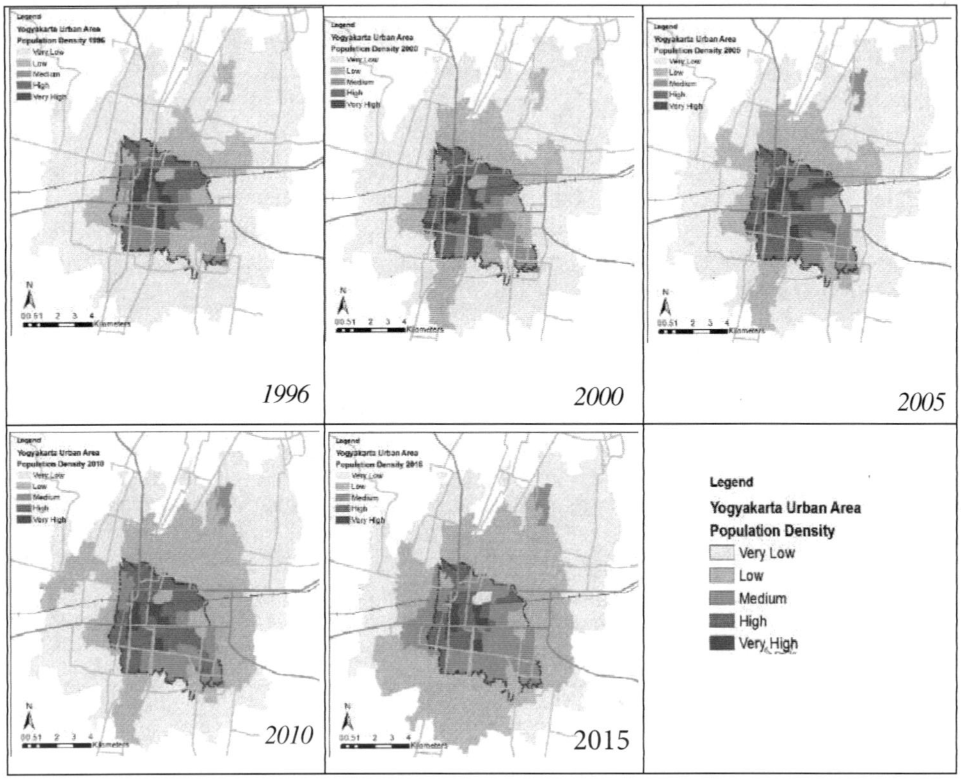

Figure 7. Population density changes 1996-2015.

As described earlier, urban form transformation results in several effects to the peri-urban area such as the loss of the non-built-up area. This area is characterized by the coexistent of agricultural and urbanized activities and functioning as supporting the inner city. Not only securing the food supply but also maintain other important environment function such as air quality and groundwater catchment area.

Some field observation and interview to key informants for a triangulation process were conducted to explain the implication of urban form transformation to the society. Triangulation is a measurement technique to put an object in space by depending on two known points in order to assess on an unknown fixed point in that space and use this concept in the validation process of social research results [20]. The in-depth interview steps are initiated with stakeholders mapping process so that it can be used as further guidance and analysis.

The peri-urbanization, as the consequences of the urban growth, realized by some community members and worrisome their daily lives. New urbanized development surrounding their neighborhood, to some extent, is a good thing resembling grow and overall economic condition improvement. They hope getting benefit from the multiplier effects of the new development, whether directly and/or indirectly benefitted. To the shop owners, the development means bigger opportunities for new potential consumers. Producers also could increase their production capacity to fulfill new demands. However, there are some people who have anxiety towards the effects this new urbanized development.

There are some negative perceptions given by the respondents towards the transformation of the urban form. The transformation, which is represented by peri-urbanization, has been blamed as one of the factors on the increase of average daily temperature. The peri-urbanization felt by some of the respondents, to some extent, contribute to the global warming by lessening the number of trees while increasing the number of vehicle and the needs of electricity and energy. Respondents believe that the presence of trees surrounding their residential area will contribute to better air quality and lower the average temperature.

Since the peri-urban area used to be dominated by agricultural activities, many roads are built without pedestrian facilities such as the sidewalk. Many respondents felt discomfort if they have to walk at roadside since their safety is threatened, despite the good traffic conditions. Their surrounding environment recently is more urbanized, but the infrastructure provided was focused on agriculture activities.

Another issue as the effects of peri-urbanization is the people's anxiety of the availability of groundwater. Almost all respondents use well as a source of clean water since they are not living in an area covered by local clean water provider agency network. The urbanized development nearby their residential, such as apartments or hotels will require plenty of clean water for their activities. Thus, since their area is not covered by the clean water network, respondents have doubtfulness towards developers in fulfilling the groundwater extraction regulation. Another reason that the development would impact the water catchment ability of the peri-urban area, since more ubiquitous impermeable surfaces.

5. Conclusions

Based on data results, analysis, and discussion in previous sections, it can be concluded that:

1. The Yogyakarta Urban Area form experienced a transformation from a relatively compact in its core in 1990 into a more sprawled and scattered urban area in 2017, followed by the increase of built-up area from 22.8% to 62.19% respectively.
2. While its form remains compact in core since 1990, Yogyakarta Urban Area scattered to its peri-urban dominated by the development of the built-up area to the Northern part of the urban area, with the speed of 1.46% per year.
3. The Yogyakarta Urban Area showed a symptom of population de-concentration at its core in 2015, while the peri-urban area population density increased gradually. It is also followed by the new development of a more urbanized land used in the peri-urban area.
4. Yogyakarta urban area as a metropolitan area required new infrastructure, i.e. new road, however, it drives the built up to increase rapidly in the peri-urban area.
5. The Yogyakarta peri-urban communities faced some negative implications due to urban form transformation.

6. References

[1] Musakwa W and van Niekerk A 2014 Monitoring urban sprawl and sustainable urban development using the moran index *Int. J. of Appl. Geo. Res* 5(3) 1–20

[2] Badan Pusat Statistik BKKBN Kemenkes and Macro International Inc. 2013 Survei demografi dan kesehatan Indonesia 2012: laporan pendahuluan Badan Pusat Statistik Jakarta

[3] Indonesian Association of Urban and Regional Planners (IAP Indonesia) 2015 Indonesia Livable City Index 2014

[4] Anderson W P Pavlos S K and Eric J M 1996 Urban form, energy, and the environment: A Review of Issues, Evidence, and Policy 33 7–35

[5] Tsai Y 2005 Quantifying urban form: compactness

[6] Varshney A 2017 Improved NDBI differencing algorithm for built-up regions change detection from remote-sensing data: an automated approach 7058

[7] Zha Y Gao J and Ni S 2003 Use of normalized difference built-up index in automatically mapping urban areas from TM imagery *Int. J. Remote Sensing* 24(3) 583-594

[8] Sinha P and Verma N K 2016 Urban built-up area extraction and change detection of adama municipal area using time-series landsat images 5(8) 1886–1895

[9] Kiyoki Y Chen X Sasaki S and Koopipat C 2016 Multidimensional semantic computing with spatial-temporal and semantic axes for multi-spectrum images in environment analysis *Inf. Modelling & Knowledge Bases* XXVII (280) 14

[10] Chander G and Markham B 2003 Revised Landsat-5 TM Radiometric Calibration Procedures and Postcalibration Dynamic Ranges *IEE Transaction on Geoscience and Remote Sensing* 41(11)
[11] Permana A S Perera R and Kumar S 2008 Understanding energy consumption pattern of households in different urban development forms: a comparative study in Bandung City, Indonesia *Energ. and Pol.* 36(11) 4287–4297
[12] Dempsey N Brown C Raman S Porta S Jenks M and Jones C 2010 Elements of urban form
[13] Besussi E Chin N Batty M and Longley P 2010 Remote sensing of urban and suburban areas 10
[14] Simon D 2008 Urban environments: issues on the peri-urban fringe *Ann. Rev. of Envi. and Res.* 33(1) 167–185
[15] Simon D McGregor D and Nsiah-Gyabaah K 2004 The changing urban-rural interface of African cities: Definitional issues and an application to Kumasi Ghana *Env. and Urbanization* 16(2) 235–248
[16] Allen A 2003 Environmental planning and management of the peri-urban interface: perspectives on an emerging field *Env. and Urbanization* 15(1) 135–148
[17] Zasada I 2012 Peri-urban agriculture and multifunctionality: urban influence, farm adaptation behavior and development perspectives 189
[18] Bertrand N 2007 Introduction: ESDP ideals and the inheritance of rural planning failures Europe's city-regions competitiveness: growth regulation and peri-urban land management 1–35
[19] Goetz A R 2011 The global economic crisis, investment in transport infrastructure, and economic development *Transportation and Economic Development Challenges* 53
[20] Mertens D M, & Hesse-Biber S 2012 Triangulation and mixed methods research *Jour. of Mixed Meth. Res.* 6(2) 75–79
[21] Tjiptoherijanto P 1999 Urbanisasi dan Pengembangan Kota di Indonesia *Populasi* 10(2) 57-72
[22] Fikri A A H S Sholeh M Baroroh K 2016 Fenomena kemiskinan perkotaan (urban poverty) di Yogyakarta: suatu kajian struktur dan respon kebijakan

IOP Conf. Series: Earth and Environmental Science **148** (2018) 012011 doi:10.1088/1755-1315/148/1/012011

Working in the global world: looking for more modern workplace overseas

Agus Joko Pitoyo[1]

[1]Faculty of Geography, Universitas Gadjah Mada, Yogyakarta, Indonesia

aguspit@ugm.ac.id

Abtract. International labor migration overseas is very complex and involves multiple dimensional issues, ranging from economic to social, political, and cultural issues. A very long history migration has affect the live of households which has been dependent very much on it, not only in economic terms but also in broader aspects.This research aim is to understand the history and process of international labor migration from the District of Ponorogo. The research was done using mixed method design, combining quantitative and qualitative approaches. Firstly, a survey was done by involving about a thousand households and followed with qualitative method applying in-depth interviews and FGDs to sub-samples households and some key informants. The results have shown that intergenerational migration has established in Ponogoro. There has been an expansion of destination countries with the establishment of many international migration routes. Migrant workers from Ponorogo have traveled to very distant continents and to more modern countries.

1. Introduction

The history of migration between regions is comprehensively described by [5]. Migration is a spatial interaction which already has been long occurred in the development of human live. As the interaction between geographical regions, population decided to move from one place to another mainly for surviving. In the history of human, people were dynamic and performed the geographical migration in order to fulfill their life necessities [8]. So, perspective of regions was more corresponding as the basic analysis of population migration [11]. Dynamic of sending migrant workers Indonesia was inseparable from the history of international population mobility in the world. That is why, it is important to understand the dynamic of Indonesian labour overseas through history.

At least there were five big phases of international migration which influenced the population dynamic of the world [9]:

- era of the slave trade
- era of indentured workers
- era of Europe to the new World
- era after World War II
- era of doors closed to Europe.

During the slavery era, geographical migration was more caused by reason of food gathering. This phase took place since people started to do hunting and fishing or food gathering which continued until at the early of the Middle Ages (about the 15th century). As the example of international migration in the Middle Ages was the invasion of Ancient Mongol and Roman in the world. The slavery was recognized as the beginning of labour migration history in the world. It was estimated that around 40 million of world population who are recently in America were from the slavery era. Even though now slavery has been eliminated, labour migration in other forms, as indentured workers, keeps on occurring.

Next era was indentured labour and the coolie system. Along the passage of time, started from the middle of the 19th Century, the slavery was gradualy turned into forced labour system. The coolie system was emerging in during the colonial era by sending many labours from China, India and Asia to British colonies, France, German and Dutch, such as to North America, Caribbean Islands, Africa, Asia and Australia. Number of collie workers to be sent at that time was estimated as 37 million people. Migration under the collie system kept on happening until the end of World War I in 1941.

Around 1840 to 1939 was the era of industrial revolution in Europe which followed by colonial period of new countries or known as from Europe to the "New World". This phase was assumed to be the beginning of modern labour migration under the reason of economic development and profit oriented. It was stated that in the period of 1891-1920 as much as 27 million of population, mainly from South and East Europe, have left Europe. This continued to happen until the First World War, but at that time the number decreased and then stopped at the Second World War. During the period of 1846-1939, 51 million of population has left Europe. Detailed according to the destination countries were United States of America (38 millions), Canada (7 millions), Argentina (7 millions), Brazil (4.6 millions) and around 2.5 million went to Australia, New Zealand and South Africa.

The next phase in the history of international migration of world population was after the Second World War (1945-1973). During that time emigration of population in the large number took place after the Second World War, such as people from England, Italy, German and Dutch headed to Australia, Canada, United States, South America and Israel. United States became the main recipient countries of migrants from Europe with 2.3 million of migrants in 1946-1963. Corresponding to that, Canada and Australia also received approximately 2 million migrants. During 1948-1963 there was a large number of Europe people migrated to Israel so at the end of 1963 around two third of Israel population were immigrants from abroad.

Other side of after the Second World War was that some countries, such as German, British and France, needed labours from the neighboring countries to work on various public facilities reconstruction, such as from Italy, Portugal, Spain, Yugoslavia and Turkey. Aside from European labours, many labours were also imported from Asia and Africa. In the reconstruction era, it was estimated that around 10 million people were entering West Europe.

Year 1973-2002 was the era for "Closing Europe for Migrants". Year 1970 was the peak of number of people heading for Europe. As for the example, British has been accepting immigrants from the former British colony which were incorporated in the British Commonwealth so many immigrants entered the UK until the end of 1960s. The economic growth has increased rapidly which reached to 5 percent each year at that time. The economic problems started to occur in 1973 when oil price shock took place. This condition has caused the migrants to start expanding their destination countries not only to Western Europe, but to Southern Europe, such as Italy, Spain and Portugal.

The closing of Europe did not just cause the economical recession, but also raised the number of refugees and/or asylum seekers caused by the collapse of communism in Eastern Europe. The collapse of communism in the middle of 1980s has caused people of Eastern Europe to be more freely in choosing their residence. It was reported that in 1989-1998 there were more that 4 million of people were seeking for asylum in Europe (43 percent from Europe, 35 percent from Asia and 19 percent from Africa). This problem then tricked the European countries to limit, even to shut the migration entering Europe. This migration happened in all over the world heading to countries which were assumed to be able to provide security (figure 1).

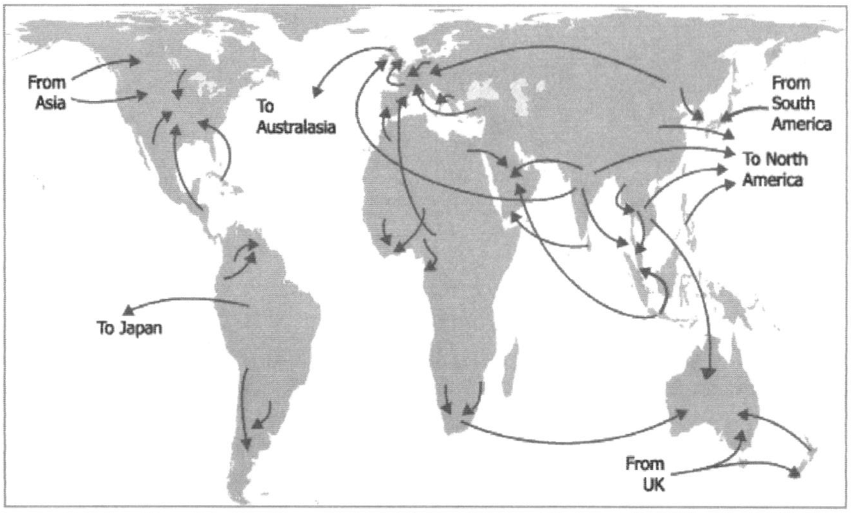

Figure 1. The direction of international Migration in the world after World War II.

The drastically increase of oil price in 1973 in Europe which followed by the closing of Europe to the immigrants has given a positive impact on oil producing countries in the Middle East. The economic growth of oil producing countries, such as as Saudi Arabian, United Arab Emirates, Quwait, Bahran and Iraq, increased rapidly during 1974-1980. This condition also caused a rapid development in many living sectors in those countries. The needs of labours was also increasing, especially in domestic sectors, such as housemaids, baby sitters and drivers, as well as contruction sectors, such as construction workers.

The history of international population migration in Europe also connected to the sending of Indonesia labours to overseas, spesifically from Kabupaten Ponorogo. People in Ponorogo also had a long history in sending their migrants workers abroad. In general, the overseas workers sending was categorised into two periods: the period before the independence and the period after the independence of Indonesia. Before the independende, the history of Ponogoro's labours sending has started. Indonesia has been a part of maritime trading in the South East Asia for a period of time along with Malaysia, Singapore and the Phillipines which used to be knows as Malay World.The trading and migration in these regions already happened in the past which started from the glory of Malacca Kingdom and involved many people in the economic activities from Sumatera, Singapore and Malaysia. The phenomenon of *merantau* (going abroad) for people from Sumatera has also supported the development of trading in Malacca and the regions around it [7].

The sending of huge number of labours during the Netherlands colonial happened when the implementation of forced cultivation in the colonial of Netherlands. The sending of labours abroad at that time was closer to the term of forced cultivation workers or contract-coolie instead of migrant workers (Hugo, 2004). Many of forced cultivation workers were sent to several Dutch East Indies colonies, such as to Malaya, Suriname and Caledonia, as construction and agriculture workers. In 1930 many of Javanese people were sent as agriculture workers, such as to Malaya (259,735 people), Suriname (31,000 people) and Caledonia (6,000 people). Asides from that, around 8,000 Javanese people also worked in Sabah dan Sarawak [1].

However, based on the data provided, in 1860-1930 Indonesia also received many migrants which dominated by Chinese and European people [3]. As recorded during 1860-1900 the number of Chinese and European people was doubled, so was during 1900-1930. The Netherlandscolonie in Indonesia at that time has opened opportunity of labours in the industry and trading sectors. The growth of Chinese and European people in Indonesia happened during 1860-1930 and in general has increased the number of the population continuously (table 1).

Table 1. The Dynamic of Indonesian Population based on Indigenous and Non-indigenous People, 1860–1930.

Year	Indonesian Indigenous People	Non-indigenous Poeple		
		European	Chinese	Others
1860	15,409,944	48,876	221,438	8,909
1900	34,666,659	91,142	537,316	27,399
1930	59,138,067	240,417	1,233,214	115,535
Average Annual Rate 1860–1900	2.05	1.86	2.23	2.85
Average Annual Rate 1900–1930	1.78	3.28	2.80	4.91

2. The dynamic of Indonesian labour working overseas

Since the beginning, Malaysia has been major destination for Indonesian migrants, and the number kept on increasing in the 1930s. Meanwhile, the dominant motives of Indonesia migrated to Malaysia for the purposes of trading and working in the plantations. The direction of migration of Indonesian natives to Malaysia was sporadic from various regions in Indonesia (figure 2). Most of them came from Java ethnic, including Ponorogo, then other ethnic groups, namely Banjar (South Kalimantan), Minangkabau (West Sumatera), Batak Mandailing (North Sumatera), Bugis (Makassar) and Bawean [1] [2] [3].

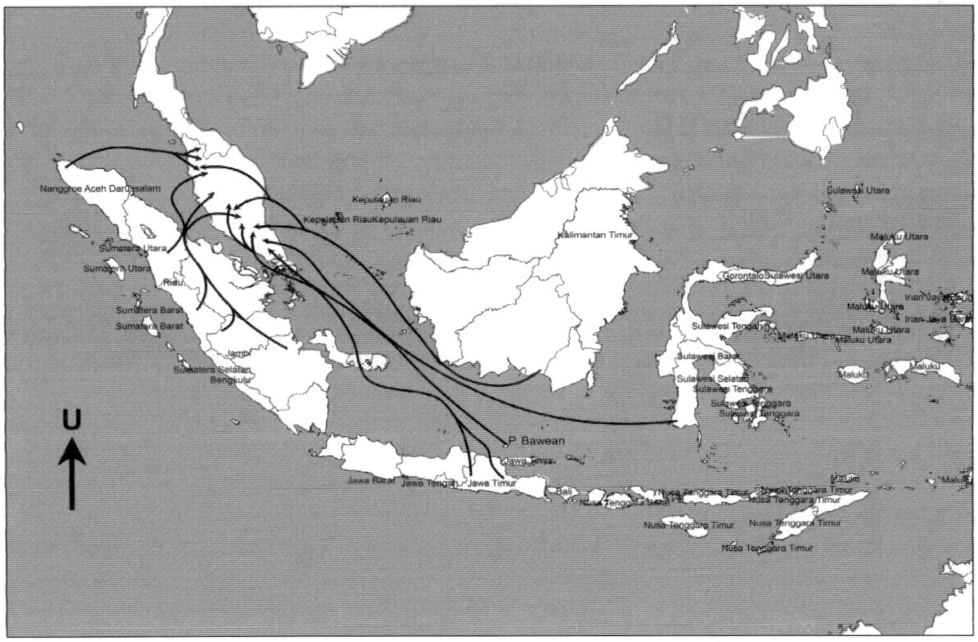

Figure 2. Direction of migration of Indonesian people to Malaysia before the independence of Indonesia.

The history of labour migrants from Ponorogo to overseas before the independence was also associated with the development of international migration in the world. There were at least three important things which related to the history of migrant workers from Ponorogo. First was the history of Ponorogo; second was religious journey for hajjs and thirdly was the colonization of the Dutch colonial era.

At the end of the Kingdom of Majapahit, migration of Ponorogo's population has occurred, primarily the migration under the purpose of trading towards the ports in the coast of Java, such as Demak, Jepara, Tuban, Gresik and Surabaya. The ports grew to be small kingdoms which had trading relations with Malacca and then separated themselves from Majapahit. The strong interaction has created the trading system known as Malay World. The interaction between Ponorogo residents and ports in the Java coastal regions, mainly with Demak and Surabaya, grown to be strong since they had a leader who was a descendant of Prabu Brawijaya [10].

Interaction between Demak as a great empire with Kadipaten Ponorogo has existed since the beginning of 16th Century. At that time Demak was led by Raden Patah who was Prabu Brawijaya's son. Meanwhile at the same time another son of Prabu Brawijaya, Raden Katong, also established Ponorogo under the assistance of Seloaji and Kiageng Mirah [6]. Asides from being the sons of Prabu Brawijaya, cooperation between Raden Patah and Raden Katong was also in form of trading between the fellow Moslems. Initially the relationship between Ponorogo's residents with ports on the north coast of Java was originally purely economic and trading as fellow descendants of Majapahit, but gradually it became religious relationship. Direction of migration of Ponorogo population during development of Demak Empire to the outside areas was done through ports of Java's coastal areas, such as Demak, Jepara, Tuban, Gresik and Surabaya, then connected heading to Malacca.

During the Dutch colonial, migration of Ponorogo's population to outside areas occurred through a mechanism of contract workers or coolie system. Migration of workers from Ponorogo in the era of colonization took place towards the plantation areas were opened by the Dutch Government. Plantation areas which were developed at that time were the plantations in Sumatra, such as in Lampung, Jambi and Metro. Around 1938 the Dutch Government brought population of Madiun, Kediri, and Ponorogo toward Lampung, Jambi and Metro. Then the sending of workers continued under the control of Dutch Government until the arrival of the Japanese army in 1942 [10].

Long before coolie system was applied, some history books described that in the early 19th century there had been a transfer of workers from Ponorogoto Malaysia. In the 19th century in the Malay Peninsula which was occupied by the British colonial, the economic expansion occurred in three sectors, namely food agriculture, export agriculture and mining, thus a large amount of labours was required. Having small number of residents in the Malay Peninsula caused British colonial had to import people from China, India and Java.

There were many of Java's population who has migrated to the Malay Peninsula. Initially most of them worked as labours on farms and exports, such as sugar cane, coffee, pepper, cinnamon and tobacco. The rubber industry in the Malay Peninsula began to grow in the last decade of the 19th century. At that time of British colonial ordered the businessmen to plant rubber since the crops had better future compared to other types of crops. Rubber prices were also high at that time. Then the British Government created policies which forced the businessmen to plant rubber. After that, they recruited many of Java's population to work as rubber plantation workers. This process continued until the beginning of the 20th century. The rapid development of plantation crops, farm exports and mining have caused many residents of Java with illegal channels migrated to the Malay Peninsula to work.

Path of migration of Javanese people (Bawean, Bali, Madura, Demak, Gresik, Tuban, Lamongan, Surabaya, Tulungagung, Ponorogo) followed the path of both legal and illegal courses. Transportation lines used was the sea pathway ship. There were at least five pathways for Ponorogo population to migrate to Malaya as follows.

1. Java–Mallaca
2. Java–Singapore–Mallaca
3. Java–Singapore–Penang–Mallaca
4. Java–Belawan Port–Singapore–Mallaca
5. Java-Jambi Port- Mallaca.

The first pathway, which was Java directly to Mallaca, was conducted under the coordination of British Government to transport workers from Java with the shipping pathway to go directly to the

Malacca. The need for labours in the food crops required the British Government to seek for cheap labours from Java. Labours were collected through several ports in Java, such as Port of Batavia, Cirebon, Demak, Semarang, Tuban, Gresik and Surabaya. Migrant workers from Ponorogo to Malacca passed the pathways in the ports of Surabaya, Gresik, Tuban, Bawean, and Demak.

The second pathway was from Java towards Malacca, but a stopped over in Singapore. As the first pathway, the second pathway was also carried out under the coordination with the British Government. Recruitment of workers was conducted by the agency in coordination with the Government of the Dutch East Indies in Java. The labours were not only heading to Malacca, but also to the Port of Singapore. The rapid grow of plantations in Singapore also required a lot of workers. Therefore, not all Javanese workers were brought to Malacca since most of them were then employed in Singapore. As in the first pathway, Ponorogo migrant workers sent through the ports in Surabaya, Gresik, Tuban, Bawean, and Demak.The third line was the route to Malacca but stopped over in Singapore then Penang and Malacca to the final. This third pathway was conducted since there was the development of export plantations, mainly in pepper and rubber. The British government opened a lot of plantations, including in Penang. Recruitment of workers was increasing and conducted by the British government with the help of the agency. Coordination with the Government of the Dutch East Indies in Java also performed. The Dutch East Indies Government permitted the placement of workers as long as there was a guarantee of protection for workers from Java. This pathway was more complex since it must go through Penang prior to Malacca. Migrant workers from Java, including from Ponorogo, were sent through the ports in Surabaya, Gresik, Tuban, Bawean and Demak.

The fourth pathway was the pathway of migrant workers from Java to Malacca, but through Belawan Port and then to Singapore and lasted in Malacca. This pathway was the result of the British Government agreement with the Dutch East Indies. Trading and transportation between the ports of Belawan, Penang, and Malacca have facilitated the delivery of labours. In addition, the Dutch also had an interest in bringing in labours from Java to work on the plantations of Sumatra.

The fifth pathway was the sending pathway of migrant workers from Java to Malacca via Port of Jambi. The fifth pathway was supported by two important things, namely the rapid export of agricultural plantations, such as rubber, and the enactment of forced cultivation system by the Dutch Government. On this pathway the Dutch East Indies Government had an interest also in sending a lot of migrants from Java to be hired cheaply in many plantations in Sumatra through Port of Jambi. Therefore, most workers from Java were dropped in Jambi while others were headed towards Malacca.

The migration process of Javanese workers from the first to fifth pathways is ongoing. The increasingly complex relationship facilitates the movement of migrant workers from Java without having to go through a formal recruitment process. As lasted until now, many migrants who worked in Malaysia plantations were through these pathways which have formed since a long time ago. The switching was done amicably, by intermediaries of relatives and friends. Those longstanding relationships coupled with the rapidly evolving information technology have been further simplifying the process of migrating workers from Ponorogo to Malaysia. The displacement of workers into Malaysia continues to grow increasingly complex, not only as a destination for workers from Java, but also workers who come from Sumatra, Kalimantan, and Sulawesi.

Sending of migrant workers in the period after the independence of Indonesia could be divided into two phases. First was after the independence until the late of 1960s (the Old Order era) and second was since the New Order era (1970 to the present). Sending of migrant workers at the Old Order era tended to be individuals and/or based on kinship using the trading pathway mechanism and the departureswere still traditional. Some studies suggested that the role of government was still low, while the presence of friends, intermediaries and brokers/*taikong* tended to be more dominant in this process. The main destination countries were Malaysia and Saudi Arabia. These two countries have become major destinations since the similarity of religion and especially for Malaysia, it was also confirmed that there were existence of a common language and cross-border relations between countries. The departure to Malaysia was usually temporary because they only worked for the provision of achieving the ultimate goal, namely to Saudi Arabia to perform the Hajj. Before and/or after the pilgrimage, many Indonesian migrants under the status of migrant workers worked in Saudi Arabia. This process went on so many Indonesian people later settled and became citizens of Saudi Arabia.

> "It has been long, Sir, long before worried period of the Communist party, I already lived in Arab. I don't know for sure, I worked there until my family, including my parents and many others came to follow me here, performed the pilgrimage, then after that they felt comfort enough to stay here (*krasan*), worked there until now. I have been assisted many people from Blembem to work in Arab" (Rjt, 2013).

> "Asides from *reog* community, we are also religious community ... to perform the pilgrimage is the special pride we hold, as NU people, that is the main thing and should be put into reality. Many people here who already performed the pilgrimage, even long before the Indonesia's independence, they went to Malaysia then went straight to Arab, many of them did not come home" (Gwn, 2014).

Both of these excerpts show that in Ponorogo, the migration process also has been going on since a long time, even before the Indonesia's independence. Indonesian migrants, including from Ponorogo, were generally taken by the party which in charge of Hajj and/or Umrah. In addition, Indonesian migrants who have been settled for a long period of time in Saudi Arabia also played a very important role. For migration to Malaysia, the condition which supported it was that he presence of cross-border migration towards Malaysia. Cultural and linguistic similarities enabled Indonesian citizens to work in Malaysia. After 1970 many Indonesian citizens have exodused to Malaysia. Most of them worked only for temporarily period and then continued the journey to Saudi Arabia. Many of those also worked for a long time and even permanently became Malaysian citizens. The journey of migrant workers from Ponorogo with Saudi Arabia as their main destination for Hajj and Umrah was done through five different pathways.

It was also implied that the form of migration which occurred in the Old Order era was a leapfrog migration and chain migration using the track of family and/or friends. Leapfrog migration was migration which was not getting to the destination directly, but through other areas in between. As an example, the migration which occurred in the Old Order era to Saudi Arabia. Initially the migration actors were headed to Malaysia in advance to work then later after they had sufficient supplies, they continued the trip to Saudi Arabia. Meanwhile it was called as chain migration because it happened gradually, bit by bit, through family and friends. It could be observed that the trip pathway of migrant workers from Ponorogo to Saudi Arabia for the purpose of pilgrimage or Umrah could be categorized into five trip pathways as below.

1. Pathway I was the departure of migrant workers directly from Ponorogo to Saudi Arabia. This pathway was usually done with the shipping lines, conducted by several people at the same time as well as involving (*kiyai*) religious leaders, and community. Among them, there was a guide who knew the path of travel of Ponorogo toward Saudi Arabia.
2. Pathway II was the departure with the main destination of Saudi Arabia through Bawean. This migration path involved migrant workers and religious leaders. As previously stated that the pathway of Ponorogo-Bawean-Malacca as a trade and shipping pathways workers have been formed as a cooperation between the British Government in Malaya with the Dutch East Indies government in Indonesia. This cooperation has also formed a travel route heading to Saudi Arabia for pilgrimage purposes. Many of the migrant workers, through an agent, worked in Saudi Arabia, both before and after the Hajj.
3. Pathway III was the departure with the main destination of Saudi Arabia through West Malaysia. This was a form of gradual migration of workers from Ponorogo: first they worked in various plantations in West Malaysia before heading to Saudi Arabia. Labor export program from Java to the plantation crops, farm exports, and mining in Malaya which have occurred since the beginning of the 19th century has been supporting the development of this third

pathway. Pathway III also supported the fact of the existence of Ponorogo community in Malaysia, such as in the area of Batu Pahat, Sri Medan, Johor and Selangor.

4. Pathway IV was the departure with the main destination of Saudi Arabia through East Malaysia (Sarawak). This form of gradual migration has similarity to the third pathway. At first migrant workers worked in various plantations in East Malaysia before heading to Saudi Arabia. This pathway was also supported by the labor export program from Java to the plantation crops, farm exports, and mining in Malaya which have occurred since the beginning of the 19th century.
5. Pathway V was the departure with the main destination of Saudi Arabia through Bawean then headed West Malaysia and East Malaysia before reached Saudi Arabia. The connection between Ponorogo and Bawean which has existed for a long periode of time supported the forming of Pathway V. Many residents of Ponorogo resided in Bawean.

Delivery of Indonesian migrant workers in large numbers which managed by the government occurred from the 1970s. The main destination countries were Middle East countries which at that time was experiencing an oil boom. This condition has caused the countries in the Middle East required a lot of cheap labours with male labours as the main target. Most of them worked on construction and processing works. The requirement of male labours in the Middle East during 1970s was also addressed by several informants from Ponorogo as follows.

> "My grandfather was an elder *Kyai* here; at first he just went there to perform the pilgrimage. There were many jobs there with high payment so he didn't go back home directly. He worked as well as performed the hajj there. At the time, many men went there to perform the Pilgrimage" (Rjt, 2013).

> "At first many men went to Arab, but now the opposite of it happened … many who went to Arab were women. Men preferred to go to Malaysia or Korea … so the number of women now are much more than men" (Ntj, 2013).

The high price of oil in the world which began in the 1970s has led to prosperity for many oil-producing countries in the Middle East, such as Saudi Arabia, Qatar, Kuwait, and United Arab Emirates. This prosperity was followed by a variety of employment opportunities which were created so that the local population (the population in the oil-producing countries) left out various menial jobs, such as construction workers, domestic work, drivers and babysitters. This condition caused job opportunities in the domestic sector and the menial work [4]. This influenced the needs of many workers in the domestic sector as housekeepers and babysitters.

An organized delivery of migrant workers by the government was based on Government Regulation No. 4/1970 regarding the program of Inter Regional labour sending (*Program Antar Kerja Antar Daerah*/AKAD) and program of Inter-State labour sending (*Antar Kerja Antar Negara*/AKAN). In the period of 1970-1980 was the starting point of the recruitment of women in large numbers to be placed in Saudi Arabia. The connection between Ponorogo-Malaya-Saudi Arabia which has been formed long ago supported the delivery program of female migrant workers.

Overseas migration pathway of migrant workers from Ponorogo during 1970 - 1980 was a migration pathway to Malaysia and Saudi Arabia. Migration of workers which has lasted for long, started since the establishment of Kadipaten Ponorogo, the sending of workers to Malaya during the colonial era, as well as through the pathway of pilgrimage, as a whole has supported the development of the number of migrant workers to Malaysia and Saudi Arabia. This fact was also supported by number of people from Ponorogo who have settled, both in Malaysia and Saudi Arabia. Migration pathway towards Malaysia and Saudi Arabia continues to evolve until today so that the two countries are considered on the top of the list in terms of the number of migrant workers. Migration path to Saudi Arabia during this period was divided into two: indirect and direct pathways.

First, the indirect pathway was not direct migration pathway of migrant workers from Ponorogo who made their transit first prior to Saudi Arabia. It hase been noted that before getting to Saudi Arabia, migrant workers transited in some places. The purpose of transitting in Malaysia was to work first before heading to Saudi Arabia for the Hajj. These indirect pathways were included Ponorogo-

Selangor-Saudi Arabia, Ponorogo-Johor-Saudi Arabia, Ponorogo-Johor-Selangor-Saudi Arabia, and Ponorogo-Bawean-Selangor-Saudi Arabia. Second, the direct pathway was the pathway of sending migrant workers from Ponorogoto Saudi Arabia. In this second pathway, migrant workers from Ponorogo were sent directly to Saudi Arabia without transit in other areas. Migrant workers from Ponorogo sent directly to Saudi Arabia and accommodated by their relatives who were first there for work. After working for a long time, those migrant workers performed the pilgrimage.

Sending of migrant workers from Ponorogo during 1980-1990 was related to the development of the various destination countries. The government intensively developed the sending programs of migrant workers by cooperating with the private sectors, such as recruitment agencies as well as by the implemention of migrant workers' placement. Service of migrant workers' placement abroad locally was carried out by Regional Office of Depnakertranskop in the provincial level and Office of Depnakertranskop II in the district level. The result of this cooperation was the increasingly extensive network with overseas destination countries which was also varied, such as Singapore. The new sending pathways of migrant workers also headed toward East Asia, such as Hong Kong, Taiwan, Korea, and Japan. Sending of migrant workers from Ponorogo even reached some European countries, such as Spain, the Netherlands, Italy, and France.

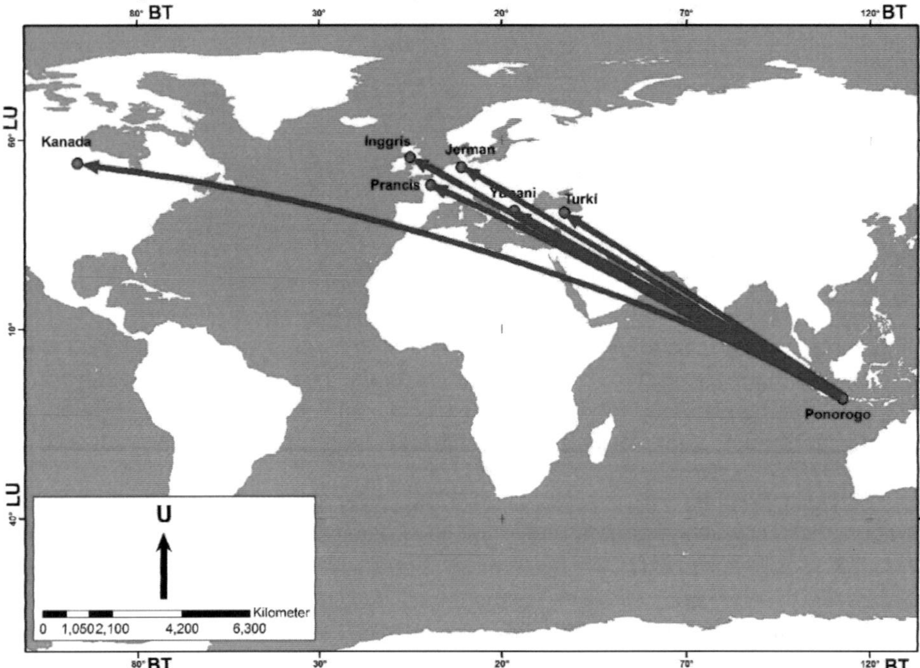

Figure 3. Pathway of Sending Ponorogo's Migrant Workers toward Europe.

Sending of Ponorogo's migrant workers has grown considerably in the period of 1990-2000. In addition to the better placement system and a wider network, the destination countries were widespread up to European countries (Turkey, Greece, Germany, France, Britain, Italy, and Spain), Canada, and the United States (Figure 3). Migrant workers during this period were definitely the heroes of development in Ponorogo. The local government has built an integrated eight-floor government building based on the contribution of those migrant workers. For the services of migrant workers, the Government of Ponorogo designed a sculpture of women workers in front of the building.

During the period of 2000 to present was the period of reform and globalization in sending migrant workers abroad. Changes in sending program were in form of very complex destination areas and the novelty of the programs launched by the government. In 2001, the Directorate General of Binapenta was disbanded and replaced by Directorate General of Manpower Placement and Protection of

Overseas (PPTKLN), whereas placement services of migrant workers at the provincial/regional level were ran by BP2TKI. In 2004 Law No. 39 of 2004 on Manpower Placement and Protection of Indonesian Overseas was published. In the operational level, Presidential Regulation No. 81 Year 2006 on the Establishment of BNP2TKI was published also.

Period of 2000-2010 was called as the era of globalization in sending migrant workers abroad. As a result of the increasingly complex international relationships, variations in destination countries were also increasingly widespread. Ponorogo's migrant workers have far reached all the continents, even to small countries, such as Cyprus, Maldives, Fiji, Mauritius, Suriname and New Zealand. Globalization of migrant workers from Ponorogo was also supported by the modernization of information, namely the use of a mobile phone. Utilization of mobile phone has cut the distance for mutual communication, both between workers, with the families, or with employment agencies. New pathways of migration which were formed was the development of migrant workers on the type of formal employment, either with the scheme of Government to Government (G to G), Government with the Company (G to C), as well as the scheme of the Company to Company (C to C).

3. Conclusion

International migration in Ponorogo has been going on since 500 years ago. There are three important things that need to be addressed. *Firstly*, the intensity of migration is higher and does not involve only one particular group or population characteristic. Migration is more general, not limited by age, sex, and marital status. *Secondly*, there has been an expansion of destination countries with the establishment of many international migration routes. This has resulted in an increasingly complex relationship between migrant workers from Ponorogo and overseas. Migrant workers from Ponorogo have traveled to distant continents, to modern countries, to global world, even to small countries such as Cyprus, Maldives, Fiji, Mauritius, Surinam and New Zaeland. *Thirdly*, the phenomenon of migration has eliminated the physical barriers of state boundaries. The development of information technology, for example with the use of mobile phones (HP) to make the barrier between Ponorogo and the destination increasingly disappeared. The use of this mobile phone has cut the distance to communicate between each worker, with family, and with employment agencies. This shows that interaction can be done at any time without limited distance. There has been a globalization of workers unrestricted by distance. Nowadays Indonesian migrant workers are facing new way of lifes, that is a digital era that futher embedded into world system of migration.

4. References

[1] Bahrin T S 1967 The Pattern of Indonesian Migration and Settlement in Malaya *Asian Studies* **5** (Cambridge: Cambridge University Press) pp 233 – 57

[2] Hugo G 1993 Indonesian Labour Migration to Malaysia: Trends and Policy Implications *South-East Asian Journal of Social Science***21**(1) pp 36 – 72

[3] Hugo G 2004 International Migration in Southeast Asia since World War II *International Migration in Southeast Asia* eds Ananta A and Arifin E N (Singapore: Institute of Southeast Asian Studies)

[4] Kritz M and Zlotnik H 1992 Global Interaction, Migration Systems, Process and Policies *International Migration System: A Global Approach.* (Oxford: Clarendon Press) pp 1 – 16

[5] Norris R E 1972 Migration As Spatial Interaction *Journal of Geography* **71** pp 294 – 301. Purwowijoyo 1985 *Babad Ponorogo I* (Ponorogo: Depdikbud)

[6] Salazar Z 1998 The Matter with Influence: Our Asian Linguistic Ties *The Malayan Connection: Ang Pilipinas sa Dunia Melayu* (Lunsod Quezon: Palimbagan) pp 59 – 80

[7] Simmons J W 1971Changing Residence in the City *Geographical Review* **58** pp 622 – 52

[8] Stalker P 2007 *Stalker's Guide to international migration* www.pstalker.com/migration/

[9] Sugiarso 2003 *History of Ponorogo Regency:A local Potency* (Ponorogo: Reksa Budaya)

[10] Wolpert J 1965 Behavioral Aspect of the Decision to Migrate *Regional Science Association***19** pp 159 – 69

IOP Conf. Series: Earth and Environmental Science **148** (2018) 012012 doi:10.1088/1755-1315/148/1/012012

User satisfaction level of parking space facility: a case of Faculty of Geography, Universitas Gadjah Mada, Indonesia

S R Budiani[1], M Iffani[2], I Novianti[2], M A F Alfana[1], R Harini[1] and A Rofi[1]

[1]Lecturer in Faculty of Geography, Universitas Gadjah Mada, Yogyakarta, Indonesia
[2]Assistant Researcher in Faculty of Geography, Universitas Gadjah Mada, Yogyakarta, Indonesia

srahayu@ugm.ac.id

Abstract. Every vehicular trip starts and finishes at a parking lot. Therefore, the availability of parking space has to meet the demand of vehicle users so that their activities can be performed smoothly and in time. The needs for parking space in Faculty of Geography tend to increase from year to year due to the increasing use of private vehicles in the faculty. This research aimed to identify the user satisfaction level of the parking space in Faculty of geography and to understand the influence of the parking facility on user satisfaction level. It was survey research because it took samples from a population and used a list of questions as a data collection tool. The data was analyzed quantitatively with descriptive statistics. The satisfaction level was calculated according to an existing regulation, namely the Decree of the Indonesian Minister of Administrative and Bureaucratic Reform No. Kep/25/M.PAN/2004 about General Guidelines for the Assessment of Community Satisfaction Index of Public Service Unit.

The results showed that the mean user satisfaction level of the parking facility and each of its elements were good. In other words, the parking facility was already good in each element. Therefore, the faculty should focus the improvement on making it better.

1. Introduction

Performing an activity requires mobilization, and, for this purpose, vehicles are essential. Without regard to the diverse use of a vehicle, parking lots are necessary for smoothly and punctually running activities in a daily basis. Its availability is inseparable from its setting, layout, and effective capacity, as well as parking service quality, all of which play a role in optimizing parking facilities [6]. Available parking facilities are urgently needed because the use of private vehicles in the Special Region of Yogyakarta increases every year [1].

The Decree of the Indonesian Minister of Transportation No. 66/1993 on Public Parking Facilities explains that:

a. Parking is a condition representing non-moving vehicle of a non-temporary nature;
b. Parking facilities outside the roadways are built specially in the form of a parking area (outdoor) and/or a parking building (indoor); and
c. Public parking facility is outside the roads and in the form of a parking building (indoor) and/or a parking area (outdoor) developed as a stand-alone business that provides public parking services.

This decree also describes the requirements for a public parking facility, namely:

a. A public parking facility can ensure the safety and smoothness of traffic;
b. It is easily accessible by users;
c. For indoor parking lot, it has to meet the construction requirements by the applicable laws and regulations;
d. For outdoor parking lot, it must have certain limits;

e. The circulation and position of parking vehicles in either indoor or outdoor parking lots are arranged using traffic signs or road markers; and
f. Every space used for parking vehicle is marked with letters or numbers so that users can easily find their vehicles.

According to [8], the service quality, as perceived by users, can be identified with a set of customer satisfaction indicators that include five dimensions of service quality. These dimensions, as well as their sub-dimensions, are as follows:

a. Tangibles (service quality in the form of physical office, computerized administration, waiting room, and information center or point). This dimension represents the modernity of the equipment, the attractiveness of the facilities, the neatness of the officers, and the completeness of the supporting equipment (pamphlet or flowchart).
b. Reliability (ability to provide reliable services). This dimension is a guarantee that the parking facility can accomplish the desired objectives, facilitate user complaints, perform the right services, provide timely services (as promised), and demand for recording errors.
c. Responsiveness (the ability to help and provide services quickly and precisely and be responsive to what customer desires). This dimension includes, for instance, a notification from the employees to users about the services provided, fast delivery of the services, and the willingness of the employees to assist users (officers are never too occupied to serve user's demand).
d. Assurance (hospitality, or the ability of employees in courteously ensuring the confidence of users in the services). This dimension is related to the behavior of the employees (confidence in providing services to users), the user's sense of security, and the ability (knowledge) of the employees to answer questions from users.
e. Empathy (firm but attentive attitude towards users). This dimension includes, for instance, individual and particular attention from employees to users, a timely service process for all users, and profound services from the employees provided through an understanding of the specific needs and desires of users.

This study focused on motor vehicle parking facilities in Faculty of Geography, Universitas Gadjah Mada, which is located in Sekip Utara, Bulaksumur, Sinduadi District, Sleman Regency. This faculty is one of the educational institutions in Universitas Gadjah Mada that are equipped with parking lots as a supporting facility for campus operation. Therefore, vehicle parking facilities are expected to efficiently help smooth the teaching-learning process, which is the core of the faculty's operation. The effectiveness of parking facilities is inseparable from the layout, capacity, and services of the parking lots provided by the faculty.

The objectives of this study were as follows:
1. To identify user satisfaction level with the parking facility in Faculty of Geography; and
2. To analyze the effect of the parking facility in Faculty of Geography on the satisfaction level

2. Method

This study was survey research because it took samples from a population. It used both quantitative and qualitative method. The analysis unit was user individuals as it fitted the context of this research, namely user satisfaction with parking service quality. The population was the users of the parking lots in Faculty of Geography UGM, which consisted of lecturer, employee, and student of the faculty. The respondents were sampled randomly according to the unit area of each parking service.

The research data were collected using two sampling techniques, namely simple random sampling and purposive sampling. These methods were employed due to the large population in this research. Purposive sampling means that the population does not have the same opportunity to be selected as a sample due to the several criteria developed according to the research objectives. In other words, it uses particular considerations that are relevant or can represent the purpose of a study [7].

IOP Conf. Series: Earth and Environmental Science **148** (2018) 012012 doi:10.1088/1755-1315/148/1/012012

The data collection tool in this study was a questionnaire, complemented by direct observations to obtain more in-depth picture of the parking lot condition. There were two sets of population acquired in this research, namely executors or managers and parking facility users. Due to the large population of the parking lot users, the respondents were selected using simple random sampling. Meanwhile, the purposive sampling on the population aimed to find more details about the implementation of parking facilities in the faculty. There were roughly 1,500 vehicle users in the faculty. Based on the Morgan table, the sample size with 95% confidence level is 306. The detail of the samples is presented in table 1.

Table 1. Number of respondents by status and type of vehicle.

Vehicle Types	Status				Total	Percentage
	Students	Lecturers and Staffs	Guests	Others		
Motorcycle	244	17	6	4	271	88.9 %
Car	14	9	0	0	23	7.5 %
Bicycle	9	1	0	0	10	3.3 %
Others	1	0	0	0	1	0.3 %
Total	268	27	6	4	305	100 %

The research data were analyzed using a descriptive statistic method. It is one of the scientific ways of collecting, presenting, and analyzing data with actual figures whose basis can be justified to find the correct conclusion. The satisfaction level was calculated according to the Decree of the Indonesian Minister of Administrative and Bureaucratic Reform No. Kep/25/M.PAN/2004 on General Guidelines for the Assessment of Community Satisfaction Index of Public Service Unit. The indicators adopted from the decree are presented in table 2.

The user satisfaction with the parking facility was calculated using the "weighted average value" of each service provided by the facility. The formula is as follows (eq. 1).

$$Weighted\ average\ value = \frac{Total\ weight}{Number\ of\ elements} = \frac{1}{17} = 0.058 \quad (1)$$

Furthermore, the Public Satisfaction Index was obtained from the weighted average value with the following formula (eq. 2).

$$Public\ Satisfaction\ Index = \frac{Total\ values\ per\ item}{Number\ of\ elements} \times Weighted\ average\ value \quad (2)$$

The Public Satisfaction Index ranges from 1 to 5.01. To facilitate its interpretation regarding the quality service and service unit performance of the parking facility (table 3), this index was multiplied by a base value of 20. The formula is as follows (eq. 3).

$$Public\ Satisfaction\ Index\ 20 \quad (3)$$

Table 2. Satisfaction level indicators.

Nr.	Indicators	Sub Indicators
1	Service procedure	• The level of information disclosure regarding the service procedure • The level of clarity on the flow in the service procedure • The level of simplicity of the service procedure
2	Terms of service	• The level of openness regarding the terms of service • The level of convenience in maintaining and meeting the service requirements • The level of clarity regarding terms of service
3	Clarity of service employees	• The level of certainty about the identity and responsibility of the service employees • The level of ease in meeting and contacting the service employees
4	The discipline of service employees	The ability of the employees in completing services promptly
5	Responsibilities of service employees	• The level of clarity of employees' responsibility • The level of assurance of employees' responsibility • The level of disclosure of employees' responsibility
6	The ability of service employees	• The level of physical ability. • The level of employees' intellectual capacity.
7	Speed of service	• Time efficiency in performing the desired services • The level of openness of service completion time
8	Service equality	• The level of service equality to all users • The extent of the coverage of service delivery
9	Courtesy and hospitality of employees	• The level of courtesy and friendliness of the service employees • The level of respect from employees for users
10	Fairness of service charge	-
11	The certainty of service charge	-
12	The certainty of the service schedule	• The clarity of the service schedule • The level of service schedule's reliability
13	Environmental comfort	• The level of cleanliness and neatness of the environment where the services are offered • The availability level of service support facility. • The level of completeness and update of service facilities and infrastructure.
14	Safety and security	• The level of safety of the facilities • The level of security in the use of facilities and infrastructure services

Table 3. Perception values, public satisfaction index, converted public satisfaction index, service quality, and service unit performance.

Perception Values	Intervals of Public Satisfaction Index	Intervals of Converted Public Satisfaction Index	Quality of Service	Service Unit Performance
1	1.00 – 2.00	20.00 – 40.00	D	Very Bad
2	2.01 – 3.00	40.01 – 60.00	C	Bad
3	3.01 – 4.00	60.01 – 80.00	B	Good
4	4.01 – 5.01	80.01 – 100.00	A	Very Good

3. Results and Discussion

3.1. User satisfaction level with the parking facility in the Faculty of Geography UGM

The user satisfaction level was calculated from the value of each element, which was adopted from the Decree of the Indonesian Minister of Administrative and Bureaucratic Reform No. Kep/25/M.PAN/2004 on General Guidelines for the Assessment of Community Satisfaction Index of Public Service Unit. Meanwhile, the value of each element was derived from users' assessment of each element of the parking facility. The results are summarized in table 4.

Table 4. User satisfaction level with each element of the parking facility in Faculty of Geography UGM.

Questions	Value Per Element	Descriptions
Disclosure of information on parking service procedure	3.06	Good
The clarity of the flow in the parking service procedure.	3.01	Good
The simplicity of parking service procedure	3.73	Good
The ease/convenience of parking service procedure	3.80	Good
Certainty about the identity and responsibilities of service employees	2.98	Bad
The ease of meeting and contacting service employees	3.66	Good
Discipline in providing the services	3.47	Good
The responsibilities of employees in providing services	3.68	Good
Equal treatment to all users (to receive services)	3.89	Good
The evenness of coverage in service delivery	3.50	Good
The ability (physical and intellectual) of employees in providing services	3.90	Good
The attitude (courtesy and friendliness) of employees in providing services	3.73	Good
the conformity between the service implementation with the established standards	3.21	Good
The cleanliness and neatness of the parking lot	3.09	Good
The availability of supporting facilities	2.80	Bad
The completeness and update of service facilities and infrastructures	2.66	Bad
The safety and security of the parking facility	3.28	Bad

IOP Conf. Series: Earth and Environmental Science **148** (2018) 012012 doi:10.1088/1755-1315/148/1/012012

Table 4 showed that 13 (out of 17) elements observed in this study were in a good category. These elements were information disclosure on parking service procedure, parking service procedure, the simplicity of parking service procedure, the ease of parking service procedure, the ease of finding and contacting the service employees, discipline in the services, the responsibility of employees to provide services, equal treatment to all users (to receive parking service), the evenness of the coverage in service implementation, the ability (physical and intellectual) of employees in providing services, the attitudes of employees in providing services, the conformity between the service implementation with the established standards, and the cleanliness and tidiness of the parking lot. On the contrary, the other four elements were categorized by the respondents as either bad or very bad. These elements were the certainty about the identity and responsibilities of service employees, the availability of supporting facilities, the completeness and update of service facilities and infrastructures, and the safety and security of the facilities.

Furthermore, the overall user satisfaction level was calculated by summing the value of each element and multiplying it by the weighted average value (0.058). This calculation yielded an index of 3.38, which was multiplied by a base value of 20. The result of this multiplication was a Converted User Satisfaction Index of 74.37, indicating a B-grade or good service quality.

3.2. The impact of parking facility on user satisfaction level

The Special Region of Yogyakarta is renowned as the city of students. This title makes people with interest in science from Indonesia and other countries choose Yogyakarta as their academic destination. As a consequence, the number of vehicles in Yogyakarta increases from year to year, as seen in figure 1.

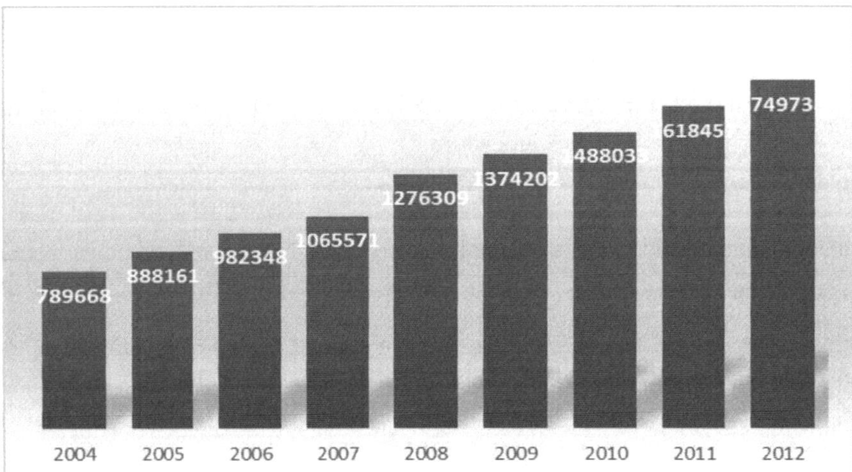

Figure 1. The number of registered motor vehicle users in the Special Region of Yogyakarta.

Universitas Gadjah Mada is one of the many educational means in the Special Region of Yogyakarta. Its Faculty of Geography also contributes to the rising trend of the number of motor vehicles. Therefore, parking facilities should be provided by every academic service that has high mobility. The manager is expected to offer parking services so that the vehicles can enter to and leave from the faculty neatly and conveniently. This study assessed the 17 elements of user satisfaction based on the users' perception of the services provided by the parking facilities in the Faculty of Geography.

IOP Conf. Series: Earth and Environmental Science **148** (2018) 012012 doi:10.1088/1755-1315/148/1/012012

The manager had prepared a plan for the development of parking facilities to be comfortable, orderly, and safe. This plan included the construction of the facilities, which, according to the manager, had currently caused some inconvenience to parking lot users in the faculty. Due to the unrealized plan to develop a single parking facility in the university, instead of one facility in every faculty, the construction of on-site parking lots in the Faculty of Geography continued. Physical improvements would likely continue in the form of road blocking, CCTV installation, plan development, and road sign and lighting installation. There was another plan to arrange the layout of parking space by type of vehicle for a more systematic and comfortable parking lot utilization by students, lecturers, employees, and guests.

The results showed that each element was perceived as good by the users, indicating a good facility as a whole. If an element is good, a further development likely emphasizes on making it better. On the contrary, if an element receives a bad rating from users, then further improvement has to analyze the cause of such assessment first. In this case, researchers can design open-ended questions for identifying the arguments or reasons behind the poor and very-poor rating of an element given by users.

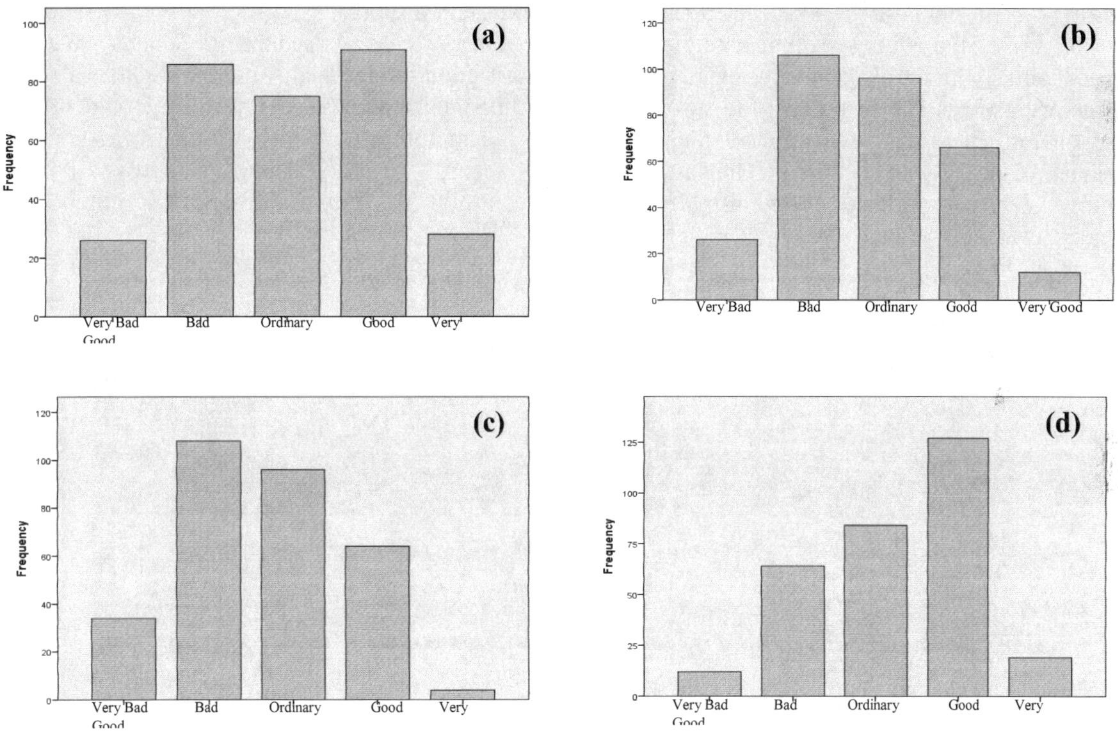

Figure 2. The frequencies of elements with averagely bad and good rating: (a) the certainty of the identity and responsibilities of employees, (b) the availability of supporting facilities, (c) the quality and completeness of facility and infrastructure, and (d) safety and security.

Regarding the identity and responsibility of the employees of the parking facility in the Faculty of Geography (figure 2a), on average the respondents gave an unsatisfactory rating with the element value of 2.98. Their identity was still difficult to distinguish from employees in other departments because there was no unique identification. Based on direct observation in the field, they did not wear uniforms unlike the security personnel in the faculty. Also, the lack of information on the shift schedule of the employees was perceived as a difficulty for parking lot users when they needed assistance. As an effort to improve the parking service in the faculty, the respondents suggested the provision of information boards containing the personal data of the employees in charge of a particular shift along with their responsibilities.

Regarding the availability of supporting facilities (figure 2b), on average the respondents gave a poor rating with the element value of 2.80, categorized as bad service quality. Some of the shortcomings expressed by the respondents were the lack of CCTV monitoring, minimum lighting, non-roof parking space, uncomfortable access to parking lots (i.e., bad drainage), frequently changed parking signs and markers, and poorly functioning parking barrier gates. The respondents expected the existing development and improvement to be able to resolve these inconveniences.

Regarding the completeness and the extent of facilities and infrastructures (figure 2c), on average the respondents gave a poor rating with the element value of 2.66, categorized as bad service quality. The respondents complained the impractical use of parking tickets that aimed to calculate the number of vehicles entering to and leaving from the faculty. Moreover, the license plate written on the ticket was never checked accurately against the vehicles passing through the barrier gates. Other complaints were the minimum of the security, modern parking facilities (the parking service was still conventional), lighting, and parking signs and markers, as well as the poorly functioned automatic barrier gates. As an effort to improve the parking services in the faculty, the respondents suggested a parking system similar to the one implemented at the UGM Central Library.

Regarding the safety and security of the parking lots (figure 2d), on average the respondents gave a good rating, with the element value of 3.28, categorized as a good service quality. According to some of the respondents, the security of the parking facility in the faculty was fair except for the outdoor parking lot. There has been rampant theft of helmets and motorcycles, which was a worrisome problem, especially to the users. The less-than-optimal security coverage could be improved by monitoring remote parking lots and installing CCTVs to control the security of the parking lots in the faculty.

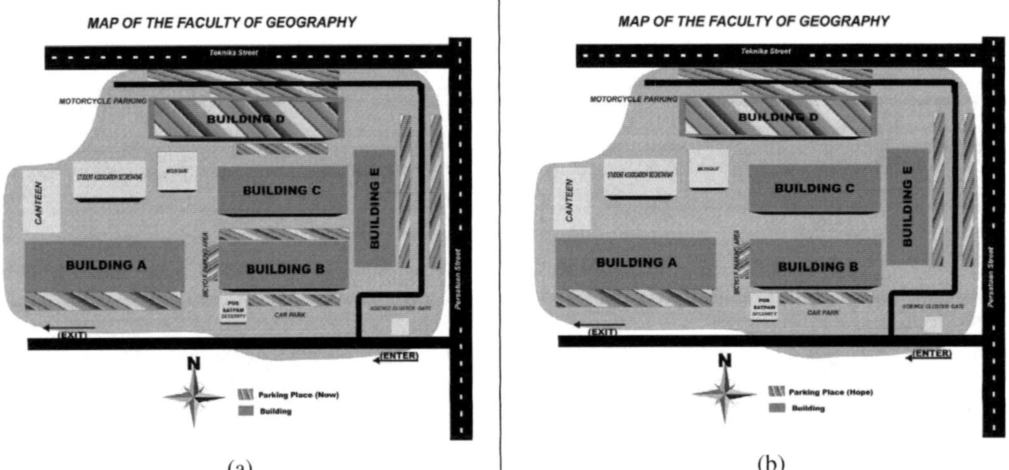

(a) (b)

Figure 3. The designs of the parking space in Faculty of Geography at the time of research (a) and based on users' expectations (b).

Figure 3 shows the difference between the current parking space design and the desired layout by the users and the manager. At the time of research, nearly all of the empty spaces in the faculty was used as a parking lot (figure 3a). The expected locations of the parking lots by the users (figure 3b) did not show significant differences. Users' expectations mostly included the reparation/improvement of the facilities, services, and systems. Based on the results of the interviews with users and the provider of the parking facilities, some expectations were described, for instance:

1. According to the plan, Building D was used as a parking lot. However, because of the current circumstances, the space on the north of this building remained unoccupied and was often used for parking motorcycles. Users expected this situation to continue in the future.
2. Users demanded regularity and clarity about the allotted space for parking motorcycles, namely the basement and on the east side of Building E. However, Building E required some improvements or a plan to arrange the parking motorcycles neatly. Meanwhile, cars were preferably parked between Buildings B and C, on the south and east sides of Building B, and in front of Building A.
3. The university's plan to develop parking lots outside the faculties and provide special vehicles like bicycles to connect the lots to the faculties was appreciated.
4. Instead of a parking lot, users would like to have open space on the south of Building D.
5. The ticketing system for vehicle entrance was still performed per cluster by the policy of the university. The entrance gate of the faculties in science cluster was planned to relocate from Faculty of Geography to Faculty of Biology in the following year.
6. The floor plan for other infrastructures had been already developed, including the addition and improvement of parking signs and facilities like CCTV. Users also demanded more lighting because the students often stayed late in the faculty and accessed the parking lots at night.

4. Conclusion

This research concluded that:

1. The average user satisfaction level with the parking facility in the Faculty of Geography was good. This level was obtained from the calculation of the users' assessment on every element of the parking services in the faculty; and
2. The results showed that each element was perceived as good by the users, indicating a good facility as a whole. If an element is good, a further development likely emphasizes on making it better. However, four element received poor rating due to the following factors: the unclear identity of the employees, minimum parking signs, bad drainage, ineffective ticket system, and the absence of CCTV installations.

5. Acknowledgment

This research was part of the Laboratory of Population and Economic Resource, Faculty of Geography UGM. Taking place in the Faculty of Geography, this research was expected to contribute to the decision making and the development of the faculty. This research was funded by BPPTN-BH in 2017 and conducted for eight months. The 2014 students of the Department of Geography and Environmental Science participated in data collection through interviews with approximately 300 respondents.

6. References

[1] Badan Pusat Statistik (BPS) 2012 *Daerah Istimewa Yogyakarta dalam Angka 2012* Yogyakarta: BPS
[2] Badan Pusat Statistik (BPS) 2013 *Daerah Istimewa Yogyakarta dalam Angka 2013* Yogyakarta: BPS
[3] Decree of the Indonesian Minister of Administrative and Bureaucratic Reform No Kep/25/M.PAN/2004 on General Guidelines for the Implementation of Public Service
[4] Decree of the Indonesian Minister of Transportation No. 66/1993 on Public Parking Facilities
[5] Moenir AS 2001 *Public Service Management in Indonesia*. Jakarta: BumiAksara
[6] Munawar A and Prasetyo HP 2000 *Parking Requirement in Master Program Building (Case Study: UGM Management Master's Building)* Proceedings of the Symposium III. Yogyakarta: FSTP Universitas Gadjah Mada
[7] Mantra I B, Kasto and Tukiran 2014 *"Penentuan Sampel"* dalam Sofian Effenfi dan Tukiran Metode Penelitian Survei Jakarta: LP3ES pp. 151-180
[8] Parasuranman A Zeithaml V A and Berry L L 1988 SERVQUAL: A Multiple-Item Scale for Measuring Consumer Perceptions of Sevice Quality. *Journal of Retailing (64)*, pp. 12-40

IOP Conf. Series: Earth and Environmental Science **148** (2018) 012013 doi:10.1088/1755-1315/148/1/012013

Coastal management strategy for small island: ecotourism potency development in Karimata Island, West Kalimantan

A W Rudiastuti[1], Munawaroh[1], I E Setyawan[1] and G H Pramono[1]

[1]Badan Informasi Geospasial
Jalan Raya Jakarta-Bogor Km 48, Cibinong, Kab. Bogor, Indonesia 16911

aninda.wisaksanti@big.go.id; anindarudiastuti@gmail.com

Abstract. Sustainable coastal management is playing an important role in coastal resources conservation, particularly on small islands. Karimata archipelago has unique characteristics and great potential to be developed as a tourism object, one of which is Karimata Island as the largest island and also reserve area. The concept of ecotourism focuses on the ecology conservation, economic benefits, and social life. Ecotourism aims to build sustainable tourism that provides economically viable and social benefits to the community. This study aims to develop coastal management strategy based on ecotourism at Karimata Island. Spatial approaching through coastal type was done. Qualitative descriptive analysis and SWOT are used to develop sustainable management strategies for the coast of Karimata Island, where the opportunities and challenges to the development of coastal ecotourism Karimata Island also included. If this potential is optimally utilized, it can be relied as an economic opportunity for local communities. Structurally shaped coast, marine depositional coast and coast build by organism are several of coastal types found at Karimata Island. Coastal ecosystems inhabited Karimata Island are mangroves, coral reefs, and macro-algae. Karimata Island have not been optimally utilized for tourist destinations. The biggest obstacle encountered is the accessibility from Kalimantan or other island at Karimata islands. Several problems related to the utilization of coastal resources were found such as mangrove and coral reef damage, also regulation that less supportive. The results of this study are expected to provide an overview of solutions for the development of coastal tourism potentials in Karimata Island.

1. Introduction

As a country with a vast coastal area, Indonesia obliged to oversee its coastal areas, especially the small islands. The coastal region has a strategic value that plays an important role in improving the national economy and people's welfare. However, coastal areas are highly vulnerable to threats of damage [1]. Smart management which involves proportion of coastal environmental protection interests and proportional economic interests is indispensable. Coastal area management strategies and policies focus on such issues as physical characteristics, resources, and resource utilization needs for the community. In the process of managing coastal areas, policy-making is directed at preservation. Goodwin (1997) in [2] suggests efforts to ensure local community participation and steps in eco-tourism activities. The need for interaction between the three parties involved, namely the government sector, private and local communities. Sustainable coastal management in the area of regional economic development and conservation of coastal natural resources, especially on small islands, is a long-term responsibility between government and society [3], [4]. Involvement and understanding of local community to the economic, social and environmental issues become the basis of shaping related policies [5].

For decades, marine tourism has grown throughout the world, especially in those who are in the tropical area with coastal assets; Reefs, Mangroves, and beaches. That coastal asset which has natural beauty represents an important touristic attraction for water sports (surfing and scuba diving), and natural tourism [6]. A row of contributions that put up by Mangrove ecosystem to protect the environment as well as local economic value. As the coastal protection from disasters [7], a place to spawn, to nurse, and to be living habitat [8], [9], as nutrient sources for organisms, as a source of seafood provision and timber [10], also for tourism. Mangrove-based tourism has been encouraged in Surabaya [11], Jakarta [12] and Bali [13]. The sustainable management approach relies on the availability of coastal habitat resources, together with conservation efforts, cultural education, and economic empowerment of local communities. Those lead to a concept of sustainable tourism of eco-tourism [14]. Eco-tourism becomes one of the alternative development that can help overcome the problems in the utilization pattern that is destructive to the sustainability of coastal resources [15]. Therefore, to be able to optimize the potential of ecotourism in Karimata Island, it is necessary to do an assessment to know the potential and existing problems.

Utilization of remote sensing imagery to provide potential data and information on coastal resources are common nowadays. Thus, it is advantageous to sense the coastal condition for sustainable development and management. Remote sensing data as one of the best source to acquire condition of mangrove areas and benthic habitat, in the manner of advantages such as synoptic coverage, capable of inaccessible area, historical and near real-time data [16]–[22].

This research intends to develop coastal management strategy of Karimata island based on coastal characteristics and resources. Both of spatial and qualitative analysis (SWOT) was carried out to get an overview of tourism management strategy in Karimata Island. Karimata Island is chosen because it is a small island that has a wealth of good quality of coastal ecosystem and have not been empowered. Development of sustainable tourism concept is needed, it is in line with the status of Karimata Island as a marine reserve area. The concept of ecotourism focuses on ecology, economic benefits, and the local community. In addition, this research is addressed to support promotion of Karimata Island to be one of Indonesia's eco-tourism destinations in accordance with sustainable development.

2. Methodology

This study focused on two processes. First, coastal habitat mapping using remote sensing technology assisted with ground-truthing results that have been carried out in 2016 by PPIT-BIG team. Second, forming the sustainable management strategies for eco-tourism in Karimata Island using SWOT analysis. Both of spatial and non-spatial data were used to deliver the strategy of coastal management in Karimata Island (table 1). Brief and clear work flow of this study were depicted in figure 1.

Table 1. Data and sources.

Types	Data	Source	Year
Spatial	World-View2	LAPAN	2015
	Free Hi-res Imagery	ESRI	2016
	RBI 1:50.000	BIG	2016
Secondary	Statistic Data of Karimata	BPS; Marine and Fisheries Agency of Kayong Utara	2015
	Marine Reserve Boundary Map	BKSDA – Kalimantan Barat	2016
	Zonation Planning Document	Marine and Fisheries Agency of Kayong Utara	2016
	Spatial Planning Document	Public Works Agency of Kayong Utara	2016

2.1. Study area

Karimata archipelago stands for a group of 83 small islands, which every island has its name registered in United Nation. Only 14 islands are inhabited. Since 1981, Karimata Islands have becoming Nature Reserve prior to Decree of the Director General of Forestry No. 2240/ DJ / I / 1981 dated June 15, 1981. The nature reserve status is continued and strengthened by Decree of the Minister of Forestry No. 381 / Kpts - II / 1985 where Karimata appointed as Marine Reserve, with extent shown in Figure 2. As it states, it is demanding people having the permit to enter the Marine Reserve area. In figure 2, Karimata

Marine Reserve area covered all the coastal and small islands in the eastern part of Karimata Island. Thus, the western part of Karimata Island can be accessed without a permit.

2.2. Methods

2.2.1 Image interpretation. For spatial data as listed in table 1, the preparatory steps which applied include geometric correction and radiometric correction. For radiometric correction, different treatments applied to the image depending on the needs related to the mapping method used. For mangrove classification and interpretation, as well as coastal typology, correction process used to attain the reflectance value, whereas the images prepared for shallow water habitat mapping further corrected by sun glint and water column correction. Digital transformation, multispectral classification, and visual interpretation are used because of their effectiveness to obtain optimal results.

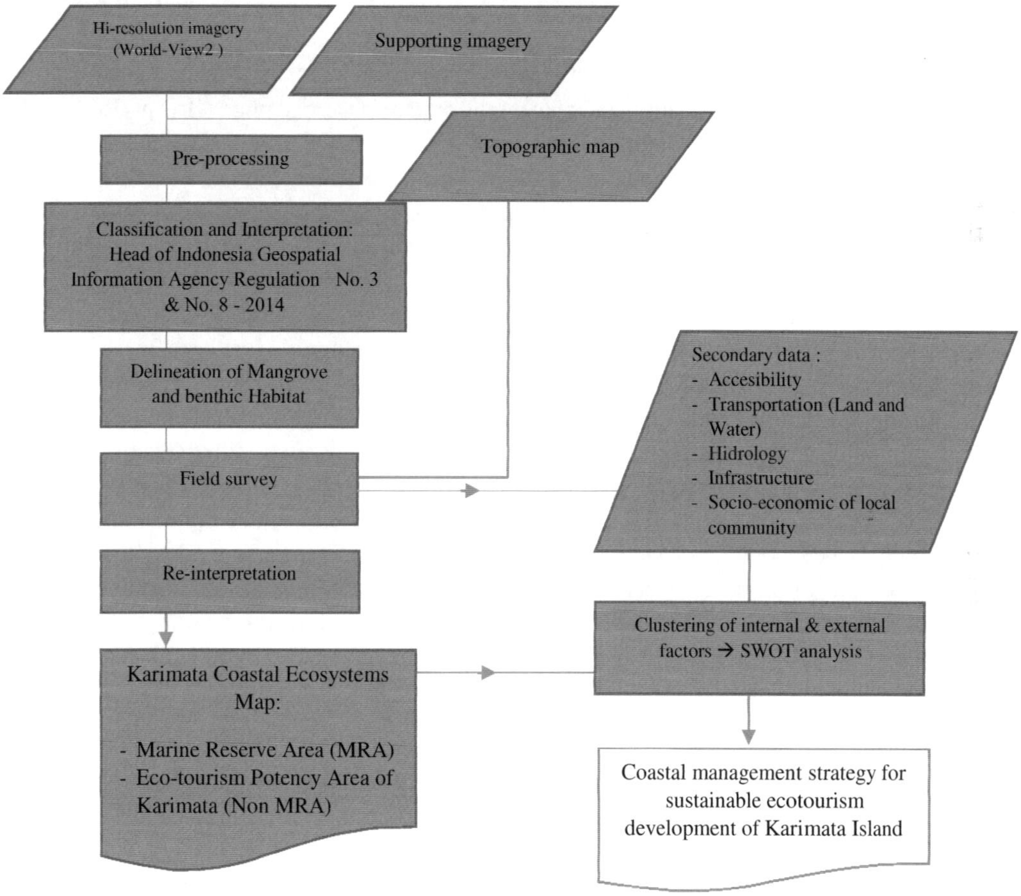

Figure 1. Flowchart of data processing and analysis.

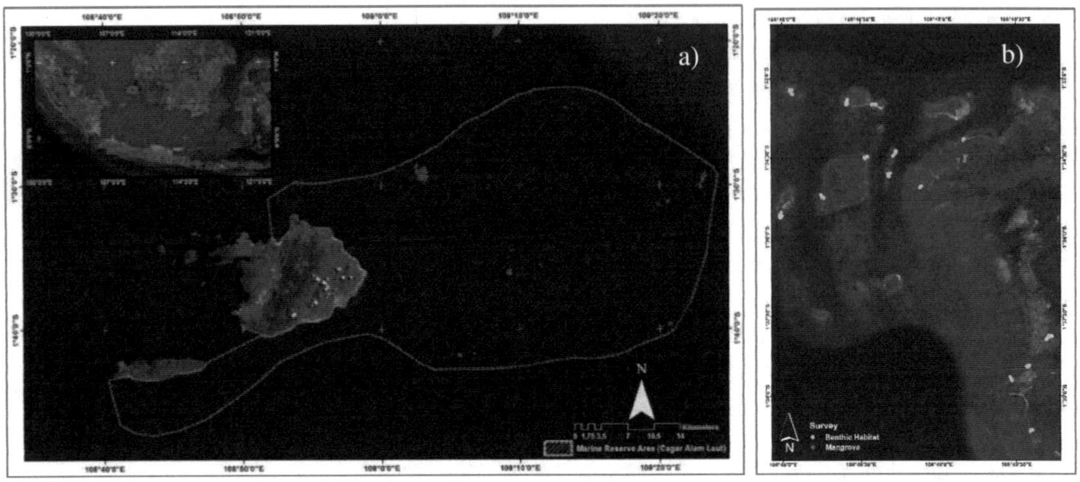

Figure 2. (a) Karimata Island and Marine Reserve Boundary (b) Survey plots in Karimata Islands.

Combination of digital classification and visual interpretation were applied. Green - Red - NIR bands combination used for on-screen digitation of mangrove areas. Canopy density classification using NDVI transformation were also implemented [23] besides the visual interpretation. Besides, delineation of mangrove extension and canopy density classes takes the advantages of hi-res imagery such as World-View2. Canopy density classes can be easily differentiated using enhanced true and false color composite. Benthic habitat interpretation of satellite imagery using several approaches such as composite band and image reinforcement with Depth Invariant Index Algorithm. This algorithm applies the correction method of the water column known as the Lyzenga Algorithm [24]. This method is effective for improving the quality of the identification and classification of shallow basin habitats thematically. In line with this, the distinction of coastal type is done by visual interpretation and then combined with field survey results. Classification of canopy density referred to the national standard for Mangrove Surveying and Mapping (SNI 7717: 2011) scale of 1: 25.000. Meanwhile, the classification for benthic habitat referred to Draft of Technical Guidelines for Collection and Processing of Geospatial Data of Shallow Benthic Habitat (scale of 1: 10,000). For mangrove, the canopy density classes are very high, high, moderate, rare, and very rare. For benthic habitat there are eight classes, and only five classes will be used according to the purpose. However, drawbacks found in some data, so that another type image is used to cover up the shortcomings from another imagery. The temporary map resulted from image interpretation becomes the primary information to conduct ground truth.

2.2.2 Field survey. Determination of sample points using stratified proportional random sampling method in each class of image data processing results. The sample size is determined proportionally to the class extent of each map as well as considering the ease of accessibility in sampling.

The land survey initiated to do the measurement of geomorphological parameters (slope, lithology, and geological structure); In situ soil parameters (pH, soil lime, organic matter, and soil drainage); and also the measurement of hydrology (measurement of groundwater level). For mangroves, the method for canopy density validation used is canopy photos using the fisheye camera. For benthic habitat, field data collection is done by two methods namely "Stop and Go" to obtain many ground truth points; and the 'transect quadrant' method to get a percent closure of the habitat class (figure 2b).

IOP Conf. Series: Earth and Environmental Science **148** (2018) 012013 doi:10.1088/1755-1315/148/1/012013

2.2.3 SWOT Analysis. In this study, the SWOT (Strength, Weakness, Opportunities, and Threats) analysis had been used to reach qualitative results. SWOT analysis is a strategic planning method used to evaluate the strength factor, weakness, opportunities (opportunities), and threats in a region [25]. These factors are based on the considerations that occur in the research location as follows:

- Strength factor, which is the internal strength owned by Karimata Island. By knowing the strength, then this region can be directed to be more resilient.
- Weakness Factor: all internal factors that are not profitable or harmful to Karimata Island
- Opportunity Factors: external factors that potentially improve the welfare of Karimata Island community
- Factor Challenges (Threats): external factors that can bring harm to Karimata Island

Qualitatively, the determination of a SWOT analysis carried out after descriptively determine and analyze the strengths, weaknesses, opportunities, and threats [25]. This qualitative model produces the SWOT matrix as shown in the following table 2.

Table 2. Matrix of SWOT Analysis.

Internal External	**Strength (S)**	**Weakness (W)**
Opportunity (O)	S-O Strategies	W-O Strategies
Threat (T)	S-T Strategies	W-T Strategies

3. Result And Discussions

3.1. Coastal ecosystems of Karimata Island

Karimata archipelago is a group of islands in West Kalimantan province which connected to the Karimata Strait and the South China Sea. Karimata Island is the largest and populated island besides Penebang Island, Serutu Island, Pelapisgenting Island, and Pelapistiang Island. [26] Informed that people in Karimata Island inhabit coastal areas tend to have maritime culture rather than agrarian. About 80% of land cover in Karimata Island is the forest where it indicates the high level of human activity that influences the pattern of land use change. Human activities in Karimata Island are still limited based on land cover variations such settlements, roads, coconut plantations, clove plantations and mixed plantations.

The existence of coastal ecosystem in Karimata Island can be found either at east or west side (figure 2a). The establishment of Marine Reserve Area on the northeast to the southern coast of Karimata Island covering an area of 77,000 ha to the southeast. This study is focused on the Non-Marine Reserve area which spread along the southwest to the northwest coast of Karimata Island. Worldview-2 imagery acquired in 2015 showed that on the west side of Karimata Island is dominated by mangroves and coral. While much of the terrestrial environment is forests and plantations, the coastal environment consists of a diverse array of habitats, including mangroves, sand, and corals. A large mangrove area is located in western Karimata (Sei Ulu Betok - Dusun Kelumpang), spanning around 12 km of the coastline (figure 2a). All at Karimata Island, it is predicted to be around 1,122.24 ha and more than 85% located outside of the Karimata nature reserve area (table 3). In the Karimata Marine Reserves, mangrove existed on the south and east coast, especially at Tanjung Ujung Pandang and Tanjung Gemuk (figure 3). Mangrove canopy density class categorized into five classes. The 'very high' canopy density has the most substantial extent (727.22 ha). Several mangrove genera were found such *Sonneratia*, *Bruguiera*, *Lumnitzera,* and *Rhizophora* sp. is dominantly appeared in this region (figure 4). The mangrove condition in Karimata is becoming a concern to develop a sustainable management strategy for mangrove ecosystems in Karimata island, which is socially beneficial to the surrounding communities [11] [27] [28].

Tabel 3. Mangrove and Benthic Habitat in Karimata.

Mangrove Density	Area (Ha) Total	Area (Ha) Non Marine Reserve	Benthic Habitat	Area (Ha) Total	Area (Ha) Non Marine Reserve
Very high	770.4890	727.2179	Coral	1106.71	162.9769
High	174.8999	132.7026	Dead Coral	1385.56	328.6137
Moderate	30.0461	11.0469	Macro Algae	488.04	127.5050
Rare	17.0395	8.0261	Sand	1625.76	288.5112
Very rare	41.6724	41.2027	Rubble	26.16	6.1129
Total	1034.1469	920.1962	Total	4632.23	920.7197

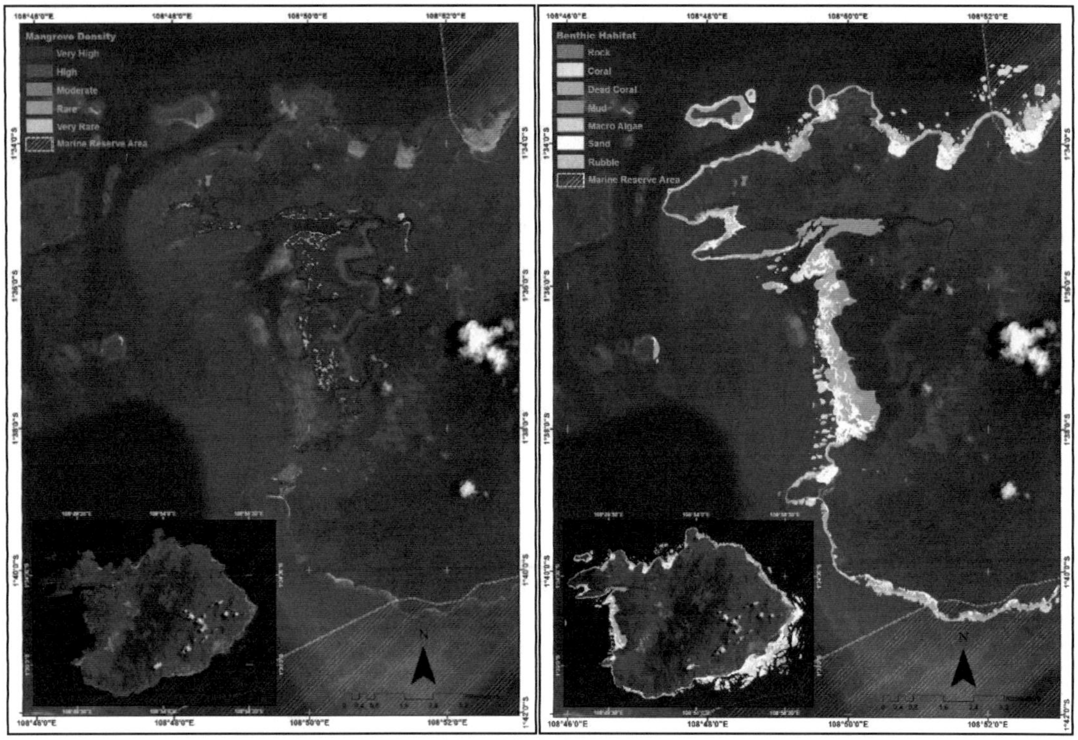

Figure 3. Mangrove and benthic habitat at non-marine reserve area.

Figure 4. Mangroves at non-marine reserve area in Karimata Island.

Shallow water habitats which owned natural beauty such coral and seagrass are great attractions that suitable for water sports such as snorkeling and scuba. Coral reefs in good condition are spread evenly in the southeast, north and south sides, which belong to the Karimata Nature Reserve. In the north coastal, coral reefs are found in small island clusters in front of Penghujan Bay and Teluk Alam (figure 3). Among the benthic habitat, coral reefs area extent to ± 1.106 Ha (21%) on a scale of 1: 10,000 (table 2). In line with the previous research results [29], Karimata Island is dominated by non-Acropora with a masive and branched life form, built between 3 - 15 m depth. Seagrass is hardly found on the coast of Karimata Island, its existence only in a small extent.

Based on the results of image interpretation, several coral spots located outside the Marine Reserve region spread on the southwestern side, northwest (Tokong Kepayang Islands - Kera Island) and north of Karimata Island (Teluk Liseng and Teluk Alam). In the southwestern part of the island (Tanjung Kalung Bay and Tanjung Kalung), living corals form separate colonies with Acropora tabulate (ACT), coral sub-massive (CS), and coral massive (CM) (figure 5). The coral health condition is poor, and visibility was limited due to wave condition [26].

On the western side of Karimata, on the Kelumpang Bay area bordering on mangroves, benthic habitat dominates by algae, ruins, mud sludge substrate and dead corals (Figure 6). Several small places of coral reefs found in the view of Dusun Kelumpang, in the southern part of Teluk Kelumpang. Based on survey results (2016), benthic habitat habitats encountered are corals with a cover range of 5 - 10% comprising Acropora Digitite (ACD), CM, Branching, encrusting, and foliose. Seagrass existence is very rare, only Cymodocea sp and Syringodium sp. found in a tiny patch (< 10% cover). Similarly, Domination of algae is coming from the Sargassum and Padina genera (figure 7). From table 2 it can be known the extent of habitat benthic residing in Non-Marine Reserve area at Karimata Island. However, dead corals and sand substrate dominated by benthic habitat. Various conditions above might show that Teluk Kelumpang can be directed to be mangrove tourism.

Figure 5. Coral at Teluk Tanjung Kalung dan Tanjung Kalung.

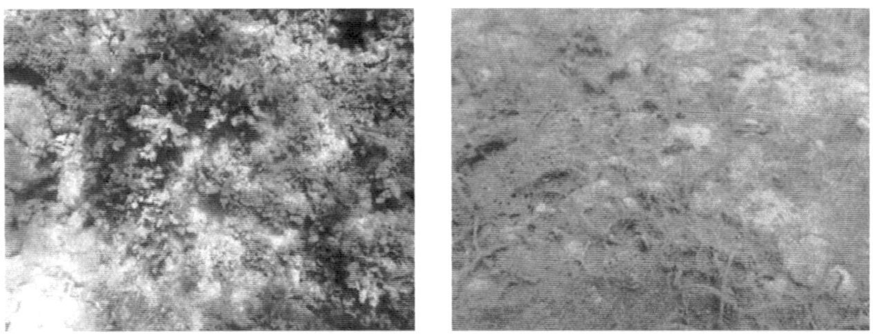

Figure 6. Benthic habitat condition at Teluk Kelumpang, in front of Mangroves area.

Figure 7. Benthic habitat at Dusun Kelumpang (southern part of Teluk Kelumpang).

3.2. Coastal Type

3.2.1 Sub aerial deposition coast. Coastal typology of the sub aerial deposition coast is found in the western part of Karimata Island, at the mouth of the Sei Ulu Betok, the Sei Are and the Sei Gelombang (figure 8; yellow box). The coastal area had sediment material from the erosion material that carried by the river and deposited at the mouth of the river. The material carried down to the downstream is a fine-sized material and becomes the substrate. The results of the image interpretation showed the existence of clay substrate connected with the existence of mangrove ecosystems along the Gulf Kelumpang that can be used as a mangrove tourism (figure 2).

3.2.2 Marine deposition coast. Marin deposition coast found in several areas such as Dusun Kelumpang, Teluk Alam, Teluk Penggaram, and Teluk Lising (figure 8; red box). Those coasts are consist of two forms; tidal flat and pocket beach. This typology formed by the tidal process of seawater that carried marine sediments. The extent of these two can vary depending on the tidal activity, and can only be distinguished by identifying the base material. The pocket beach has a bright white quartz sand material, while the tidal flat has a finer clay-sand material. Sandbanks also sometimes formed in the tidal plains. This marine sediment beach typology has the potential to be developed into tourism.

3.2.3 Coast built by organism. Coastal built by organisms can be found almost around Karimata Island forming a fringing reef, especially on Tokong Kepayang Island, Kepayang Island and Teluk Begunyai (figure 8, blue box). These coastal typologies include reef flat, reef patches such as in northern of Teluk Liseng and Teluk Alam, and reef crests. Those coastal has potential for snorkeling or diving sites, due to the coral and seagrass. Another coastal type found in Karimata Island is erosional coast (figure 8, green box) which also has potency for sunbathing locations and panoramas seeker [30].

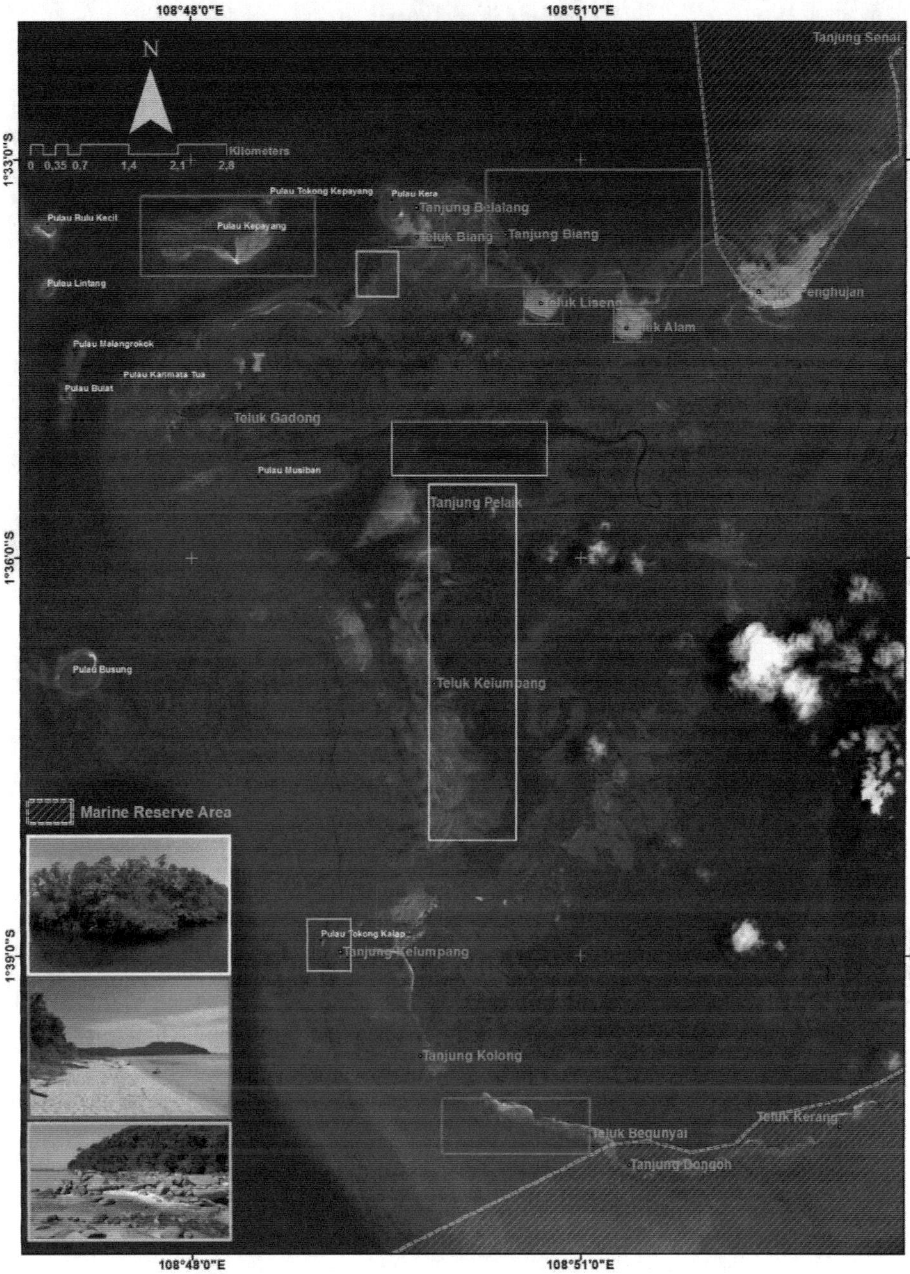

Figure 8. Coastal types at Karimata Island.

3.3. Strength, Weakness, Opportunities, and Threats (SWOT) Analysis

Sustainable tourism can be developed and also aware of its impact by incorporating environmental resource management and monitoring in an integrated plan [2]. Cultural and physical aspect are an integrated unity that supports each other as a coastal and marine tourism area. All characteristics and conditions of Karimata Island translated as strengths, weaknesses, threats, and opportunities. Those are needed to deliver an excellent strategy for sustainable management of Karimata island's natural resources (table 3).

Table 3. SWOT Analysis.

Internal External	Strength (s)	Weakness (W)
	1. Karimata has mangrove forest area outside the marine reserve with good condition 2. Marine depositional coastal typology as a tourist attraction 3. Having historical value as an ancient shipping lane becomes a region's marketing potential 4. Abundant fishery potential (seasonal) 5. Local people are maritime community	1. Residents are still dependent on marine resources 2. Inadequate public transport facilities and infrastructure 3. Inadequate electricity and communications infrastructure 4. Not complete benthic habitat, Sea-grass ecosystems are found very rarely 5. Conditions of coral reefs mostly are damaged 6. Unavailability of facilities and infrastructure of marine product management 7. The fulfillment of logistics needs still depends on the supply of the mainland
Opportunity (O)	S-O Strategies	W-O Strategies
1. The location is strategic because it is the archipelagic sea lanes of Indonesia (ALKI 1) and international shipping lanes 2. Good hydrological condition to sustain the water needs 3. Natural condition that support the development of ecotourism and marine tourism 4. Potential development of micro hydro power plant for energy independence Karimata Island 5. Inter-regional	1. Ecotourism development in mangrove forest area and Optimizing the potency of marine tourism sector in coastal areas 2. Development of marine tourism based on local wisdom and cultural heritage 3. Development of facilities and infrastructure to support the independency of energy, food and tourism development 4. Development of fishery-based aquaculture sector as an alternative to regional development 5. Regulation framing for eco-tourism management at Karimata Island	1. Enhancing the quality of human resources (entrepreneur training) 2. Optimization of cooperation between regions to develop facilities and infrastructure of marine product management 3. Development of facilities and infrastructure to support energy independence and tourism 4. Coastal Ecosystem rehabilitation by involving the community

cooperation to develop fisheries and tourism potential 6. The status of Karimata Island Region as Marine Reserve based on the Decree of the Minister of Forestry no. 381 / Kpts - II / 1985 control the utilization of natural resources		
Threats (T) 1. Natural and man-made hazard (Sea level rise, oil spills, coastal erosion) 2. Natural resource management has not been implemented optimally and sustainably 3. Unavailability of environmental sanitation 4. IUU Fishing	S-T Strategies 1. Development of tourism areas based on local wisdom 2. Community empowerment to participate in the natural resources management of Karimata region. 3. Development of environmental sanitation facilities 4. Regulation framing to keep down the impact of man-made hazard (oil-loaded ship lanes).	W-T Strategies 1. Increasing the capacity and quality of facilities and infrastructure of transportation, infrastructure and energy 2. Regulation enforcement of Marine Reserve Area (CAL) 3. Government support for investment and integration of development programs in tourism and fisheries sectors 4. Development of engineering techniques to reduce the potential for coastal disaster risk in Karimata Island

4. Conclusion

The existence of Karimata Island as a small island in Indonesia has its attractiveness. However, the development of sustainable tourism utilizing coastal ecosystem resources has not been initiated by the local government. Most of the eastern coastal area of Karimata Island belong to the Karimata Marine Reserve area. Thus the western part of the island can be utilized for tourism activities. Based on the mapping results, Karimata Island has the coastal ecosystem resources like mangroves and coral reefs that scattered almost throughout the island. More than 80% of the mangroves area is concentrated on the western side of the island, ie, in Kelumpang Bay. The condition becomes a significant base in the development of mangrove ecotourism. The study result of coastal types, Karimata Island, has several coastal types that support marine tourism activities.

In the preparation of sustainable management strategies, it is important to assess the characteristics of the strengths, weaknesses, and opportunities and threats. The development of tourism considerations for the region of the economy is also an important step.

Karimata Island has an eco-tourism and marine tourism attraction. The availability of qualified human resources for tourism management is essential. Thus, local society resources enhancement is important. The adequacy of training in the aspect of entrepreneurial for the development of eco-tourism strongly required. Accordingly, the local government should follow an aggressive growth strategy. While all the tourists are the target, the facilities and infrastructure to support the tourism development must be implemented. Also, there is the high need for regulation, promotion, and public relations.

Acknowledgement
We would like to send our appreciation and honor to Thematic Mapping and Integration Centre – Badan Informasi Geospasial, especially Dr. Sri Hartini and her team for giving material and information to support this paper.

References

[1] B Koddeng 2011 Zonasi Kawasan Pesisir Pantai Makassar Berbasis Mitigasi Bencana (Studi Kasus Pantai Barambong-Celebes Convention Centre) Pros. 2011, vol. 5, *Group Teknik Arsitektur*, p. TA17-1-14.
[2] S H Djoeffan C Chusharini and E S Sundari 2010 Strategi pengelolaan kawasan wisata cagar budaya karangkamulyan di kabupaten ciamis.
[3] N A Sloan and A S Ugandhy 1994 An overview of Indonesian coastal environmental management *Coast. Manag.*, vol. 22, no. 3, pp. 215–233, Jan. 1994.
[4] P Paul T K Kar and A Ghorai 2016 Ecotourism and fishing in a common ground of two interacting species *Ecol. Modell.*, vol. 328, pp. 1–13.
[5] M M Masud A M Aldakhil A A Nassani and M N Azam 2017 Community-based ecotourism management for sustainable development of marine protected areas in Malaysia *Ocean Coast. Manag.*, vol. 136, pp. 104–112.
[6] M S Rhormens A de G Pedrini and N P Ghilardi-Lopes 2017 Implementation feasibility of a marine ecotourism product on the reef environments of the marine protected areas of Tinharé and Boipeba Islands (Cairu, Bahia, Brazil) *Ocean Coast. Manag.*, vol. 139, pp. 1–11.
[7] H Yanagisawa S Koshimura T Miyag and F Imamura 2010 Tsunami damage reduction performance of a mangrove forest in Banda Aceh, Indonesia inferred from field data and a numerical model *J. Geophys. Res. Ocean.*, vol. 115, no. 6, pp. 1–11.
[8] C Kuenzer A Bluemel S Gebhardt T V Quoc and S Dech 2011 Remote sensing of mangrove ecosystems: A review *Remote Sensing*, vol. 3, no. 5. pp. 878–928.
[9] I Nagelkerken S J M Blaber S Bouillon P Green M Haywood L G Kirton J O Meynecke J Pawlik H M Penrose Sasekumar and P J Somerfield 2008 The habitat function of mangroves for terrestrial and marine fauna: A review *Aquatic Botany*, vol. 89, no. 2. pp. 155–185.
[10] B B Walters 2005 Ecological effects of small-scale cutting of Philippine mangrove forests *For. Ecol. Manage.*, vol. 206, no. 1–3, pp. 331–348.
[11] K Umam S Sudiyarto and S Tj Winarno 2015 Strategi Pengembangan Ekowisata Mangrove Wonorejo Surabaya *Agrar. J. Agribus. Rural Dev. Res.*, vol. 1, no. 1, pp. 38–42.
[12] A C Putra S Anggoro and Kismartini 2015 Strategi pengembangan ekowisata melalui kajian ekosistem mangrove di pulau pramuka , kepulauan seribu Ecotourism Development Strategy with Mangrove Ecosystem Studies *J. Saintek Perikan.*, vol. 10, no. 2, pp. 91–97.
[13] Y R Fitriana 2005 Diversity and abundance of macrozoobenthos in mangrove rehabilitation forest in Great Garden Forest Ngurah Rai Bali *Biodiversitas*, vol. 7, no. 1, pp. 67–72.
[14] D Satria 2009 Strategi Pengembangan Ekowisata Berbasis Ekonomi Lokal Dalam Rangka Program Pengentasan Kemiskinan Di Wilayah Kabupaten Malang *J. Indones. Appl. Econ.*, vol. 3, no. 1, pp. 1–2009.
[15] A Tuwo 2011 Pengelolaan ekowisata pesisir dan laut: pendekatan ekologi, sosial-ekonomi,

kelembagaan, dan sarana wilayah. *Brilian Internasional.*
[16] S Chauvaud C Bouchon and R Maniere 1998 Remote sensing techniques adapted to high resolution mapping of tropical coastal marine ecosystems (coral reefs, seagrass beds and mangrove) *Int. J. Remote Sens.*, vol. 19, no. 18, pp. 3625–3639.
[17] P J Mumby E P Green A J Edwards and C D Clark 1999 The cost-effectiveness of remote sensing for tropical coastal resources assessment and management *J. Environ. Manage.*, vol. 55, no. 3, pp. 157–166.
[18] C Warren J Dupont M Abdel-Moati S Hobeichi D Palandro and S Purkis 2016 Remote sensing of Qatar nearshore habitats with perspectives for coastal management *Mar. Pollut. Bull.,* vol. 105, no. 2, pp. 641–653.
[19] A Aslan A F Rahman M W Warren and S M Robeson 2016 Mapping spatial distribution and biomass of coastal wetland vegetation in Indonesian Papua by combining active and passive remotely sensed data *Remote Sens. Environ.*, vol. 183, no. September, pp. 65–81.
[20] N A Ibharim M A Mustapha T Lihan and A G Mazlan 2015 Mapping mangrove changes in the Matang Mangrove Forest using multi temporal satellite imageries *Ocean Coast. Manag.*, vol. 114, pp. 64–76.
[21] J Long D Napton C Giri and J Graesser 2014 A Mapping and Monitoring Assessment of the Philippines' Mangrove Forests from 1990 to 2010 *J. Coast. Res.*, vol. 294, no. 2, pp. 260–271.
[22] J B Long and C Giri 2011 Mapping the Philippines' mangrove forests using Landsat imagery *Sensors*, vol. 11, no. 3, pp. 2972–2981.
[23] J M Kovacs F Flores-Verdugo J Wang and L P Aspden 2004 Estimating leaf area index of a degraded mangrove forest using high spatial resolution satellite data *Aquat. Bot.*, vol. 80, no. 1, pp. 13–2
[24] P J Mumby E P Green A J Edwards and C D Clark 1997 Coral reef habitat-mapping: How much detail can remote sensing provide? *Mar. Biol.*, vol. 130, no. 2, pp. 193–202.
[25] L Muta'ali 2015 *Teknik Analisis Regional Untuk Perencanaan Wilayah, Tata Ruang dan Lingkungan* Yogyakarta: Badan Penerbit Fakultas Geografi (BPFG).
[26] I E Setyawan M S Wijaya D M Yuwono R Nugroho and I Pujawati 2016 Karimata - Bentang Alam dan kondisi Sumber Daya. Cbinong p. 160.
[27] F Purwantiv 2001 Tourism and conservation management options for karimunjawa marine national park (Case Study and Reviews) *J. Coast. Dev.*, vol. 4, no. 2, pp. 99–106.
[28] I P A Citra 2017 Strategi pemberdayaan masyarakat untuk pengembangan ekowisata wilayah pesisir di kabupaten buleleng *J. Ilmu Sos. dan Hum.*, vol. 6, no. 1, pp. 31–41.
[29] N C D Aryanto and Y Permanawati 2009 Types and distribution of coral reef on the karimata coast , west kalimantan *Bull. Mar. Geol.*, vol. 24, no. 1, pp. 25–35.
[30] T Gunawan L W Santosa L Muta'ali and S H M B Santosa 2007 Pedoman survei cepat terintegrasi wilayah kepesisiran. Badan Penerbit dan Percetakan Fakultas Geografi.

The geomorphology and hydrogeology of the karstic Islands Maratua, East Kalimantan, Indonesia: the potential and constraints for tourist destination development

E Haryono[1], M H D Sasongko[1], D H Barianto[2], J B Setiawan[3], A A Hakim[1] and A Zaenuri[1]

[1]Karst Research Group, Faculty of Geography, Universitas Gadjah Mada, Indonesia
[2]Geological Engineering Department, Universitas Gadjah Mada, Indonesia
[3]Research Centre for Marine Science and Technology, Universitas Gadjah Mada, Indonesia

e.haryono@ugm.ac.id

Abstract. Maratua Island is one of the islands of Berau District, East Kalimantan which has great potential of natural beauty for tourism development. The area currently is one of famous tourist destination in East Kalimantan which is a carbonate reef built-up or so-called karst island. This paper is an endeavor 1) to unveil geomorphological and hydrogeological characteristics of the island, and 2) to recommend Island development as a tourist destination. Maratua Island is a V shape atoll with the open lagoon. Six geomorphological units were found on the island, i.e., fringing reef, beach, marine terrace, karst ridge, structural valley, and lagoon. Caves are also found in the karst ridge and the coast as an inundated passage. Three structural depressions in the karst ridge are other unique geomorphological feature in the area of which a marine lake environment with jellyfish is inhabited. The island is typified by two different aquifer units, i.e., porous media and fractured media aquifer. Porous aquifer lies on the beach of Boibukut area. Fractured-aquifer characterizes the other geomorphological units in the area. Freshwater accordingly is found in the beach area with a limited amount. Unfortunately, the groundwater in the marine terrace and karst ridge are saline. Maratua Island has enormous potential for tourism destination development. The major tourist activities in the area based on the geomorphological unit are snorkeling and diving (in fringing reef and lagoo n), hiking, cave exploration and marine lake exploration and cave diving (in karst ridge and structural valley); recreation and picnic (beach). The major limitation in the area is a shortage of freshwater resource and land. Limited water supply should be extracted from the beach area of Bohe Bukut village. Groundwater extraction from the beach area of Bohe Bukut must be for drinking water only. Supply of drinking water should be substituted from collected rainwater or desalination from sea water and water in the cave. Restrictions in the number of visitors and lodging development should also be considered.

1. Introduction

Indonesia is the biggest archipelagos countries in the world. The number of islands in Indonesia is 16.056 islands (Berita Geospasial BIG, 2017). As such, Indonesia islands have various physical characteristics, including geomorphology and hydrogeology. Geomorphology is the study of landforms and processes taking place on the earth surface, and the relationship between these landforms and processes in terms of spatial context [1]. The primary object of geomorphological studies is landforms. The geomorphological characteristics of a surface will affect the hydrogeological conditions of an area [2].

Karstic island is one of the islands found in some Indonesian Island. Karst landform is considered unique, where it has typical morphological and hydrological conditions due to the development of secondary porosity [3]. Regulation of the Minister of Energy and Mineral Resources (ESDM) No 17 of 2012 specifies that karst is a landscape that is formed as a result of the dissolution of limestone or dolomite.

Maratua Island is one of the small islands of the Derawan-Maratua complex islands, situated in Berau Regency, of the East Kalimantan Provence. The island is one of the carbonate islands and karst islands in Indonesia [4]. Maratua Island has great potential of natural beauty for tourism development. Presently, tourism in Maratua Island has also been growing and attracted a more significant number of both domestic and foreign tourists. This development in tourism is also responded by investment and infrastructure development as well. A carpet of a white-sand beach and, karst caves, reef, and a saltwater lake constitute the attraction of tourism in Maratua Island.

Becoming one of the leading national and regional tourist destinations makes the development in Maratua Island is growing very fast. Tourism facilities, i.e., roads, inns, airport, and other public facilities have been built in the Island. The fast development in a small island of Maratua obviously will affect its environmental conditions. The development, therefore, must consider its carrying capacity on the island. The carrying capacity of the environment is defined as the number of people or conditions that can be supported by resources in the area or ecosystem [5].

Identifying the potential and the existing limiting condition of an area is an important step as a framework for appropriate and sustainable development. The Maratua Island which has high potential for tourist destination also has its limitation in term of geomorphological and geohydrological aspects. This paper accordingly aims at determining the geomorphological and geohydrological characteristics of Maratua Island, and giving recommendations for the development based on the existing potential and constrains.

2. Study Area

Maratua Island is the easternmost island of East Kalimantan, bordering the Indonesia Archipelago with the Malaysia and Philippines. It lays at the 2°15′12″ North Latitude and 118°38′41″ East Longitude. Maratuais considered as a small island of which the land area is less than 1,000 km^2 and has a permanent population of fewer than 100,000 inhabitants [6]. The population number of the Maratuaisland is 3,402, and the area is 4,119.54 km^2. Administratively, the island belongs to Berau Regency. The Maratua Island Subdistrict has four villages, namely BoheSilian, Payung-Payung, BoheBukut, and TelukAlulu. Maratua Island consists of a main island called Maratua and a group of small isles situated within its lagoon, i.e. Semut Island, Siddau Island, Kokok Island, Pabahanan Island, Bulingisan Island, and Andongabu Island.

Physiographically, Maratua Island is situated in the northeast part of Kalimantan which is in the tip of Mangkalihat high. The island is 60 km away from the mainland of Kalimantan separated by shallow see [7]. The eastern part of Maratua Island is bordered by the slopes of the Makasar Strait. Geologically, Maratua Island is composed of limestone and located at the fringe of the NW-SE bearing Maratua Fault (figure 1.). The Maratua Island morphology, therefore, is governed by this fault to which its shoreline extends parallel to the fault orientation.

3. Methods

The main methods used in this research were remote sensing interpretation and field survey. A field survey was conducted in 2012 up to 2015. Mapping of landforms in Maratua Island was undertaken using remote sensing image interpretation which was verified by field checking. The remote sensing images used for the interpretation were Landsat TM 7 and ALOS. A field survey was conducted through geomorphological and hydrogeological surveys; including landform observation and measurement, cave mapping, groundwater level measurement, plotting of potential tourist attraction. Most of these potential tourist destination attraction was obtained from local people information. Qualitative measurement of water quality conditions were carried out in several caves and resident's wells by tasting the water.

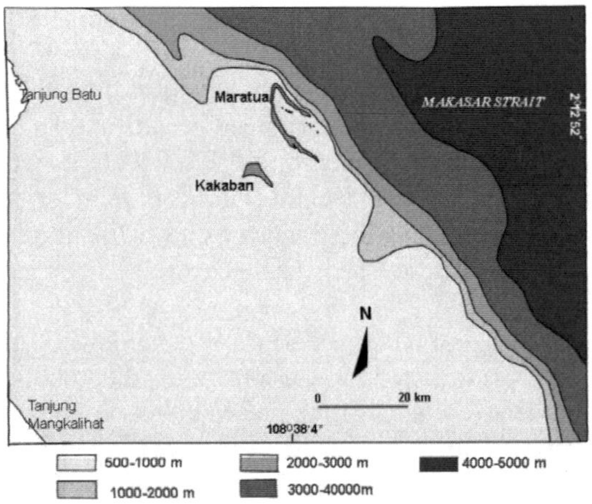

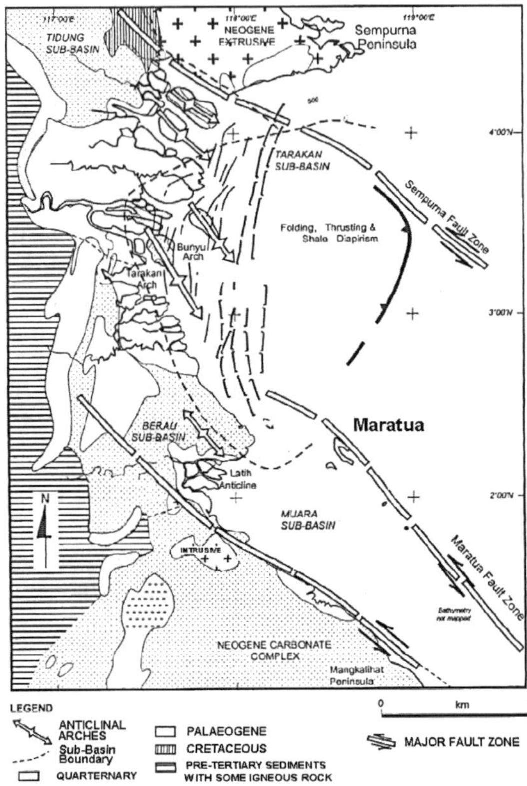

Figure 1. A Map illustrating bathymetry and the tectonic setting around Maratua Island [7], [27].

4. Result and Discussions

4.1. Geomorphology

The tectonic setting which primarily controls the formation of Maratua Island has resulted in a large number of structural features to be found in the island. The island formation is governed by NW-SE fault, the rifting of Eocene-Quarter Makasar Strait, and uplifting [8] [11]. Exogenic processes in the area further modify the structural form into different landforms. The landforms of Maratua Island, hence, consists of structural, solution, marine, and organic form (see figure 2).

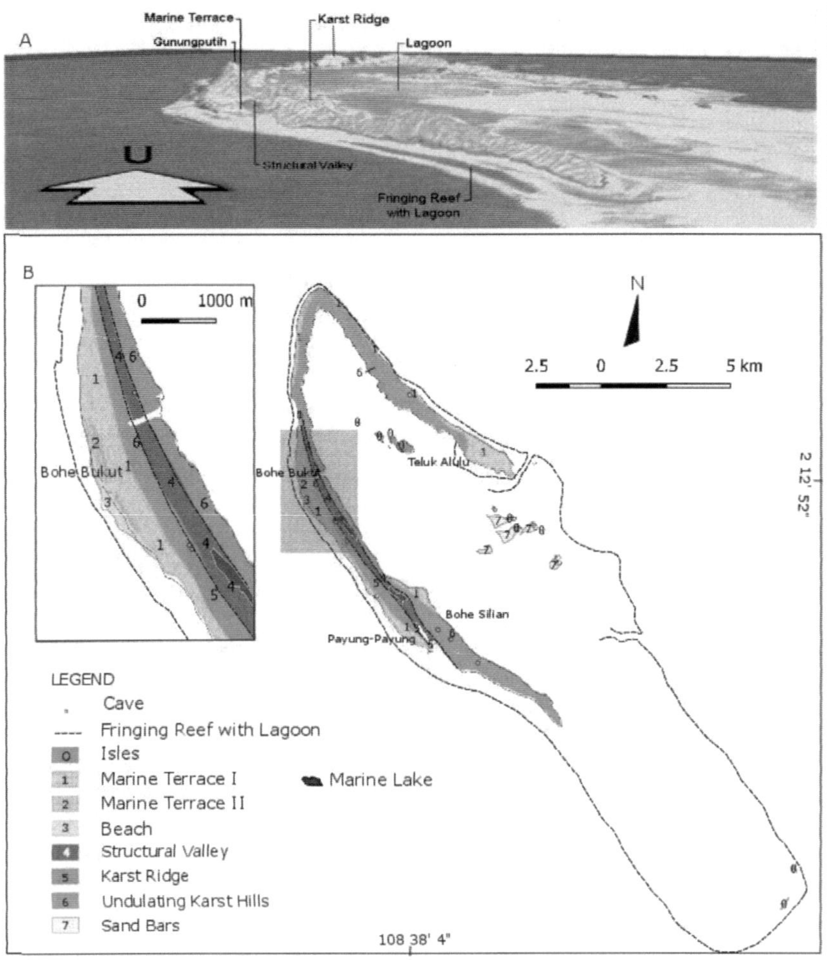

Figure 2.
(a) morphological features of the Maratua Island in 3D view and
(b) landforms of the Maratua Island.

Structural landform found in the Maratua Island is a structural valley which is associated with the NW-SE structural trend. The structural valleys in Maratua Island are filled with salt water forming a marine lake, namely Hajibuang, Tanabamban, Haji Mangku, and Siku Peninsula lake. The formation of marine lakes is suggested occurring during the last interglacial periods [12] [13]. The marine lake is comparable to the marine lake of the Misol Island complex described [28]. The planar shape of the marine lakes tends to extend with a ratio of their maximum width to their maximum doline length of about 1:3.

Table 1. Peaks of karst ridge in Maratua Island.

Nr.	Hills	Height (m from MSL)	Slope (degree)	Slope Form
1	Bohe Silian	80	25 -35	Concave
2	Gunung Putih	111 m	35-45	Concave
3	Bohe Bukut-Teluk Alulu	65 m	20-35	Concave

Since the Maratua Island is a karstic island, solution form is the most extensive landform in the Island. Karst ridge is one of the major positive karst forms. It has several peaks from the southwest to the northeast. The highest peak in Maratua Island is Gunung Putih protruding up to 111 m from MSL.

Single doline in Maratua Island develops in the middle slope of the karst ridge along with structural opening with the azimuth of 290 up to 310, coinciding with the dominant structural pattern.

In addition to the exokarst landform, caves are also found in the Island as an endokarst landform (table 2). The caves found in Maratua Island have unique and distinctive morphology. Most of the caves are structural controlled (fault or joint) caves whose narrow passage with a high cave ceiling. Few caves have a large chamber,i.e., Tangkapa Cave and Angkal-Angkal Cave.

Table 2. The Caves in Maratua Island.

Nr.	Cave Name	X	Y
1	Sembat	680678	241568
2	Tangkapa	679625	242479
3	Angkal-Angkal	679136	242826
4	Silandayan	676935	244428
5	Sipbig	677085	244183
6	Sukur	674758	247565
7	Mayat-Mayat	674554	248089
8	Pogah	675221	246891
9	Hapit Pogah	675187	246947
10	Organ	674445	248955
11	Jellau	676660	253437
12	Pahang	677594	251903
13	Penggunting	676793	244087

Conical forms such as those are found in most karst areas in Java [14]–[17], Sulawesi [18], [19], Papua [20] do not exist in Maratua Island. The geological structure governs the elongated shape found in in the island. Joints and faults are controlling factors that cause elongated hills found on the Maratua Island. Karst hills, in general, belong to the asymmetric form. Another major karst form found in Maratua Island is doline. Doline is one of the karst landform hallmarks resulted from concentrated dissolution. Dolines in Maratua Island, according to Cvijic category [21]-[22] has a funnel-shape with the ratio of width to depth is about 1:2.

Marine landforms found in Maratua Island are marine terraces, beaches, and lagoons.Two marine terraces are found on the island. The first marine terrace elevates from 8 to 10 m MSL with a relatively flat to rolling topography. The marine terrace composed of uplifted Quaternary Reef. The width of the reef terrace varies from 200 m to 2 km. In the west, the marine terrace is wider than that in the eastern part of the island. The second terrace and beach composed of white sand of coral fragments and sand. Most of the beach in Maratua Island is very narrow (less than 50 m width). In the nearshore zone, Maratua island is characterized by an organic form of fringing reef morphology. Fringing reef development is more extensive in the western part and inner lagoon than that in the eastern part.

4.2. Hydrogeology

The aquifers in Maratua Island are composed of three rocks, i.e., Miocene limestone, quarternary coral limestone, and unconsolidated sand (figure 3). The three rocks then form different aquifers. Miocene limestone occupies the central part of the mainland of Maratua, forming karst ridge and undulated karst hills extending from Bohe Silian Village to Teluk Alulu Village, forming a V shape island. The ridge whose bedrocks are Miocene limestone is often found in a caving system.

The karst aquifers in Maratua Island are characterized by high fracture porosity resulting from joints and faults. Therefore, freshwater lenses as occurs in an Atoll [23]–[26] does not develop in

Maratua Island. An extremely high permeability typifies karst aquifers in the area. The water level ranges from 2 to 5 meters above sea level. The water level in the western arm of Maratua Island is higher compared to the eastern arm such as in Teluk Alulu. The water level in karst aquifers is highly fluctuated and is controlled by tide oscillation. The fluctuations of groundwater level in karst aquifers can reach 1.5 meters. The fluctuating groundwater in karst aquifers occurs because the wet caves in Maratua Island are connected to the sea through fissures/cracks and joints. Table 3 shows caves with underground water found in Maratua Island. Water found in most of the caves and are brackish to saline.

Quarternary coral limestone occupies the marine terrace located in food slop of karst ridge either on the outer side overlooking the open sea or on the inner side that overlooks the inner lagoon. Marin terraces are characterized by rocks dissolution cavities and coral structure cavities. The rocks in the marine terrace less hard compared to the Miocene limestone of the karst ridge and undulated karst hills. The groundwater level in this aquifer ranges from 2 to 3 meters. The depth of groundwater from the ground surface ranges from 5 to 8 meters. The groundwater level in terrace aquifer is also affected by tides. Aquifers of reef terraces have high productivity with water quality similar to that of Miocene limestone aquifer.

The third aquifer is composed of loose sand occupies the lower marine terrace (marine terrace II). The aquifers in marine terrace II and beach has medium permeability. Porosity is formed by the cavities between the loose grains of fine to coarse sand. The water level is very shallow, ranging from 1.5 to 2 meters. There are many dug wells were made by the local people in this area which serve as the primary source of clean water. The water quality is considerably good. The water is colorless, odorless, and tasteless. The water level is slightly affected by tides. Productivity of these sandy aquifers is low. Marine terraces with loose sand sediments are found more extensive in the western part of Maratua Island overlooking the mainland of Kalimantan Island, especially in Bohebukut Village with a width of about 300 m and a length of more than 2 km. In the northern part of the island and on the inner lagoon, marine terraces with loose sand sediment are not formed. Bohe Bukut Beach can form wide loose sand because this area is a deposition area. The existence of undisturbed reefs in front of Bohebukut Beach also serves as a factor that protects the existence of sandy marine terraces that do not experience wave erosion. Therefore, aquifers in Maratua Island can be classified into three groups, namely karst aquifer, reef marine terrace aquifer, and sandy marine terrace aquifer (figure 3).

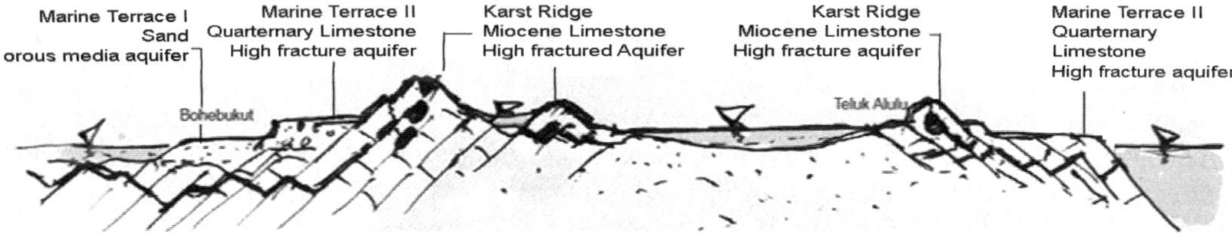

Figure 3. A Cross-Section of Maratua Island Showing a the Rocks Forming Aquifers (not to scale).

4.3. Potential for Tourist Destination Development

As a karstic island with unique landforms as described previously, the Maratua Island has enormous potential to be designated as a tourist destination (table 4.). The fringing reef situated in the outer side and inner side of the island is suitable for snorkeling, diving, swimming, and boat riding. The characteristics of the calm waves is suitable for underwater sightseeing, especially to enjoy turtle and other colourfull fish.

Table 3. Caves with Water in Maratua Island.

Nr	Cave Name	Coordinate			Water Depth (m)	Fluctuation (m)
		X	Y	Z		
1	Sembat	68067	24156	5	> 7	1.2
2	Tangkapa	67962	24247	6	< 2	0.3
3	Angkal-	67913	24282	6	-	-
4	Silandayan	67693	24442	7	> 13	1.5
5	Sipbig	67708	24418	6	> 6	1.4
6	Sukur	67475	24756	3	> 9	1.5
7	Mayat-	67455	24808	4	1	0.2
8	Pogah	67522	24689	5	< 2	0.3
9	Hapit Pogah	67518	24694	4	<2	0.3
10	Organ	67444	24895	4	< 4	0.6
11	Jellau	67666	25343	4	> 4	1.2
12	Pahang	677,59	25190	4	> 16	-
13	Penggunting	67679	24408	4	< 3	0.5

Table 4. Tourism Activities Potential in the Landforms of Maratua Island.

No	Landforms	Tourism Activity Potential	Type of
1	Fringing Reef	Snorkeling, swimming, and diving raiding	Recreation
2	Beach	Sunbathing, sightseeing, beach sport activity	Recreation
3	Karst Ridge and Undulating karst hills	Tracking, cave exploring, rock climbing, sunset or sunrise watching	Special-Interest Tourism
4	Marine Lake	Panoramic sightseeing, rowboat	Recreation
5	Lagoon	Snorkeling, swimming, and diving, and water Games	Recreation
6	Cave	Cave exploration	Special-Interest

A beach is a landform which is closely associated with tourism. Beaches on the Maratua Island are also well known among tourists.White sand with coconut three give a breathtaking scenery along the island. Beaches can be designated as a place for sunbathing, enjoying the panorama, doing beach sports such as beach volley ball and football.

Karst ridge and undulated karst hills can be designated as a new tourist attraction. The karst right in Maratua Island are situated in the middle of the island covered by forest. Such a condition can be used to develop several types of unique interest tourism such as nature exploration/tracking or rock climbing. Steep slope of karst ridge in Maratua Island is suitable for stimulating adrenaline through out door activities such as rock climbing. Other enjoyable activity for tourists and local people in the karst hills is watching a sunrise from the top of Gunung Putih Hill.

Another special interest tourism activity that can be experienced in the island is cave exploration. The caves in Maratua Island are unique and challenging to be explored. Some of the caves are found to be watery and have a submerged passage. This situation is good for special interest tourism development in the form of cave exploration. Tourists who wish to enjoy the nature exploration in Maratua Island should have special skills in cave exploration. However, there are some caves that can be enjoyed without having to own special skills, such as Halo Tabung Cave (cenote), Sembat Cave, and Angkal-Angkal Cave.

4.4. Constrains for the Development
Maratua Island which is considered as a small island makes it has limiting conditions. Tourism development in this island should also consider those limiting factors. The main limiting factors in Maratua Island are clean water supply and limited land. Development plan should be based on the quantity and quality of these limitation. Clean water with good quality in Maratua Island is found only in aquifers in sandy marine terrace of BoheBukut.

Fulfillment of the need for clean water should get serious attention in the management of Maratua Island for tourism purposes. There are some alternatives to meet the need for water resources. One of them is by utilizing brackish water in caves and sea water distillation. The need for clean water for consumption can be met from wells in aquifers in sandy terraces. Besides, the need for clean drinking water can also be met by using rain water. The need for water for bathing and washing which do not necessarily require the use of good quality water can be met by using brackish water contained in the cave. Distilled water obtained from brackish water in the caves will reduce production coast of water purification compared to that from sea water.

Restrictions in the number of visitors and lodging development should also be considered. This has something to do with the limitation of land and the level of pollution that can be generated as a result of such development and the number of tourists that exceeds the capacity.

5. Conclusion
Maratua Island has high potential tourism development. The uniqueness of the island results in different natural attraction for tourism destination. Among the important geomorphological features are fringing reef, lagoon, beach, and caves. The potential tourism activities in the area are diving, snorkeling and boating, canoeing, caving, trekking, as well as panoramic sightseeing. However the island has limited fresh water sources and. Therefore, the island must be managed with care for tourism development. Water supply extracted from the beach area of Bohe Bukut should be limited for drinking water only. Substitutive water supply from collected rainy water or desalination from sea water should be given priority for further water supply development. The other possible water supply is from caves. Cave water can be considered for domestic water supply. Cave water can also be used as drinking water after desalinated. Desalination of cave water will be more economical than that from sea water. Restrictions in the number of visitors and lodging development should also be considered.

6. Acknowledgments
Thanks due to LPPM UGM that made the deployment and grand of KKN-Tematik Unit Maratua possible. The Directorate of Higher Education for the funding through PUPT scheme. The Berau Office of Mineral Resources and Energy and PUSTEK UniversitasGadjahMada for making possible the groundwater investigation of Maratua Island, all the students of KKN Unit Maratua who help data collection.

References

[1] Zuidam V 1983 *Guide to Geomorphologic Aerial Photographic Interpretation and Mapping* The Netherlands: ITC.

[2] Todd D 1980 *Groundwater Hydrology* New York: Wiley.

[3] Ford DC Williams PW 2007 *Karst Hydrogeology and Geomorphology* Chicester: John Wiley & Sons, Ltd.

[4] Sasongko M 2015 Kajian Morfologi Karst Pulau Maratua Kalimantan Timur *Sekripsi* Yogyakarta: Fakultas UGM.

[5] Rees W 1990 *Sustainable development and biosphere* Teilhard studies 23.

[6] Dahuri R 2003 *Keanekaragaman Hayati Laut: Aset Pembangunan Berkelanjutan Indonesia* Jakarta: Gramedia.

[7] Van Bemmelen RW 1970 *The Geology of Indonesia*, 2nd ed The Hague: Martinus Nijhoff.
[8] Moss S Holloway R 2015 New observations on the sedimentary and tectonic evolution of the Tertiary Kutai Basin , East Kalimantan *Petroleum Geology of Southeast Asia, Geological Society Special Publication* 16 393-416.
[9] Hall R and Wilson M E J 2000 Neogene sutures in eastern Indonesia. 18 781–808.
[10] Wilson M E J Chambers J L C Evans M J Moss S J and Nas D S 1999 Cenozoic carbonates in Borneo: case studies from northeast Kalimantan *Journal of Asian Earth Sciences* 17 183-201.
[11] Witts D Davies L Witts D Davies L Morley RJ Anderson L 2015 Neogene Deformation of East Kalimantan : A Regional Perspective *Proceedings Indonesian Petroleum Association*, May.
[12] Becking LE Renema W 2011 Recently discovered landlocked basins in Indonesia reveal high habitat diversity in anchialine systems. Hydrobiologia. 677 89–105.
[13] Tomascik T Mah AJ Nontji A Moosa MK 1997 *The Ecology of the Indonesian â€™ Seas*, vol. VIII (Singapore: Periplus Editions (HK) Ltd).
[14] Haryono E Day MJ 2004 Landform differentiation within the Gunung Kidul Karst, Java-Indonesia *J. Cave Karst Stud.* 66 (2) 62–684.
[15] Haryono E Putro ST Suratman Sutikno 2017 Polygonal Karst of Karangbolong Area, Java-Indonesia *Acta Carsologica* 46 (1) 63–72.
[16] Haryono E 2013 Uplift Evidence from Karst Morphology Preliminary Evidence from Blambangan Peninsula Karst in *Proceeding of 16th International Congress of Speleology, Brno* Cech Republic.
[17] Zylshal Z Danoedoro P Haryono E 2013 And Object-Based Image Analysis Approach to semi-Automated Karst Morphology Extraction in *Proceeding of 34th Asian Conference on Remote Sensing* 4750.
[18] McDonald C 1976 Limestone morphology in South Sulawesi Indonesia" *Z. Geomorphol. N.F*, Bd 26 79–91.
[19] Sunatadirdja MA and Lehmann H 1955 Der tropiche Karst von Maros und Nord-Bone in SW-Celebes (Sulawesi) *Z. Geomorphol. N.F*, Bd 2 50–65.
[20] Verstappen HTh 1964 Karst Morphology in the Star Mountains (in Central New Guinea) and Its Relation to Lithology and Climate *Z. Geomorphol* 8 (1) 40–49.
[21] Haryono E Adji TN 2004 *Pengantar Geomorfologi dan Hidrologi Karst* Yogyakarta: Fakultas Geografi UGM).
[22] Haryono E (Ed) 2016 *Pedoman Praktis Survei Terintegrasi Kawasan Karst* (Yogyakarta: Badan Penerbit Fakultas Geografi).
[23] Werner AD Sharp HK Galvis SC Post VEA and Sinclair P 2017 Hydrogeology and management of freshwater lenses on atoll islands : Review of current knowledge and research needs *J. Hydrol* 551 819–844.
[24] Comte JC Banton O Join J L and Cabioch G 2010 Evaluation of effective groundwater recharge of freshwater lens in small islands by the combined modeling of geoelectrical data and water heads *Water Resources Research.* 46 1-10.
[25] Post VEA Houben G J 2017 Density-driven vertical transport of saltwater through the freshwater lens on the island of Baltrum Germany) following the 1962 storm flood *J. Hydrol.* 551 689–702.
[26] Cahyadi A 2015 Analisis Potensi Sumberaya Air Pulau Koral sangat Kecil: Studi Kasus di Pulau Koral Pramuka, Kepulauan Seribu, DKI Jakarta *Thesis* (Yogyakarta: Universitas Gadjah Mada).

IOP Conf. Series: Earth and Environmental Science **148** (2018) 012014 doi:10.1088/1755-1315/148/1/012014

[27] Lentini H Daman 1996 *Geological Map of Tarakan Basin* (Jakarta).
[28] Purba GYS Haryono E Sunarto Manan J Rumenta L Purwanto,Becking LE 2018 Jellyfish Lakes at Misool Islands, Raja Ampat, West Papua Indonesia *Biodiversitas* 19 172-182.

Food security assessment in the coastal area of Demak Regency

R Harini[1], H N Handayani[1,2] and F R Ramdani[1,2]

[1]Department of Environmental Geography, Faculty of Geography, Gadjah Mada University, Yogyakarta, Indonesia
[2]Undergraduate Program of Environmental Geography, Faculty of Geography, Gadjah Mada University, Yogyakarta, Indonesia

rikaharini@ugm.ac.id

Abstract. Food security is an issue of national and global level. Food is a basic human need to live.Without food will threaten human life. This research was conducted in coastal area of Demak Regency. This research is to understand the potential of human resources, natural resources and assess the level of food security of coastal communities. The data used are primary data through interviews with the local community.Also used secondary data from government agencies. Data analysis used qualitative and quantitative descriptive methods through graphs, tables and maps. The results showed that potential of human resources in Demak coastal area (Wedung, Purworejo, and Sidogemah villages) is low. It can be indicated from education level of respondents in Demak are elementary school and junior high school. Beside, total households income are about 2-4 million. This study found that the households sampled are 90% is food insecured households. The most of households in Demak coastal area are about 90% households include on insecurity food category.

1. Introduction

Coastal is an area that stretches in the interior of the sea, especially the first topographic changes on the surface of the mainland (1). Defines coastal areas as a land-and-sea encounter zone in which there is a relationship between human activity and the land and sea environment (2). The main features of the coastal area is (3) :

1. Coastal areas include land and sea
2. Land and sea boundaries determined by the level of land influence on the sea and the influence of the sea on land
3. Does not have uniform width, depth, and altitude

The problem that often arises in coastal areas is flood (4). The presence of tidal floods directly has an impact to human activity because it can damage the dwelling, roads, and other infrastructure (5). Tidal floods disaster has an impact on the socioeconomic life of the community. The presence of the tidal floods causes disruption of human economic activity. One of the sectors affected by tidal floods is the agriculture sector such as rice fields and ponds due to tidal floods that cause the decline in people's incomes.

In this regard food security is an interesting issue to study. Food security is everyone's access at all times in the foods sufficient for healthy living (6). Food security according to UU No.7 of 1996 is the condition of the fulfilled of food needs for the household which is reflected in the availability of adequate food, both quantity and quality, safe, equitable and affordable.

According to the definition of food security in Food Law. 7 of 1996, food security can also be interpreted as:

IOP Conf. Series: Earth and Environmental Science **148** (2018) 012015 doi:10.1088/1755-1315/148/1/012015

a. Fulfillment of food with adequate conditions, good food derived from plants, livestock and fish to meet the needs of carbohydrates, proteins, fats, vitamins and minerals that are beneficial to human growth and health
b. Fulfillment of food with safe conditions in which the food is free of biological, chemical, and other contaminants that can interfere harm human health
c. Fulfillment of food evenly throughout the country
d. Fulfillment of food with affordable conditions or easily obtainable at an affordable price

Based on the definition of food security, there are 5 element which is must be met :
a. Oriented on household and individual
b. Time (food is available and accessible at all time)
c. Emphasize on food access of household and individual on physical, economic, and social
d. Oriented on nutrition fulfillment
e. Intended for healthy and productive life

Food insecurity is a condition of society in an area that has the inability to obtain adequate and appropriate food for healthy living include the poor, affected communities by disaster, and communities that are live on inaccessible places (*UU Republik Indonesia No. 18 Th. 2012 Tentang Pangan*). Food insecurity condition can be seen through the measurement of food sufficiency. Food sufficiency at the household level is a condition where every household has a sufficient quantity of food to meet the nutritional needs required by all family members (9).

2. The Method

Demak Regency, Central Java is an area bordering the Java Sea and is affected by the tidal floods that result in social activities of the community. This study was conducted in 3 villages in a coastal area of Demak. Wedung, Purworejo, and Sidogemah villages were chosen because these three areas are affected by tidal floods. The data used in this study consist of primary data and secondary data. Primary data obtained by interview method. Secondary data are obtained from related institutions.

Respondents in this study are households located in coastal areas of Demak. Sampling is done by using sampling quota. Fulfillment of the number of samples used in this study is based on expert assessment of researchers. The study used 30 randomly selected respondents from each village so the total sample was 90.

Data analysis method in this research is descriptive statistic. Descriptive statistic used to analyze data by describing data that has been collected without intending to make conclusions that apply to the public or generalization.

The interview data were used to assess socio-economic characteristics. Secondary data were used to describe potential of natural resources. Food security analysis was conducted with secondary data and primary data. Secondary data is used to determine the adequacy of food in the study area. Food sufficiency is known from the ratio between food availability and food needs (11).

<u>Food Availability</u>

Rice

$$Pnet = (Px(1-(B+Pk+T)))xC \quad (1)$$

Description:

P : Rice production
B : the value of seeds (0.0088)
Pk : Feed (0.02)
T : wasted (0,054)
C : coefficient of paddy into rice (0.632) (10)

Corn, cassava, sweet potato

$$M\ net = P \times (1-(B+Pk+T)) \quad (2)$$
$$C\ net = P \times (1-(B+Pk+T)) \quad (3)$$
$$SPnet = P \times (1-(B+Pk+T)) \quad (4)$$

Description:
B : value of the seeds (corn (0.009), cassava (0), sweet potato (0))
Pk : Feed (corn (0.06), cassava (0.02), sweet potato (0.02))
T : Wasted (corn (0.05), cassava (0.0213), sweet potato (0.1))

$$Tnet = 1/3 \times (Cnet + SPnet) \quad (5)$$

Description:
Tnet : net production of cassava and sweet potatoes
Cnet : net production of cassava
Spnet : net production of sweet potato

$$Pfood = Rnet + Mnet + Tnet \quad (6)$$

Description:
Pfood : Food production cereals
Rnet : Net production of rice
Mnet : Net production of corn
Tnet : Net production of cassava and sweet potato

Food Cereals Availability

$$F = \frac{P_{food}}{t_{pop} x\ 365} \quad (7)$$

Description:
F : Food cereals availability
Pfood : Food production cereals
tpop : Midyear population

Ratio of Food Availability

$$I_{AV} = \frac{F}{C_{normatif}} \quad (8)$$

Description:
I_{AV} : Food availability ratio
Cnormatif : Consumption normative (300 gram)
F : The availability of food cereals

Table 1. Classification Ratio of Food Availability [11].

Ratio of Food Availability	Classification
R < 0,90	Prone to Food
0,90 ≤ r ≤ 1,14	Enough Food
R > 1,14	Surplus Food

Analysis food security conducted with primary data obtained from the proportion indicator of food expenditure on total household expenditure. Proportion of food expenditure can be calculate by [11]:

$$\%\ \text{food expenditure} = \frac{P\ food\ expenditure}{Total\ household\ expenditure}\ x\ 100 \quad (9)$$

Status of food insecurity indicators
Insecurity food : If presentation of food expenditure > 60%
Security food : If presentation of food expenditure ≤ 60 %

3. Result and Discussion

3.1. Socio-economic Characteristic

The socio-economic condition of a society characterizes the resource potential of people residing in a region. Better socio-economic condition of a society reflects higher quality or potential of human resources. In this study, the benchmark of human resource potential as characterized by the quality of human resources is education level, employment, and purchasing power of the community can be assessed from the average level of household income. The higher education level a person affecting to higher quality of human resources he has.

Table 2. The Education Level of the Communities.

No	Education Level human	Wedung (%)	Purworejo (%)	Sidogemah (%)
1	Didn't/never School	0	0	6.7
2	Didn't Pass from Elementary School	0	17.2	6.7
3	Elementary School	70	55.2	46.7
4	Junior High School	16.7	17.2	20
5	Senior High School	13.3	6.9	20
6	College	0	3.4	0

Table 2 shows that the average of education level of the communities in Purworejo, Wedung, and Sidogemah Villages are elementary school. This condition is in accordance with the results of research showing that the average (63.3%) of coastal community education in North Minahasa region is primary school. This is indicating that education participation of the communities in Demak is still relatively low. Coastal communities prefer work to add household income rather than pursuing education.

Table 3. The Employment of Households Members.

No	Main Activity Households Members	Wedung		Purworejo		Sidogemah	
		N	(%)	N	(%)	N	(%)
1	Fishery	15	14.02	29	27.10	2	1.96
2	Commerce	6	5.61	14	13.08	6	5.88
3	Private employees	2	1.87	4	3.74	8	7.84
4	PNS/TNI/Polri/pensioners	0	0	0	0	1	0.98
5	Local honorarium employee	1	0.93	0	0	0	0
6	Labor	18	16.82	3	2.80	37	36.27
7	Service/self employee	1	0.93	4	3.74	5	4.90
8	Industry	0	0	2	1.87	0	0
9	House wife	26	24.30	21	19.63	19	18.63
10	Student	33	30.84	28	26.17	18	17.65
11	Unemployed	5	4.67	2	1.87	6	5.88
	Total	107	100	107	100	102	100

The type of work of community members in coastal areas of Demak is known from the main types of household members' activities. The main activities of household members in this study site show varied activities. In the Wedung village most household member activities are as housewife as much as 24.30% and student as much as 30.84%. In Purworejo village the main activities of household members are dominated in fisheries and student, respectively 27.10% and 26.17%.

Different with Sidogemah village, the main activities of household members are dominated as a laborer as much as 36.27% and housewife respectively 18.63%. The more members of the household work the more income will be generated.

Table 4. Average Household Income.

No	Village	Average Income (in Indonesian Rupiah/IDR)
1	Wedung	2,774,833.33
2	Purworejo	3,848,833.33
3	Sidogemah	3,075,333.33

Household income is an indicator of household economic level. Table 4 show the average total income household in Demak coastal area. In each village, has different condition. Average household income in Wedung village is the lowest value IDR 2,774,633.33. The highest average total households income in Purworejo village is IDR 3,848,833.33. While average total household income in Sidogemah village is IDR 3,075,333.33.

The fulfillment of the needs of the community is known from proportion household expenditure divided into food expenditure and non food expenditure.

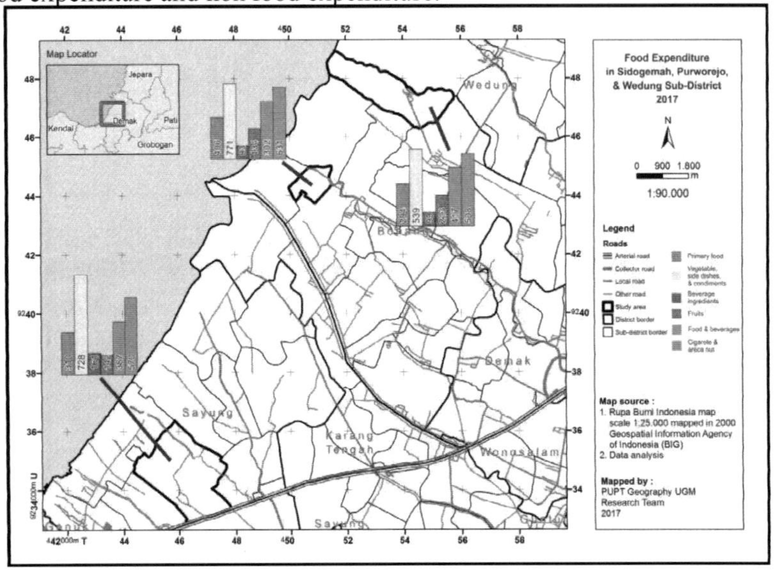

Figure 1. Map of Proportion of Food Expenditure in Demak Coastal Area.

Figure 1 shows proportion of food expenditure in the Wedung, Purworejo, and Sidogemah village. The highest proportion of food expenditure is for vegetables, side dishes & condiments in each village. The interesting thing that found in this study is consumption of cigarette is high (58.89% of the total respondents). Even some respondent who is a smoker saying that better didn't eat than didn't smoke. This condition causes expenditure for cigarette is more than expenditure of primary food.

Table 5. Average Non-Food Expenditure in Demak Coastal Area.

No	Non-Food Expenditure	Average		
		Wedung (IDR/year)	Sidogemah (IDR/year)	Purworejo (IDR/year)
1	Rent / Contract	43,263	41,933	34,333
2	Account Electricity, gas, kerosene, firewood, and others.	884,207	1,235,466	1,045,933
3	Home maintenance and repair	0	135,000	554,800
4	Drinking water, minerals	367,266	538,000	539,600
5	Various Goods and Services	559,466	852,133	775,200
6	Cost of education	2,216,433	1,406,133	1,946,733
7	Health Fee	134,136	208,333	227,866
8	Household equipment	156,000	18,600	338,433
9	Communication	497,666	583,333	709,666

Non-food expenditure in three villages showed a uniform condition, where the highest non-food expenditure is the cost of education.

3.2. Food Security

To assess the food security is done by using secondary data (serelia productivity data and population). The scope of the study in 3 sub-districts in Demak District is Sayung (Sidogemah), Bonang (Purworejo) and Wedung (Wedung) sub-districts. Based on the calculation of food availability and population in the study area indicates that the three areas are food insecure. This means that agricultural production is not able to meet the needs of people living in the area. An area is said to be food insecure if the food sufficiency index is less than 0.90. Table 6 shows the highest level of food insecurity in Sayung District, while the lowest is in Wedung Sub-district.

Table 6. Food Sufficiency Level in Demak Area.

Sub-district	F (ton)	F (gram)	C	I_{AV} (Food Availability Ratio)	Classification
Sayung (Sidogemah)	0.00006	55.13	300	0.18	prone to Food
Bonang (Purworejo)	0.00016	164.36	300	0.55	prone to Food
Wedung (Wedung)	0.00024	240.10	300	0.80	prone to Food

Table 7 shows the average household expenditure in the Wedung, Purworejo, and Sidogemah village. The average monthly household expenditure is about IDR 2 million which is divided into food and non-food expenditure. The proportion for food expenditure is more than non food expenditure. The proportion for food expenditure in the three villages are above 70%, while the non food expenditure in Wedung village is 21.76%, in Purworejo village is 21.48%, and in Sidogemah village is 19.63%. Food insecurity condition was presented by proportion of food expenditure per total expenditure. The results show that Wedung, Purworejo, and Sidogemah villages are included in the category food insecurity.

Table 7. Expenditure of Food and Non Food.

Description	Wedung	Purworejo	Sidogemah
Expenditure (IDR/month)	2,006,212	2,478,634	2,241,133
a. Food	1,571,433	1,946,120	1,801,300
b. Non Food	434,799	532,514	439,833
Proportion of food expenditure to total expenditure (%)	78.33	78.52	80.37
Proportion of non food expenditure to total expenditure (%)	21.67	21.48	19.63
Criteria	Food insecurity	Food insecurity	Food insecurity

Figure 2 show the percentage of household in the Wedung, Purworejo, and Sidogemah villages are in insecurity food. In the Wedung and Purworejo villages the households are in insecurity food as much as 90%, while 10% in security food. Different in Sidogemah village the households are in insecurity food as much as 93%, while 6.7% in security food.

Access to household economy in obtaining food on the coast of Demak is low. This is because the coastal conditions are often hit by flood tides. So that the fields and ponds are intrusive so as not to be productive. The impact is that people have to supply food from other areas with difficult accessibility. This condition indicates that government attention is needed to help communities in coastal areas of Demak to meet food needs.

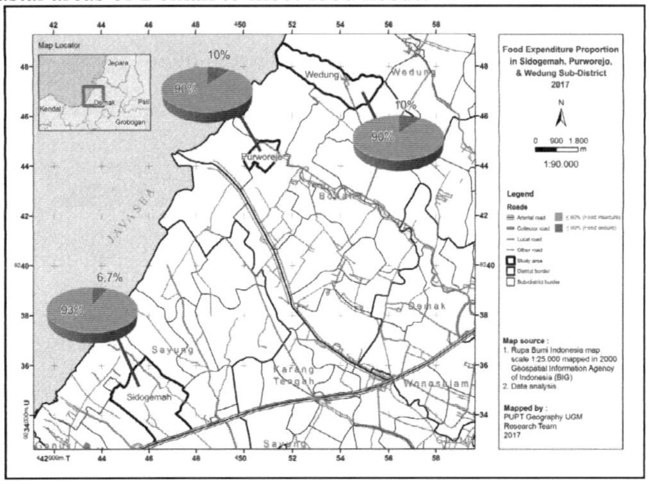

Figure 2. Map of Food Security in Demak Coastal Area.

4. Conclusion
Wedung, Purworejo, and Sidogemah are sample of villages in the Demak coastal area affected by tidal floods. The socio-economic characteristic of people in three villages have similar characteristic. The education level of the community is elementary school. Total household income are about IDR 2-4 million per month. The result of the research show is about 90% households in Wedung, Purworejo, and Sidogemah are in insecurity food.

5. Acknowledgment
The authors would like to express gratitude to Faculty of Geography UGM through Lecturer research grants through public funds and people in Demak coastal areas, especially in Wedung, Purworejo and Sidogemah villages.

References
[1] Muhsoni dan Firman F 2009 Arahan pemanfaatan ruang wilayah pesisir untuk budidaya dengan memanfaatkan citra satelit dan sistem informasi geografis di sebagian bali selatan. *J. Embryo.* **6** 67-76

[2] Cicin-Sain B and Knecht R W 1998 *Intergrated Coastal Area and Ocean Management : Concepsts and Practices* (Washington DC: Island Press)

[3] Kay R and Alder J 1999 *Coastal Planning and Management* (London: E & FN SPON)

[4] Singh M and Devkota L 2015 *People's Perspective on Flood Hazard and Adaptation Strategies in the Koshi River Basin of Nepal* (Nepal: Development Research Institute)

[5] Marfai MA and L King 2008 Potential vulnerability implication if coastal inundation due sea level rise for the coastal zone of semarang city. *Environmental Geology.* **54** 1235-45.

[6] Maxwell S and Frankenberger T R 1992 *Household Food Security: Concepts, Indicators, Measurements, A Technical Review* (Rome: International Fund for Agricultural Development, United Nations Children Fund)

[7] Undang-Undang Republik Indonesia Nomor 7 Tahun 1996 Pangan. 4 November 1996. LN 99/1996

[8] Undang-Undang Republik Indonesia Nomor 18 Tahun 2012 Pangan. 16 November 2012. Lembaran Negara Republik Indonesia Tahun 2012 Nomor 227. Jakarta.

[9] Aswatini, Romdiati A, Setiawan B, Latifa A, Fitranita dan Noveria M 2004 *Ketahanan Pangan, Kemiskinan dan Sosial Demografi Rumah Tangga* (Jakarta : Pusat Penelitian Kependudukan LIPI)

[10] United Nation 2004 *5th report on the world nutrition situation* (Geneva: UN-SCN)

[11] Smith L C and Subandoro A 2005 *Measuring Food Security Using Household Expenditure Surveys* (Washington DC: IFPRI)

[12] Wasak Martha 2012 Keadaan sosial ekonomi masyarakat nelayan di desa kinibuhutan kecamatan likupang barat kabupaten minahasa utara sulawesi utara. *Pacific. J.* **1** 1339-42.

IOP Conf. Series: Earth and Environmental Science **148** (2018) 012016 doi:10.1088/1755-1315/148/1/012016

Preliminary study of *coastal circulation cells* in the coastal area of Kendal, Indonesia

M A Marfai[1], E Trihatmoko[1], Sunarto[1], Wulandari[1], A A Risanti[1], I A Kurniawan[1]

[1]Departement of Environmental Geography, Faculty of Geography, Universitas Gadjah Mada, Yogyakarta, Indonesia, 55281

arismarfai@ugm.ac.id

Abstract. The objective of this study was to recognise the hydrodynamic pattern of sea surface current to find the potential of coastal circulation cells in Kendal coastal area. The initial identification of coastal circulation cells in this study used the modification of the band ratio method. Further identification related to the availability of coastal circulation cells utilise the results of aerial photography as a process of detailed validation and has never been done before. The results showed that there was a similar potential of coastal circulation cells in different seasons. In detail, the coastal circulation cells obtained the suspension of sediment, with also the potential of nutrient and alga, namely Rhodophyta Sp. Considered by the colour shown in the photograph of UAV.

1. Introduction

Land, atmosphere, and sea are the three primary systems on Earth's surface [1]. These three systems have different processes and affect each other with complex performance. Coastal area is the area that became the meeting between those three primary systems which include the hydrosphere, pedosphere and lithosphere, atmosphere, and anthroposphere [2].

The coastal dynamic is complicated. Wind, waves, currents, and tides are the four major forces that cause the complexity of the coastal dynamics [3]. The primary current is the sea surface currents with a depth of fewer than 100 m below the sea level [4]. The influence of these four forces as a form of hydrodynamics can occur either directly or indirectly.

The hydrodynamic aspect of the coastal area is measured by the presence of wind, waves, currents, and tides [3]. Discussion of issues of hydrodynamics in coastal areas is also much mentioned about sea level rise. However, this event is explained by [5] as the impact of land subsidence. Thus, it can be interpreted sea level rise is not a major study of the hydrodynamic aspects of the coastal areas.

Coastal circulation cells are mass of water circulation movements occurring both seasonally and yearly [6], with their forming forces being wind or current [7] [8] [9] [10]. The earlier concept of coastal circulation cell [7] said that it is formed through a combination of normal current (onshore-offshore current) with long-shore current then caused rip current. It was confirmed by another research [8], that the location of coastal circulation cell is nearshore zone. The concepts that exist so far do not discuss the coastal circulation cell firmly, with the meaning of the concept is a patent concept. One of the meanings of coastal circulation cells can be defined as the accumulation of nutrients [10] or only limited to sediments in the absence of nutrients.

The study of coastal areas is closely related to the process of sedimentation and erosion as the primary processes of coastal dynamics [11] [12]. Two processes in the coastal area occur due to the direct influence of winds that induce waves and currents movement [13] [14] [15] [16]. The direct influence of waves directly affect the coastal configuration [17] and then the sediment material will be carried by sea surface currents [18]. Based on the fact that there is a study of sea hydrodynamics aspects related to the winds, waves, and sea surface currents as the main trigger of the coastal dynamics becomes important to be done.

The existence of the coastal dynamics certainly gives its effects for the areas that receive different processes and lead to the occurrence of disasters. The variability of the effects of the coastal dynamic is very clearly indicated in the northern region of Central Java. This can be explained by the results of previous research [19] [20], it is explained that Tegal, Pekalongan and Demak Regencies are areas of accelerated erosion. Based on those research, there are intensive erosion effects recorded in Pekalongan Regency with erosion that continues to occur throughout the year with the addition of 1 m each year. The effects of other coastal dynamics also occurred in the Demak with tidal inundation events that forced hundreds of families to move from the previous residence which means they lose their homes [20]. This loss can be categorised as a disaster event.

Disaster perceived from the dynamics of the coastal area in the northern part of Central Java is reinforced by high population growth, such as the one that occurred in the coastal area of Kendal Regency. This is due to the presence of the most densely populated national transportation routes in Java Island, namely Jalur Pantai Utara [21]. The optimisation of industrial activities, especially in the area of Kendal Regency [22] also contribute the population growth. Apart from aspects of the population, the dominance of the fluvial process leads to thick and widespread sediment deposition throughout the northern part of Central Java, the effect being the loose soil material that tends to lose and has a high degree of erosivity. Based on the mentioned facts, the northern region of Central Java is mainly for Kendal Regency to be an area worth examining in this study.

This research will be obtained data of field result and secondary data related to sea surface current data as main media forming coastal circulation cells. It is said so because the wind and waves tend to have a laminar movement and direction [23] [8] [3] [24]. So that, sea surface current is considered as the greatest potential for driving the formation of coastal circulation cells. The coastal circulation cells are necessary to be studied as previous studies [7] [8] [9], explaining the potentials contained in circulation cells such as marine nutrients, sediments, and potential of hazards.

The high intensity of the interaction between the coastal dynamic and coastal circulation cells is an interesting fact since both are triggered by the same factor, which is sea surface current as part of the ocean hydrodynamics aspect [7] [8]. Bodri River as a major river in Kendal gives two possibilities which have the potential to either provide sedimentation effects or become nutritional contributors [10] or even other toxic waste and suspension for marine life in Kendal. Based on previous studies, sedimentation in Kendal is high with sedimentation up to 1.9 km^2 during ten years of observation [25]. The sedimentation comes from the Bodri River. Sedimentation occurring in the Kendal coastal area has potential to carry nutrients or even waste along the Kendal coastal area. Even so this condition can certainly cause problems. The important preventive steps were taken including monitoring the availability of coastal circulation cells that can determine the accumulation of suspension in the coastal area.

The objective of this study is to assess the coastal circulation cells in the coastal area of Kendal, Central Java. This objective will be done by modification of band ratio as the initial identification of coastal circulation cells and conduct field validation of the availability of coastal circulation cells.

2. Methodology

The research area of this work set in District Patebon with Pidodo Kulon Village as the only one area occupied the main river mouth. Current measurements are focused on sea surface currents. Hydrodynamic analyses were performed in the nearshore zone behind the breaker zone with the assumption that sea surface current energy is stabled from the reduction of the depth.

This study analyzed using ENVI software on secondary data of Landsat 5 TM and 8 OLI / TIRS Image for 2008-2017 Path 120-121 and Row 65 [26]. The result of ENVI software analysis was validated by using primary data in the form of Unmanned Aerial Vehicle (UAV) photograph. Primary data collection requires tools such as GPS, UAV, and motorboat.

Determination of the sample for the study area is done with the evidence of the satellite image interpretation to obtain the potential of coastal circulation cells. Sample determination was achieved by using Landsat 5 TM and 8 OLI / TIRS images in 2008-2017 with paths 120 and 121 and row 65 or covering all areas of Central Java. The observations from 2008-2017 were included in the medium-term variations scale. The scale is one of the scaling [27] which is defined as the systematic trend line of shoreline variations. Other scales are long-term variations (100-1000 years) and short-term fluctuations (short-term fluctuations). The acquisition of the sample area considered by the location which is semi-closed (bay) or open water. This consideration as revealed [28] those tidal factors and other physical factors are more influential on the semi-closed seas (bay) that mean there is an opposite condition for open water. This interpretation is also used to obtain locations with potential coastal circulation cells. The areas with potential coastal circulation cells or anomalous findings in the nearshore zone may be the location of the sample.

Envi software was used to process the image of Landsat with modification of band ratio method. Band ratio method was used from the beginning of site selection to the end of the analysis to identify potential locations related to the availability of coastal circulation cells. The band ratio method was chosen because it is an effective method of displaying certain objects and will be more easily detected due to differences in them [29]. Data analysis was performed in the descriptive analysis of hydrodynamic model results with field data validation, coastal circulation cell findings and suspension analysis. The research flow diagram is shown in figure 1.

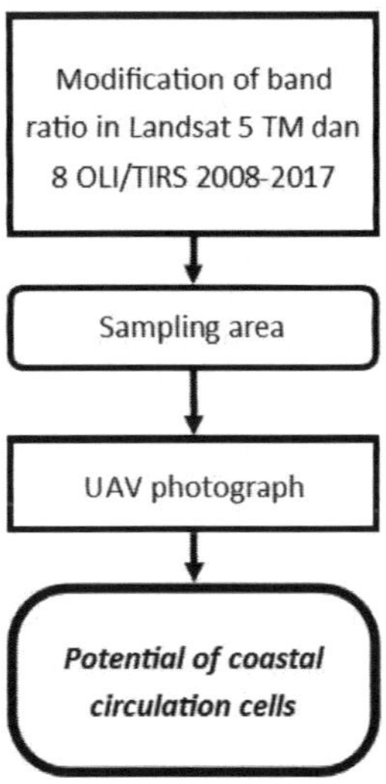

Information: process, result.

Figure 1. Research flow diagram.

3. Result and Discussion

The result of band ratio modification in band 3/2 indicated a tendency of sediment suspension accumulation at the end of river mouth of Bodri River (figure 2). Several circular movement appearances were also shown in the modified results. These were shown in Figure 6.7 and 6.10 on 21st November 2015 and 25th May 2017. Both sequentially are visible on months that tend to have entered the western and eastern seasons. Circulation patterns that occur more directed to the northeast from the river mouth of Bodri River both in the trend of the western and eastern seasons. This same characteristic could be interpreted that the potential of coastal circulation cells based on two distinct seasons has the same suspension accumulation site, i.e. a radius of less than 5 km from the tip of the river mouth of the Bodri River.

The result of band ratio modification in band 3/2 Landsat OLI/TIRS showed the existence of suspension in the potential coastal circulation cells. The suspension creates a different pattern. Figure 2 is the result of the band ratio modification of Landsat 8 OLI/TIRS dated July 28th, 2008. This figure showed the suspension tends not to be seen by the image during recording. That happened because July is a month with low rainfall intensity so that fewer sediment were brought into the sea. This condition caused the nutrients in the sea to decrease so that algae growth also reduced. So that was not recorded the suspension on Landsat 8 OLI/TIRS recording on July 28th, 2008.

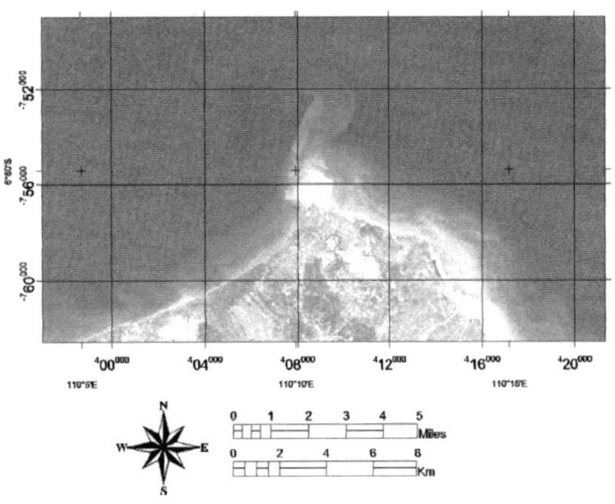

Figure 2. The result of band ratio modification in band 3/2 Landsat 8 OLI / TIRS on July 28th, 2008.

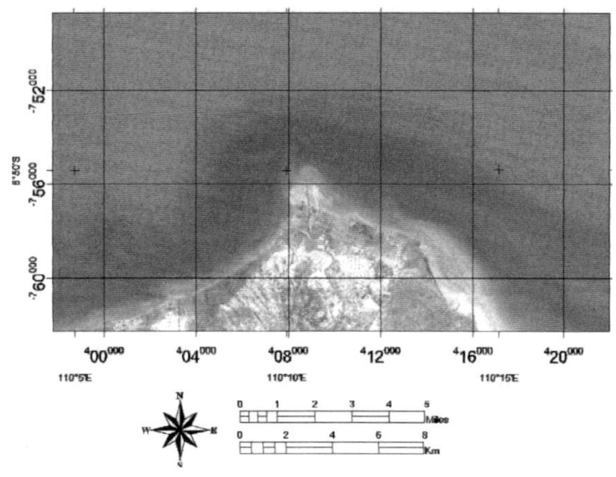

Figure 3. The result of band ratio modification in band 3/2 Landsat 8 OLI / TIRS on October 19th, 2009.

October is the beginning of the west season with the intensity of moderate rain. Figure 4 showed a suspension though not very clear. This happened because the sediment that carried nutrients began to carry into the sea, so the suspension of algae also began to grow. Conditions in November 2011 showed almost the same conditions, namely the suspension that was starting to look although still vague.

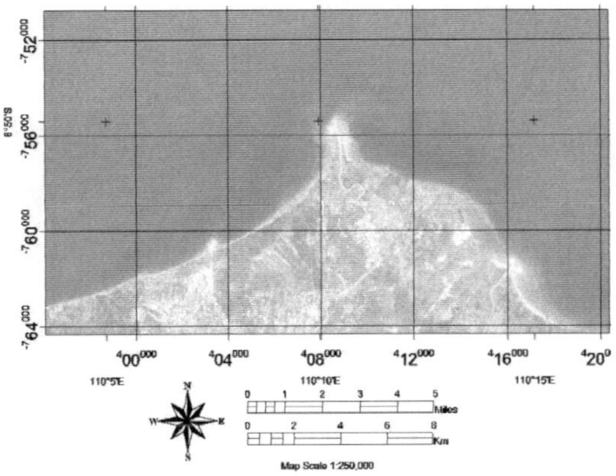

Figure 4. The result of band ratio modification in band 3/2 Landsat 8 OLI/TIRS on November 2nd, 2011.

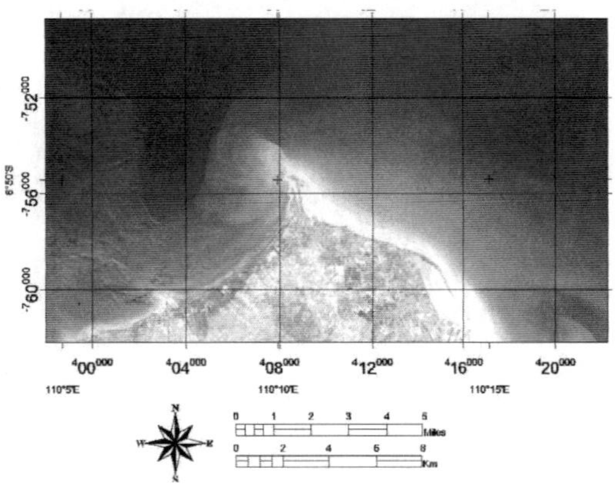

Figure 5. The result of band ratio modification in band 3/2 Landsat 8 OLI / TIRS on September 12th, 2013.

Figure 5 was the result of the band ratio modification in band 3/2 Landsat 8 OLI/TIRS dated September 12th, 2013. This figure showed a visible suspension in the nearshore zone. That happened because the intensity of rain had tended from medium to high. So that suspended sediment that carried nutrients were accumulated in the nearshore zone. This condition allowed the algae in the ocean to quickly grown and developed. This suspension of sediment, nutrients and algae were captured by Landsat 8 OLI/TIRS on September 12th, 2013. The same conditions were also recorded on the result of band ratio modification of Landsat images in figure 6, figure 7, figure 8, figure 9, and figure 10.

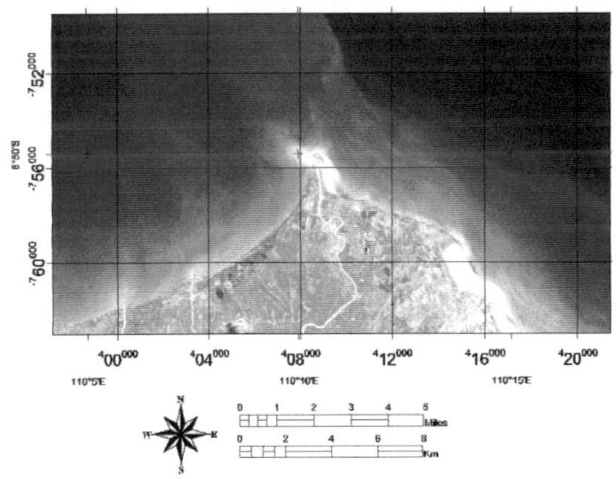

Figure 6. The results of band ratio modification in band 3/2 Landsat 8 OLI/TIRS on May 10th, 2014.

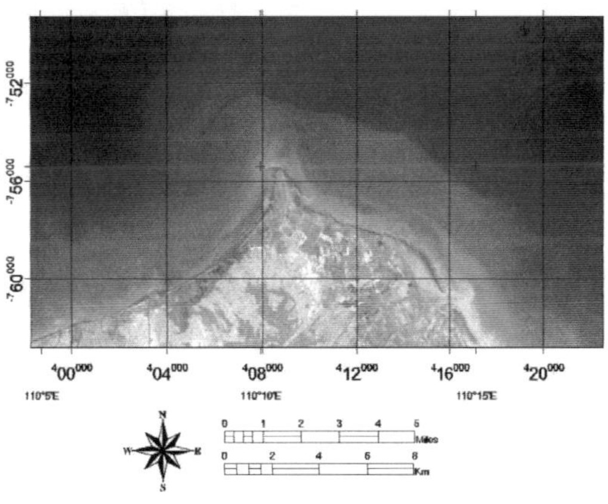

Figure 7. The results of band ratio modification in band 3/2 Landsat 8 OLI/TIRS on November 2nd, 2014.

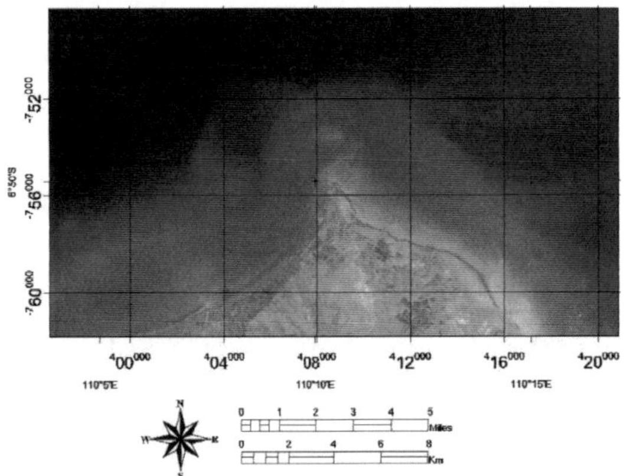

Figure 8. The results of band ratio modification in band 3/2 Landsat 8 OLI/TIRS on November 21st, 2015.

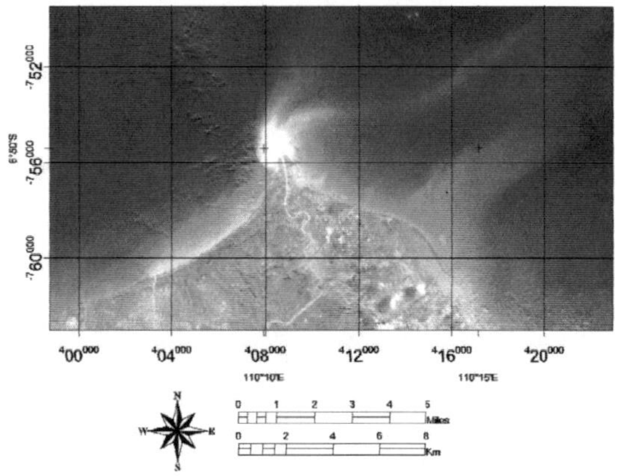

Figure 9. The results of band ratio modification in band 3/2 Landsat 8 OLI/TIRS on March 28th, 2016.

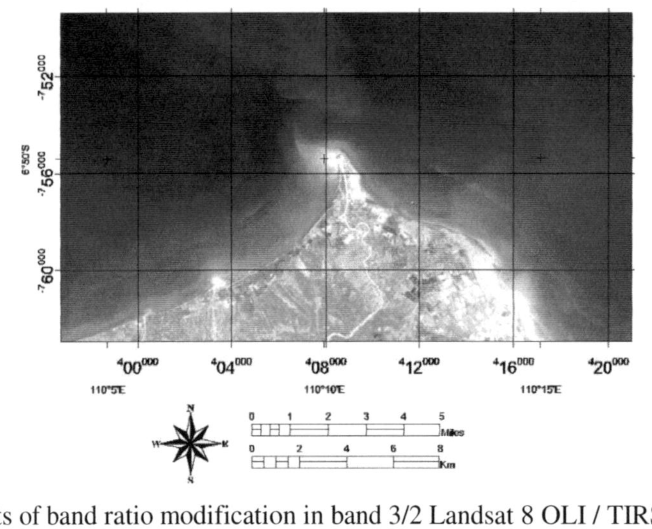

Figure 10. The results of band ratio modification in band 3/2 Landsat 8 OLI / TIRS on May 1st, 2017.

Field validation by UAV photograph showed part of the more detailed estimate of coastal circulation cells with laminar current patterns. Figure 12 shows the laminar current movement with an accumulation of brown suspension. This accumulation could be a sediment suspension that allowed the presence of nutrients carried by sea surface currents. It was said so because the sediment suspension was formed, alongside the pattern of purple laminar sea surface current flow (figure 12) which could be estimated as *Rhodophyta Sp* [30]. The purple colour of the laminar current flow was not said to be a waste as it should be. If the flow of the laminar stream was a waste that should be mixed with the brown sediment suspension. But the purple suspension naturally separated from brown suspensions, so it was certain that the purple suspension was a marine biota that colonizes the sea surface current. Thus the sediment suspension carried by surface currents is a sediment with nutrient accumulation [10]. The accumulation of nutrients was carried by the Bodri River flow into a single unity of potential coastal circulation cells in the sea near the mouth of the Bodri River.

Figure 11. Images of sea-surface currents carrying possible sediment suspensions containing nutrients.

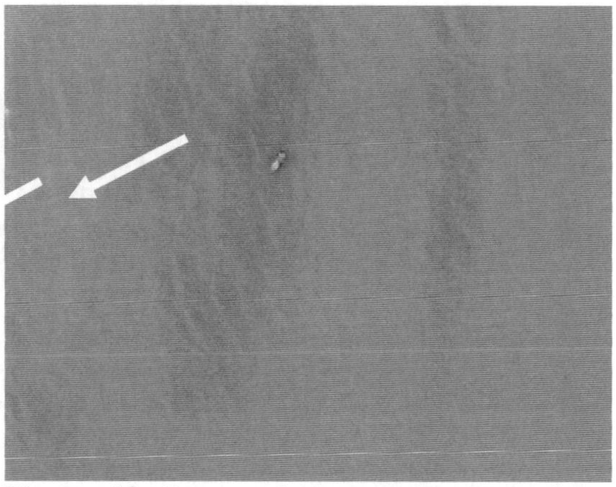

Figure 12. Image of the purple sea surface current movement, it is probably *Rhodophyta Sp*.

4. Conclusion

The potential of coastal circulation cells in Kendal coastal area showed the tendency of occurrence in the tip of the mouth of the Bodri River. The same characteristic pattern of circulation occurs in both east and west seasons with an inclination up to 22° to the northeast. Field validation results using a UAV for circulation analysis in which there was a difference of suspension on each surface current flow, which is possible as the accumulation of sediment and nutrient followed by red algae or *Rhodophyta Sp*.

5. Acknowledgement

This reseach was granted by The Reseach Grant of Lecturer and Laboratory Faculty of Geography Universitas Gadjah Mada in 2017, Grant of Pendidikan Magister menuju Doktor untuk Siswa Unggul in second batch, Grant of International Reseacrh Collaboration (Hibah KLN) Number 2244/UN1.P.III/DIT-LIT/LT/2017. We also thank to the resident of Pidodo Kulon, Kendal.

References

[1] Davidson R and Arnott 2010 *Introduction to Coastal Processes and Geomorphology*, Cambridge University Press, Cambrigde.

[2] Young R and Norby L 2009 *Geological Monitoring: Boulder*, Geological Society of America, Colorado.

[3] Sunarto 2004 *Geomorfologi Pantai: Dinamika Pantai*, Fakultas Geografi UGM, Yogyakarta.

[4] National Oceanic and Atmospheric Administration 2017 Current http://oceanservice.noaa.gov/education/pd/tidescurrents/tidescurrents_currents.html. Accessed on February 14th, 2017.

[5] Ritter D F Kochel R C and Miller Jr 1995 *Process Geomorphology,* Third Edition, Wm. C. Brown Publ., Dubuque.

[6] Dever M D Hebert D Greenan B J W Sheng J and Smith P C 2016 Hydrography and Coastal Circulation along the Halifax Line and the Connections with the Gulf of St. Lawrence, *Atmosphere-Ocean, 54(3),* 199-217.

[7] Komar P D 1976 *Beach Processes and Sedimentation.* Prentice Hall, New Jersey.

[8] Pethick J 1984 *An Introduction to Coastal Geomorphology.* Edward Arnold Ltd., Maryland.

[9] Ponte A L Velasco G D D Valle-Levinson A Winters K B and Winant C D 2012 Wind-Driven Subinertial Circulation inside a Semienclosed Bay in the Gulf of California, *Journal of Physical Oceanography, 42*, 940-953.

[10] Yuan D Li Y Wang B He L and Hirose N 2017 Coastal Circulation in the Southwestern Yellow Sea in the Summers of 2008 and 2009, *Continantal Shelf Research, http://dx.doi.org/10.1016/j.csr.2017.01.022*

[11] Junus N Hartono and Sukwardjono 2009 Penginderaan Jauh dan Sistem Informasi Geografis untuk Monitoring Perkembangan Morfologi Delta Sungai Jeneberang Menggunakan Citra Landsat Multi Temporal, *Majalah Geografi Indonesia, 23(1),* 40-56.

[12] Marfai M A 2011 The Hazards of Coastal Erosion in Central Java, Indonesia: An Overview, *Malaysian Journal of Society and Space, 3*, 1-9.

[13] Cochin V 2008 Sea Surface Current–Wave–Wind Interactions Measured by Coastal Ground Wave HF Radars, *IEEE Geoscience and Remote Sensing Letters, 5(2),* 227-230.

[14] Ghalenoei E Sharifi M A and Hasanlou M 2015 Estimating and Fusing Optical Flow, Geostrophic Currents and Sea Surface Wind in the Waters around Kish Island, *The International Archives of the Photogrammetry, Remote Sensing and Spatial Information Sciences, 40(1),* 221-226.

[15] Lee H J 2010 Preliminary Results on Suspended Sediment Transport by Tidal Currents in Gomso Bay, Korea, *Ocean Science Journal. 45(3)*, 187-195.

[16] Moriya T and Murakami K 2014 Study on the Effect of Ocean Wind on Sea Surface Current in Semi-Enclosed Basin. *Journal of Japan Society of Civil Engineers, 70(2),* 406-410.

[17] Kasim F 2011 Koreksi Pasang Surut dalam Pemetaan Perubahan Garis Pantai Menggunakan Data Inderaja dan SIG, *Jurnal Ilmiah Agrosains Tropis, 6(3),* 180- 188.

[18] Kumaat J C Kandoli S T B and Moningkey A T 2014 The Coastal Platform Morphodynamics Characteristics Belang Bay, North Sulawesi Province, *Indonesian Journal of Geospatial, 2(3),* 34-41.

[19] Marfai M A Pratomoatmojo N A Hidayatullah T Nirwansyah A W and Gomareuzzaman M 2011 *Model Kerentanan Wilayah Pesisir berdasarkan Perubahan Garis Pantai dan Banjir Pasang,* Magister Pengelolaan Pesisir dan Daerah Aliran Sungai, Yogyakarta.

[20] Marfai M A 2012 Preliminary Assessment of Coastal Erosion and Local Community Adaptation in Sayung Coastal Area, Central Java – Indonesia, *Quaestiones Geographicae, 31(3)*, 47-55.

[21] Marfai M A Hizbaron D R and Mardiatno D 2015 *Research Report International Research Collaboration and Scientific Publication (Third Year): Coastal Geomorphological Hazard in North Central Java-Indonesia*, UGM and Vrije Univ, Yogyakarta and Amsterdam.

[22] Tahapan, Tata Cara Penyusunan, Pengendalian, Dan Evaluasi Pelaksanaan Rencana Pembangunan Desa Di Kabupaten Kendal No. 2 tahun 2011. LD Nomor 2 Seri E No. 2 Tahun 2011.

[23] Duxbury A B Duxbury A C and Sverdrup K.A 2002 *Fundamentals of Oceanography* New York: McGraw-Hill.

[24] Baharuddin Pariwono J I Nurjaya I W 2009 Pola Transformasi Gelombang dengan Menggunakan Model RCPWave pada Pantai Bau-Bau, Provinsi Sulawesi Tenggara. *E. Jurnal Ilmu dan Teknologi Kelautan Tropis*, *1(2)*, 60-71.

[25] Trihatmoko E 2017 Proses dan Dampak Dinamika Wilayah Kepesisiran Jawa Tengah dan Daerah Isimewa Yogyakarta, *Thesis,* Fakultas Geografi, Universitas Gadjah Mada, Yogyakarta.
[26] USGS 2016 *Landsat 8* landsat.usgs.gov/landsat8.php. Accessed on April 3rd, 2017.
[27] Rijn L V 2010 Coastal Erosion Control based on the Concept of Sediment Cells. *Project of Concepts and Science for Coastal Erosion Management,* Deltares, Netherlands.
[28] Bishop J M 1984 *Applied Oceanography*, John Willey and Son, New York.
[29] Danoedoro P 2012 *Pengantar Penginderaan Jauh Digital.* Penerbit Andi, Yogyakarta.
[30] Nilssen I 2016 Integrated Environmental Mapping and Monitoring: A methodological approach for end-users. NTNU Grafisk senter, Trondheim.

IOP Conf. Series: Earth and Environmental Science **148** (2018) 012017 doi:10.1088/1755-1315/148/1/012017

Land use change analysis using spectral similarity and vegetation indices and its effect on runoff and sediment yield in tropical environment

N Christanto[1], J Sartohadi[1], M A Setiawan[1], D B P Shrestha[2,] and V G Jetten[2]

[1]Department of Environmental Geography, Faculty of Geography, Universitas Gadjah Mada, Indonesia
[2]Faculty of Geo-Information Science and Earth Observation (ITC), Twente University, The Netherlands

n.christanto@gmail.com

Abstract. Land use change influences the hydrological as well as landscape processes such as runoff and sediment yields. The main objectives of this study are to assess the land use change and its impact on the runoff and sediment yield of the upper Serayu Catchment. Land use changes of 1991 to 2014 have been analyzed. Spectral similarity and vegetation indices were used to classify the old image. Therefore, the present and the past images are comparable. The influence of the past and present land use on runoff and sediment yield has been compared with field measurement. The effect of land use changes shows the increased surface runoff which is the result of change in the curve number (CN) values. The study shows that it is possible to classify previously obtained image based on spectral characteristics and indices of major land cover types derived from recently obtained image. This avoids the necessity of having training samples which will be difficult to obtain. On the other hand, it also demonstrates that it is possible to link land cover changes with land degradation processes and finally to sedimentation in the reservoir. The only condition is the requirement for having the comparable dataset which should not be difficult to generate. Any variation inherent in the data which are other than surface reflectance has to be corrected.

1. Introduction

Land degradation is a common problem threatening food security and deterioration of the environment [8][9]. In the degradation processes bio-physical factors play an important role while land cover/land use changes often work in triggering or accelerating the process[7][10]. The main cause of land cover change which occurred in various parts of the world is often due to increase of population which forces the clearing of natural forest in order to create land for cultivation[15][17][26][29][36]. Other causes for deforestation leading to land cover changes are shifting cultivation [1][20][23], commercial farming (rubber, cotton, plantation of fast growing trees for fuel wood), governmental policy change[6][13][30] or due to conflicts or political cause [28].

When land cover/land use changes take place on marginal lands it can have detrimental effects on the land [14][16][19][39]. In order to study the impact of land use changes on land degradation it is imperative to study change pattern for which remote sensing techniques have played an important role. A variety of techniques starting from comparison of visual interpretation or classification of images obtained in different dates[5][25] to complex change detection techniques such as cross-correlogram spectral matching, change vector analysis, object oriented technique, neural network or principal component analysis have been applied by various researchers[4][18][22][24][31][32][33][38].

In order to analyze the negative impact of land use changes it is important to know which land cover type has undergone which changes. Finding this answer is often not straight forward. Various change detection techniques are available[38] but information on only changed or unchanged-pixels or classes is not sufficient if we want to analyze the effect of land cover or land use changes on land degradation. On the other hand, if we want to assess which land cover has underdone which changes and if we have only remote sensing data from the past, generating land cover information in the absence of trainings samples will not be so straight forward. In the present paper a technique is described to derive land cover information from previously obtained remote sensing data based on spectral characteristics of recently obtained image.

2. The Methods

The main objectives of this study are to assess the land use change and its impact on the runoff and sediment yield of the upper Serayu Catchment. In this paper we describe a method to derive land cover/land use information from a watershed in central Java for the year 1991 based on the analysis of recent data (Landsat data from 1989 to 2015) and field work. The result of the land use change analysis will be compared with data of sediment measurement to see the relation between land use change ad sediment transports.

2.1. *Study Area*

The study area is located in the upper Serayu basin, Central Java, Indonesia (Figure 1) covering a surface area of about 95,174 Ha. The east to west axis is about 43 km and north to south corner about 29 km. The elevation varies from 225 m above sea level in the western part to more than 2000 m in the north eastern part, the highest mountain being Sumbing with an elevation of 3325 m. The area is volcanic with several active craters (Sindoro and Sumbing volcano). [37] described the geomorphology of the area as strongly eroded volcanic terrain. Geomorphologically the area is characterized by having volcanic cones, plateau (Dieng Plateau), strongly dissected mountain slopes and alluvial lowlands consisting of terraces and flood plain. The area is drained by several rivers (Merawu, Tulis, Begaluh) which join the main drainage system, the Serayu river which feeds into a reservoir which is located in the south-west lowland. The climate is equatorial tropical with mean annual rainfall varying from 1700 mm up to 4200 mm per year and mean annual temperature varying from 14 to 27^0 C.

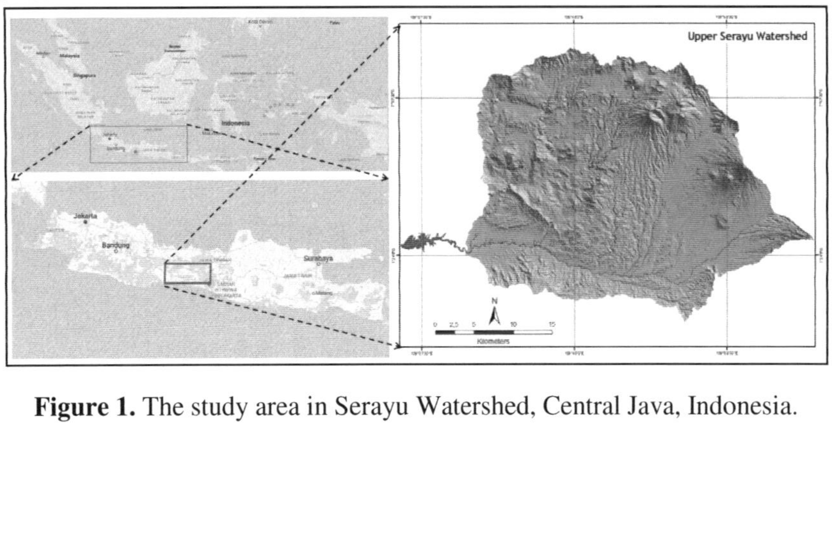

Figure 1. The study area in Serayu Watershed, Central Java, Indonesia.

IOP Conf. Series: Earth and Environmental Science **148** (2018) 012017 doi:10.1088/1755-1315/148/1/012017

2.2. Materials used and data collection

Two sets of Landsat ETM data (data acquisition 2014 and 1991) were acquired from the Serayu basin, Central Java, Indonesia. The data was geo-referenced and assigned UTM coordinates. Field work was carried out during 2014. Field observations were made using stratified random method: the first stratification is based on major land cover/land use classes. The main land cover classes identified were forest, plantation forest, dry land cultivation, paddy fields, bare soil, water body and settlements. For each land cover/land use class a number of samples points were taken which were randomly distributed, their locations were recorded using a GARMIN GPS receiver. The sample points were separated in 2 groups: one for training samples and one for validating the classification results. In addition, other data related to land degradation were also collected.

2.3. Making data comparable

In order to do land cover change analysis using remotely sensed data from different dates, it is essential to make the different datasets comparable. A data set obtained in one date has to match the radiometric and geometric parameters of other data set. Other option is to bring both the data set to a common base. Directly obtained remotely sensed image is known to have not only surface reflectance data but also artifacts caused by various factors which might be inherent in the data e.g. variations due to scattering of light, solar zenith angle, seasonal effect or due to topographic variation. Even if the two dataset are obtained in the same month there could be some difference in solar azimuth due to different acquisition dates and this could have some influence in the data.

Thus the data had to be corrected for atmospheric haze and the sensor induced gain effect which was then converted to at-sensor radiance. Finally, the at sensor radiance data were converted into top of atmosphere reflectance taking into account the solar zenith and azimuth angles based on the data acquisition date and the time. Converting the at sensor radiance value into top of atmosphere reflectance was carried out using equation and calibration data as explained in [3]. In the resulting data set pixel values represent per cent reflectance and are in the range of 0 to 100 in the corresponding spectral bands which ensure that both the data sets (1991 and 2014) are comparable and ready for further analysis.

2.4. Defining reference spectra for the main land cover classes

In order to make land cover/land use map, it is important to define reference spectra for the main land cover types. Based on the sample points taken from each land cover/land use class in the fieldwork carried out during 2014 the corresponding reflectance values were collected from the corrected Landsat TM data of 2014.

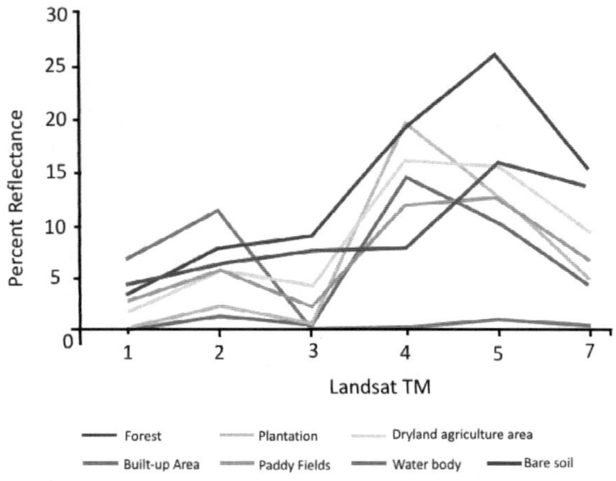

Figure 2. Spectra of various land cover/land use classes.

2.5. Generating NDVI and DVI

The reference spectral for each land cover class is derived based on the top of atmosphere reflectance data generated on Landsat of 2014. In figure 2 and in table 1 the spectra representing the main land cover/land use classes in the area are shown.

For land cover/land use classification and change detection analysis, spectral reflectance data alone may not be sufficient. The basic question on whether the land surface is covered by vegetation or not can be easily answered by using the normalized difference vegetation index (NDVI). For deriving NDVI the commonly used algorithm for Landsat Thematic Mapper bands (tm3 and tm4) is used. The resulting NDVI values are in the range of -1.0 to +1.0, the positive and higher NDVI values indicates the presence of vegetation cover while the values close to zero may mean water bodies or bare surface. For further image analysis the data with negative value is not convenient, thus it is converted into positive digital numbers (DNNDVI) as follows:

$$DN_{\mathrm{NDVI}} = 25\,NDVI + 25 \qquad (1)$$

The NDVI is multiplied by 25 and an additional 25 is added to avoid negative values which means that any pixel value of less than 25 means that it is not covered by vegetation. Thus the resulting index (DNNDVI) can be used to separate between vegetation cover from bare surfaces. But it cannot be used to differentiate between different vegetation types since it is not sensitive to variations of the difference between the reflectance in near infrared and red. For an example plantation forest and natural forest cover may show similar DNNDVI value (table 1) but the plantation forest cover has relatively higher reflectance in the near infrared band because of different tree species (AlbisiaFalcataria) as compared to that of natural forest. Moreover in the plantation forest area the fast growing tree species are often intercropped with durian (*Durio zibethinus*) or salak (*Salacca zalacca*, snake skin fruit) or other field crops thus increasing the canopy cover. In order to map these differences the difference vegetation index (DVI) would be handy. The DVI is generated simply by subtracting TM3 from TM4 as follows:

$$DVI = TM4 - TM3 \qquad (2)$$

IOP Conf. Series: Earth and Environmental Science **148** (2018) 012017 doi:10.1088/1755-1315/148/1/012017

The magnitude of the DVI could then be used to differentiate the different forest types. For an example the mean DVI value of forest is 11 and that of plantation forest is 19, showing a clear difference between them, whereas their DNNVDI values shows 50 and 48 respectively which can be considered similar (table 1). For the generation of DVI the output pixel value is assigned 0 if the value of tm3 is larger than tm4. This is done in order not to have negative values in the resulting DVI image.

2.6. Deriving classification parameters

Once the reference spectra are defined and the mean DNNDVI and the DVI are generated for each cover class (table 1), the remote sensing data obtained in earlier date can be classified. The mean values in the corresponding spectral bands in each cluster are considered as being the reference for that class. Similarly, the mean DN and DVI values are generated for each land cover/land use class.

Table 1. Reference spectra for major land cover/land use types.

	Forest	Plantation forest	Dryland cultivatio n	Bare soil	Built up area	Paddy field	Water body
TM1	0	0	2	3	4	3	5
TM2	1	2	6	8	6	6	9
TM3	0	1	4	9	8	2	0
TM4	11	20	16	19	8	12	0
TM5	8	13	16	26	16	12	1
TM7	3	5	10	15	14	6	0
DN[illegible]	50	48	40	34	25	43	0
DVI	11	19	12	10	0	10	0

For land cover classification similarity index is applied. Similarity can be defined by Euclidian distance and the spectral angle between the test pixel and the reference spectra. If the Euclidian distance is within the threshold, then it is considered similar. The similarity vector ($\vec{S}$) can be computed as follows:

$$\vec{S} = \begin{pmatrix} t_1 - c_{11} \\ t_2 - c_{12} \\ t_n - c_{1n} \end{pmatrix} = \begin{pmatrix} \delta s_1 \\ \delta s_2 \\ \delta s_n \end{pmatrix} \quad (3)$$

Where t1- tn are the reflectance values in the corresponding bands of the test pixel, and c11 - c1n are the corresponding reflectance values of a reference class. $^{\delta}s1..^{\delta}sn$ represent the elements of the similarity vector $\overrightarrow{\Delta S}$. From the test pixel distances are calculated to each class center. The magnitude of the similarity vector being:

$$|\vec{S}| = \sqrt{{}^{\delta}s_1{}^{2+}\,{}^{\delta}s_2{}^{2+\cdots\cdot}{}^{\delta}s_n{}^{2}} \quad (4)$$

The test pixel is classified and class label is assigned according to nearest distance if that distance is within the threshold limit for that class. Threshold values are determined using statistics of the main land cover classes. The training samples collected in the field are used to find the distribution pattern (the mean and standard deviation). The class mean and standard deviation are used to derive the spread of the class. In table 2 the statistics of various cover types are given.

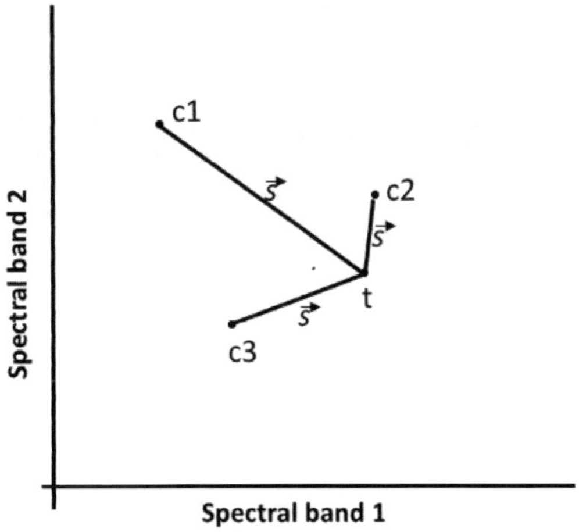

Figure 3. From the test pixel t, the similarity distances can be estimated and the one with the shortest distance to a class centre can be considered similar if it is within the threshold value from that class centre.

So, spread of reflectance value (95% confidence interval) from its mean in any spectral band for a given class can be estimated using its mean ($\bar{x}c11$) and the standard deviation ($sc11$) as follows:

$$Spread_{c11} = (\bar{x}_{c11} \pm 1.96\, s_{c11}) \quad (5)$$

The magnitude of the class spread from its center location can be estimated as:

$$|C_1| = \sqrt{((1.96\, s_{c11} - \bar{x}_{c11})^2 + (1.96\, s_{c12} - \bar{x}_{c12})^2 \ldots .(1.96\, s_{c1n} - \bar{x}_{c1n})^2)} \quad (6)$$

Similarly, spectral angle is also checked if the test pixel is similar to a class. Spectral angles are computed between test pixel and the reference spectra which is given by (Kruse, Lefkoff et al. 1993) as follows:

$$Spectral\ angle = cos^{-1}\left(\frac{\sum_{i=1}^{n} t_i\, c1_i}{(\sum_{i=0}^{n} t_i)^{0.5}(\sum_{i=0}^{n} c1_i)^{0.5}}\right) \quad (7)$$

Final assignment of the class label is done by checking the corresponding DNNDVI and the DVI values if they are also within the threshold limits for that class.

2.7. Classification

It is cumbersome to compute classification on a pixel by pixel basis thus it is performed in two steps: (i) first grouping of spectrally similar classes is done by unsupervised clustering, and (ii) assigning class label to the generated cluster based on similarity measures. The iterative self-organized unsupervised clustering algorithm (ISODATA) of ERDAS Imagine software was used to do grouping of spectrally similar classes. A separate dataset was created which consists of 6 bands of Landsat TM, DNNDVI and DVI. A series of unsupervised classification runs was implemented on this new dataset with a predefined number of classes starting from 5 with an increase of one in every step and with maximum number of iterations of 50. The convergence threshold was set to 1.0. The divergence statistical measure of distance between generated cluster signatures by each run was used to select the optimal run [11].

Table 2. Statistics of the reflectance values of the major land use classes.

Forest	min	mean	max	stdev	no
b1	0	0	1	0.5	607
b2	0	1.4	5	1.232	607
b3	0	0.6	6	1.203	607
B4	1	14.6	33	6.642	607
b5	1	10.2	24	4.368	607
b7	0	4.2	11	1.94	607

Bare soil	min	mean	max	stdev	no
b1	1	3.6	6	0.933	176
b2	4	7.9	12	1.222	176
b3	3	9.1	16	2.127	176
B4	13	19.1	24	2.371	176
b5	19	26.1	33	2.473	176
b7	8	15.3	23	2.556	176

Dryland cultivation	min	mean	max	stdev	no
b1	0	1.7	6	1.02	434
b2	2	5.6	10	1.316	434
b3	1	4.2	11	1.88	434
B4	10	16.2	28	3.064	434
b5	11	15.6	21	1.766	434
b7	5	9.5	18	2.36	434

Builtup area	min	mean	max	stdev	no
b1	1	4.3	16	1.581	260
b2	3	6.4	18	1.426	260
b3	1	7.7	28	2.494	260
B4	4	8	19	2.096	260
b5	10	15.9	37	2.926	260
b7	6	13.9	38	2.962	260

Plantation forest	min	mean	max	stdev	no.
b1	0	0.2	1	0.325	360
b2	1	2.3	5	0.71	360
b3	0	0.8	4	0.708	360
B4	13	19.7	28	2.572	360
b5	8	12.8	19	1.814	360
b7	3	4.8	9	0.918	360

Water body	min	mean	max	stdev	no
b1	0	6.6	18	3.039	271
b2	0	11.4	23	3.909	271
b3	0	0	0	0	271
B4	0	0.3	10	1.254	271
b5	0	0.8	8	1.033	271
b7	0	0.3	4	0.623	271

Paddy	min	mean	max	stdev	no
b1	0	2.8	9	1.033	642
b2	3	5.9	12	1.15	642
b3	0	2.3	11	1.91	642
B4	5	11.9	22	2.829	642
b5	2	12.8	27	3.96	642
b7	1	6.8	17	2.459	642

The optimal run with a distinguished peak in divergence separability explains the optimal size of spectrally similar groups [12]. In case of 1991 data, the highest divergence value of 3645 was obtained for 28 classes (figure 4). Once the optimal number of groupings is made, they can be linked to class label using the similarity parameters as explained in section 3.5. For assigning the final class label to the spectrally similar groups, the following procedure was followed.

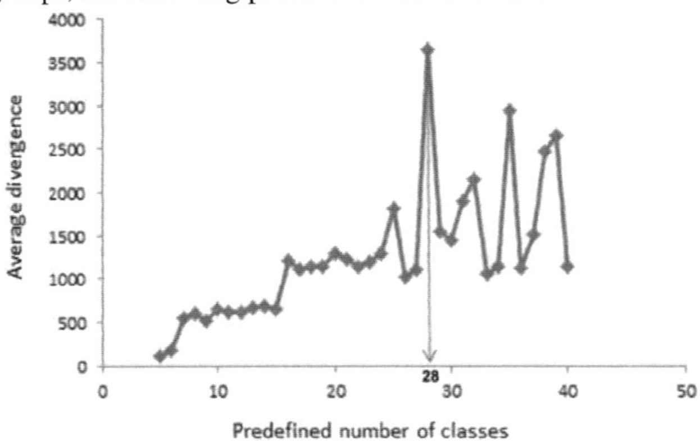

Figure 4. Average divergence values versus number of predefined number of classes in unsupervised classification.

Condition 1

- Euclidian distance and the spectral angle to a class under consideration has to be minimum of all the distances and that distance should be within the limit set by the class spread for the corresponding class (Table 3.). If there is discrepancy, then the spectral angle with the lowest value will be considered.
- In addition to this the DNNDVI and the DVI values are checked if they are within the threshold limits for that class.
- If all the above criteria are fulfilled the cluster will be class labeled or else go to condition 2.

Condition 2

- The minimum Euclidian distance and the spectral angle are larger than the threshold limit for any land cover class.
- If this is the case then the class label "Unknown" will be assigned to the cluster.

IOP Conf. Series: Earth and Environmental Science **148** (2018) 012017 doi:10.1088/1755-1315/148/1/012017

- In addition to the classification using similarity measures, the maximum likelihood classification algorithm using training samples collected during the fieldwork was also applied for 2014 image. This was done to compare the results of two classification methods.

Table 3. Euclidian distances from the resulting classes of unsupervised classification results to established land cover classes, the one with minimum distance will be class labelled.

Unsupervised Classes	Forest	Plantation forest	Dry land cultivation	Bare soil	Built up area	Paddy fields	Water body
1	15.34	8.56	**5.68**	13.82	13.96	10.18	26.69
2	18.74	16.65	8.97	11.79	**6.56**	12.02	26.08
3	13.86	**5.89**	7.78	17.26	16.87	10.47	26.19
4	14.35	**5.83**	10.99	20.39	20.26	12.69	27.12
5	19.91	13.99	**7.6**	8.73	12.15	13.53	29.98
6	19.53	**9.89**	13.88	20.25	23.5	17.27	32.23
7	20.12	11.38	**9.41**	12.48	17.71	15.16	31.91
8	18.34	**8.89**	10.84	17.05	20.23	14.94	30.75

2.8. Classification Accuracy Assessment

A set of field sample points was set aside for testing the classification accuracy for 2014 image.

3. Results and Discussions

3.1. Classification of the recent image

Result of the land cover classification for 2014 image using similarity measures is given in figure 5. The result shows that the watershed covers highest surface cover by plantation forest (54%) which is followed by primary forest (21%) and dry land cultivation (13%) (table 6). Overall accuracy of classification which is obtained by the total correctly classified pixels over total number of test pixels is 77 %t with 75 % of reliability (table 4).

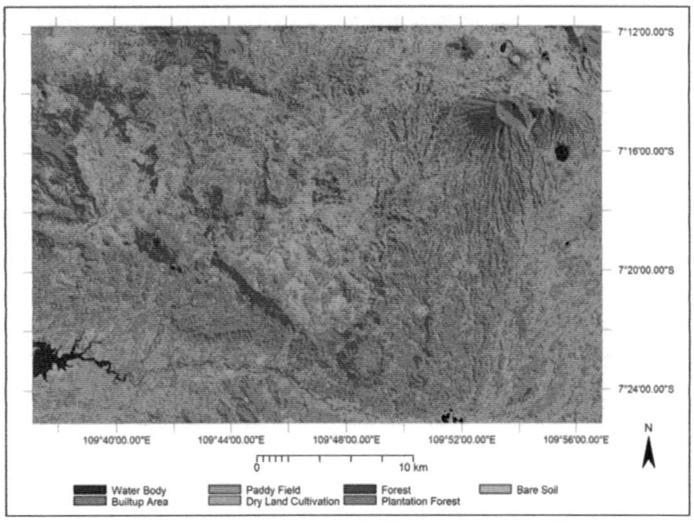

Figure 5. Land cover in 2014.

The result of applying the maximum likelihood algorithm on the same data set shows overall accuracy of 68 percent which is lower than that of applying similarity measures (table 5). Also maximum likelihood classification resulted in lot of pixels being undefined or unclassified (5 % of the total area which include also water bodies being classified as unknown). In both the classifications paddy field scored the lowest accuracy (35 %).

Table 4. Accuracy assessment of similarity measures.

	Test pixels	Water Body	Builtup Area	Paddy Field	Dry Land Cultivation	Forest	Plantation Forest	Bare Soil	Accuracy
Water body	409	391	0	0	0	18	0	0	0.96
Builtup area	135	0	115	12	0	0	0	8	0.85
Paddy	165	0	1	58	10	137	5	0	0.35
Dry land cult.	117	0	0	16	67	12	15	7	0.57
Forest	967	0	0	0	23	601	343	0	0.62
Pl. forest	882	0	0	1	9	32	840	0	0.95
Bare soil	71	0	0	0	35	0	0	36	0.51
	Reliability	1.00	0.99	0.67	0.47	0.75	0.70	0.71	

Table 5. Accuracy assessment of maximum likelihood.

	Test pixels	Water Body	Builtup Area	Paddy Field	Dry Land Cultivation	Forest	Plantation Forest	Bare Soil	Accuracy
Water body	409	393	0	0	0	1	3	0	0.96
Builtup area	135	0	75	0	0	0	0	11	0.56
Paddy	165	0	0	56	9	12	4	0	0.34
Dry land cult.	117	0	0	1	65	14	1	5	0.75
Forest	967	0	0	0	2	572	376	0	0.59
Pl. forest	882	0	0	0	14	233	635	0	0.72
Bare soil	71	0	1	0	7	0	0	63	0.89
		1.00	0.99	0.98	0.67	0.69	0.62	0.80	

3.2. *Classification of 1991image*

The reference spectra based on 2014 data (table 1) was applied to classify 1991 image following similarity measures as explained in section 2.6. The result shows the highest surface cover by plantation forest (45 %) and dry land cultivation (21 %) which is followed by primary forest (18 %) (figure 6).

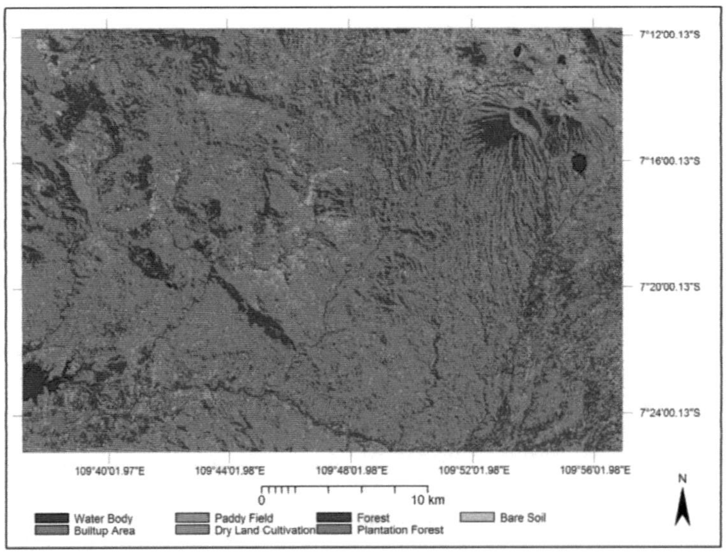

Figure 6. Lands cover classification of 1991 image.

Land cover map is not available from early nineties to validate the accuracy of the applied method but wide spread land use change in the form of deforestation to increase land for cultivation and to implement the government sponsored introduction of fast growing trees for increasing the production of the area is reported in the Serayu watershed [27]. The widespread deforestation in the early nineties in the watershed resulted in very high soil erosion in the steep slopes with the consequent deposit of high amount of suspended sediment in the Mrica reservoir, located in the southwest lowlands of the Serayu basin. The annual deposit of suspended sediment in the reservoir in 1991 (6,018,471 m^3) is considered to be the year with one of the highest recorded annual sediment deposition in the reservoir (figure 7).

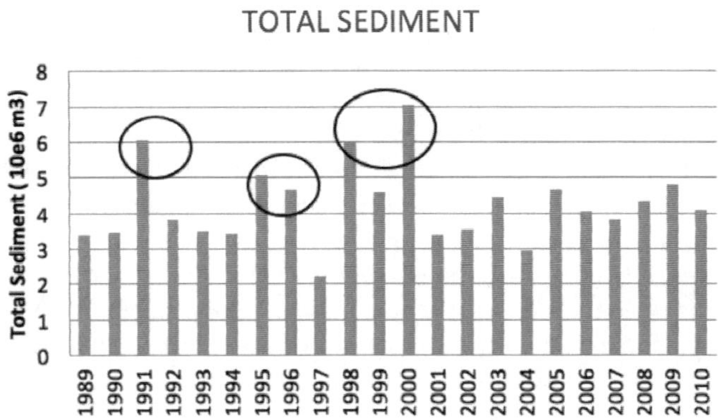

Figure 7. Suspended sediment deposition in the reservoir.

This can be used as an evident to validate the land cover classification performed in Landsat TM 1991 data. Estimation of the yearly deposit of suspended sediment is based on daily measurement of sediments at the intake of the reservoir. This data then calibrated with bathymetry measurements which also done on a yearly basis. The data is available for the year 1989 – 2010. From figure 7 we can understand that high sedimentation in 1991, 1998 and 2000 and relatively lower sediment deposition in the year 2009.

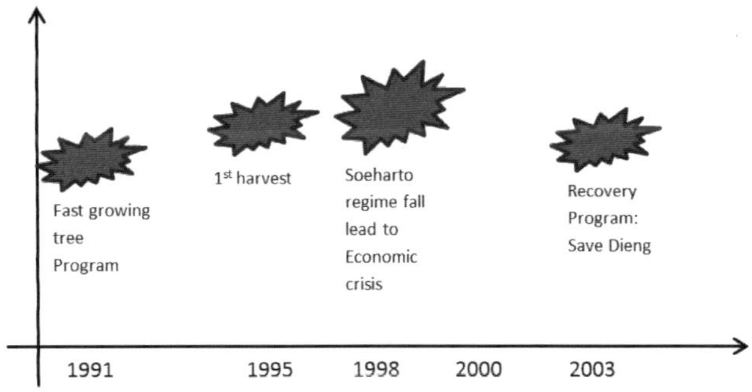

Figure 8. Major land use change policies in Indonesia.

By comparing land-use/cover in 1991 and 2014, we can see that in 1991 image there are significant bare lands in the study area. Most of the bare land is belong to Merawu catchment, and if we compare from figure 7 we can see that in 1991 the erosion rate from the Merawu Catchment is significantly high. Using this evident, we can claim that most of the erosion is come from the bare land. In the other hands, if we take a look at figure 8, we can see that in early 90's, the government has a program for a community forest. Therefore, to meet the land needs for the community forest, some land clearing was done. This fact is another evident that the image in the 1991 is qualitatively correct.

3.3. Land covers change from 1991 to 2014

The results of the land cover classification of 1991 and 2014 shows the highest changes in plantation forest and dry land cultivation in the area. Plantation forest cover is increased in 2014 by 8391 ha (9.27 % of the whole area) and the dry land cultivation area is decreased by 7348 ha (8.11 %) (table 6). Paddy field area is increased by 1.49 %. High soil erosion in the Serayu watershed resulting in high sedimentation in the Mrica reservoir is very much related to government sponsored land use policies in the area. In early nineties community forest program (called *sengonisasi* in Indonesian) was introduced to increase the economic condition of the Serayu basin [27]. Land was cleared to implement this program. The government introduced fast growing trees called Sengon laut (*Paraserianthesfalcataria*) in the area which grows fast enough to be ready for harvest after 4-5 years thus increasing the income of the people living there.

The result of the first harvest of the fast growing trees in 1995 had led to high soil losses and high sedimentation in the reservoir (figure 5). In 1998 there was again economic crisis in Indonesia following the fall of the Soeharto regime and the global economic crisis which led to high unemployment problem (figure 6). Since the education level in Dieng plateau and its surrounding area is low [2], agriculture is the only sector that could absorb large number of labours [34]. The increase of agricultural activity in the area has led to more forest clearing which increases soil erosion significantly. This high rate of erosion not only creates a major problem in the agricultural area, but also sedimentation problem in the reservoir built for hydro power generation (figure 7).

Table 6. Land cover change from 1991 to 2014.

Land cover/land use class	1991		2014		Land cover change from 1991 to 2014
	Area (ha)	Area (%)	Area (ha)	Area (%)	Area (ha)
Water body	713	0.79	368	0.41	-345
Built up area	2403	2.65	2606	2.88	+203
Paddy field	2900	3.2	4253	4.7	+1353
Dry land cultivation	19312	21.34	11964	13.22	-7348
Forest	16169	17.86	19375	21.44	+3206
Plantation forest	40337	44.57	48728	53.84	+8391
Bare soil	8593	9.49	3214	3.55	-5379
unknown/unclassified	81	0.09	0	0	Note:
Total	90508	100.00	90508	100.00	+ increase - decrease

4. Conclusion

The main objective of this study was to assess the land use change and its impact on the runoff and sediment yield of the upper Serayu Catchment. Two period of land use analysis was done from 1991 to 2014 using Landsat imagery reveals that there is a significant land use changes. Largest increase on the area was found on plantation forest followed by forest, paddy fields and built up area. The changes were 8391 ha, 3206, 1353 and 203 respectively. Dry land cultivation and bare soil were reduce significantly during the last decade by -7348 ha and -5379 ha. The result then compared with the sediment measurement. Taking into account the results of the land use analysis, we can conclude that the sediment was decrease significantly due to land use changes.

The study shows that it is possible to classify previously obtained image based on spectral characteristics and indices of major land cover types derived from recently obtained image. This avoids the necessity of having training samples which will be difficult to obtain. On the other hand, it also demonstrates that it is possible to link land cover changes with land degradation processes and finally to sedimentation in the reservoir. The only condition is the requirement for having the comparable dataset which should not be difficult to generate. Any variation inherent in the data which are other than surface reflectance has to be corrected.

References

[1] Angelsen A 1995 Shifting cultivation and "deforestation": A study from Indonesia. *World Development*, *23*(10), 1713–1729. http://doi.org/10.1016/0305-750X(95)00070-S.

[2] BPS 1998 Wonosobo Dalam Angka. Wonosobo, Badan Pusat Statistik.

[3] Chander G Markham B L and Helder D L 2009 Summary of current radiometric calibration coefficients for Landsat MSS, TM, ETM+, and EO-1 ALI sensors. *Remote Sensing of Environment*, *113*(5), 893–903. http://doi.org/10.1016/J.RSE.2009.01.007.

[4] Chen X Chen J Shen M and Yang W 2008 Land-use/land-cover change detection using change-vector analysis in posterior probability space. In L. Liu, X. Li, K. Liu, X. Zhang, & X. Wang (Eds.) (Vol. 7144, p. 714405). *International Society for Optics and Photonics*. http://doi.org/10.1117/12.812671.

[5] Chen Z and Wang J 2010 Land use and land cover change detection using satellite remote sensing techniques in the mountainous Three Gorges Area, China. *International Journal of Remote Sensing*, *31*(6), 1519–1542. http://doi.org/10.1080/01431160903475381.

[6] Chipika J T and Kowero G 2000 Deforestation of woodlands in communal areas of Zimbabwe: is it due to agricultural policies? *Agriculture, Ecosystems & Environment*, *79*(2–3), 175–185. http://doi.org/10.1016/S0167-8809(99)00156-5.

[7] Christanto N 2008 *Hydrological – Slope Stability Modeling for Landslide Hazard Assessment by means of GIS and Remote Sensing Data in Geo-Information for Spatial Planning and Risk Management*. Retrieved from https://www.itc.nl/library/papers_2008/msc/ugm/nugroho.pdf.

[8] Christanto N Hadmoko D S Westen C J Lavigne F Sartohadi J and Setiawan M A 2009 Characteristic and Behavior of Rainfall Induced Landslides in Java Island, Indonesia : an Overview. *EGU General Assembly 2009, Held 19-24 April, 2009 in Vienna, Austria http://meetings.copernicus.org/egu2009, p.4069, 11*, 4069. Retrieved from http://adsabs.harvard.edu/abs/2009EGUGA..11.4069C.

[9] Christanto N Sartohadi J Setiawan M A Hadi M Jetten V G and Shrestha D P 2017 Investigating The Effect of Conservation Techniques on the Land Degradation of Tropical Catchment Prone to Landslide. *Jurnal Geografi, UNNES*, *14*(2), 1–10.

[10] Christanto N Shrestha D P Jetten V G and Setiawan M A 2012 Modeling the effect of terraces on land degradation in tropical upland agricultural area. *EGU General Assembly 2012, Held 22-27 April, 2012 in Vienna, Austria., p.1075, 14*, 1075. Retrieved from http://adsabs.harvard.edu/abs/2012EGUGA..14.1075C.

[11] Davis S M Landgrebe D A Phillips T L Swain P H Hoffer R M Lindenlaub J C and Silva L F 1978 Remote sensing: The quantitative approach. *New York, McGraw-Hill International Book Co., 1978. 405 P.* Retrieved from http://adsabs.harvard.edu/abs/1978mhi..book.....D.

[12] de Bie C A J M Khan M R Smakhtin V U Venus V Weir M J C and Smaling E M A 2011 Analysis of multi-temporal SPOT NDVI images for small-scale land-use mapping. *International Journal of Remote Sensing*, *32*(21), 6673–6693. http://doi.org/10.1080/01431161.2010.512939.

[13] Deacon R T 1995 Assessing the Relationship between Government Policy and Deforestation. *Journal of Environmental Economics and Management*, *28*(1), 1–18. http://doi.org/10.1006/JEEM.1995.1001.

[14] Hartemink E Veldkamp T and Bao Z 2008 Property rights, land conflicts and deforestation in the Eastern Amazon. *Turk Journal Agriculture*. Retrieved from http://citeseerx.ist.psu.edu/viewdoc/download?doi=10.1.1.367.3151&rep=rep1&type=pdf.

[15] Entwisle B Rindfuss R R Walsh S J and Page P H 2008 Population growth and its spatial distribution as factors in the deforestation of Nang Rong, Thailand. *Geoforum*, *39*(2), 879–897. http://doi.org/10.1016/j.geoforum.2006.09.008.

[16] Gebresamuel G Singh B R and Dick Ø 2010 Land-use changes and their impacts on soil degradation and surface runoff of two catchments of Northern Ethiopia. *Acta Agriculturae Scandinavica, Section B - Plant Soil Science*, *60*(3), 211–226. http://doi.org/10.1080/09064710902821741.

[17] Hamandawana H Nkambwe M Chanda R and Eckardt F 2005 Population driven changes in land use in Zimbabwe's Gutu district of Masvingo province: Some lessons from recent history. *Applied Geography*, *25*(3), 248–270. http://doi.org/10.1016/J.APGEOG.2005.03.005.

[18] Jensen J R Rutchey K Koch M S and Natumalani S n.d. lnland Wetland Ghange Detection in the Everglades Water Conservation Area 2A Using a Time Series of Normalized Remotely Sensed Data. Retrieved from https://www.asprs.org/wp-content/uploads/pers/1995journal/feb/1995_feb_199-209.pdf.

[19] Kangalawe R Y M 2010 Changing land use/cover patterns and implications for sustainable environmental management in the Irangi Hills, central Tanzania. *Environment, Development and Sustainability*, *12*(4), 449–461. http://doi.org/10.1007/s10668-009-9204-5.

[20] Kingwell-Banham E and Fuller D Q 2012 Shifting cultivators in South Asia: Expansion, marginalisation and specialisation over the long term. *Quaternary International*, *249*, 84–95. http://doi.org/10.1016/J.QUAINT.2011.05.025.

[21] Kontoes C C 2008 Operational land cover change detection using change vector analysis. *International Journal of Remote Sensing*, *29*(16), 4757–4779. http://doi.org/10.1080/01431160801961367.

[22] Li X and Yeh A G O 2002 Neural-network-based cellular automata for simulating multiple land use changes using GIS. *International Journal of Geographical Information Science*, *16*(4), 323–343. http://doi.org/10.1080/13658810210137004.

[23] Liang L Shen L Yang W Yang X and Zhang Y 2009 Building on traditional shifting cultivation for rotational agroforestry: Experiences from Yunnan, China. *Forest Ecology and Management*, *257*(10), 1989–1994. http://doi.org/10.1016/J.FORECO.2008.11.032.

[24] Maldonado F D Santos J R Dos and De Carvalho V C 2002 Land use dynamics in the semi-arid region of Brazil (Quixaba, PE): Characterization by principal component analysis (PCA). *International Journal of Remote Sensing*, *23*(23), 5005–5013. http://doi.org/10.1080/01431160210000133l3.

[25] Narumalani S Mishra D and Rotwell R 2004 Analyzing landscape structural change using image interpretation and spatial pattern metrics. *GIScience and Remote Sensing*.

[26] Ningal T Hartemink A E and Bregt A K 2008 Land use change and population growth in the Morobe Province of Papua New Guinea between 1975 and 2000. *Journal of Environmental Management*, *87*(1), 117–124. http://doi.org/10.1016/j.jenvman.2007.01.006.

[27] Prastowo H 1995 Pemanfaatan Produk Hutan Rakyat Sebagai Pemasok Bahan Baku Industri. Seminar Pengembangan Hutan Rakyat. D. P. d. P. S. D. Kehutanan. Bandung, Indonesia.

[28] Puppim de Oliveira J A 2008 Property rights, land conflicts and deforestation in the Eastern Amazon. *Forest Policy and Economics*, *10*(5), 303–315. http://doi.org/10.1016/J.FORPOL.2007.11.008.

[29] Rasul G Thapa G B and Zoebisch M A 2004 Determinants of land-use changes in the Chittagong Hill Tracts of Bangladesh. *Applied Geography*, *24*(3), 217–240.

http://doi.org/10.1016/J.APGEOG.2004.03.004.
[30] Redo D Millington A C and Hindery D 2011 Deforestation dynamics and policy changes in Bolivia's post-neoliberal era. *Land Use Policy*, *28*(1), 227–241. http://doi.org/10.1016/J.LANDUSEPOL.2010.06.004.
[31] Salmon B P Olivier J C Kleynhans W Wessels K J and van den Bergh F 2010 Automated land cover change detection: the quest for meaningful high temporal time series extraction. In *2010 IEEE International Geoscience and Remote Sensing Symposium* (pp. 1968–1971). IEEE. http://doi.org/10.1109/IGARSS.2010.5653723.
[32] Servello E L Kuplich T M and Shimabukuro Y E 2010 Tropical land cover change detection with polarimetric SAR data. In *2010 IEEE International Geoscience and Remote Sensing Symposium* (pp. 1477–1480). IEEE. http://doi.org/10.1109/IGARSS.2010.5653215.
[33] Su X Wu W Li H and Han Y 2011 Land-Use and Land-Cover Change Detection Based on Object-Oriented Theory. In *2011 International Symposium on Image and Data Fusion* (pp. 1–4). IEEE. http://doi.org/10.1109/ISIDF.2011.6024300.
[34] Sunderlin W D Angelsen A Resosudarmo D P Dermawan A and Rianto E 2001 Economic Crisis, Small Farmer Well-Being, and Forest Cover Change in Indonesia. *World Development*, *29*(5), 767–782. http://doi.org/10.1016/S0305-750X(01)00009-2.
[35] Svoray T and Ben-Said S 2009 Soil loss, water ponding and sediment deposition variations as a consequence of rainfall intensity and land use: a multi-criteria analysis. *Earth Surface Processes and Landforms*, *35*(2), n/a-n/a. http://doi.org/10.1002/esp.1901.
[36] Turkelboom F Poesen J and Trébuil G 2008 The multiple land degradation effects caused by land-use intensification in tropical steeplands: A catchment study from northern Thailand. *CATENA*, *75*(1), 102–116. http://doi.org/10.1016/J.CATENA.2008.04.012.
[37] Verstappen H T (Herman T & International Institute for Aerospace Survey and Earth Sciences 2000 *Outline of the geomorphology of Indonesia : a case study on tropical geomorphology of a tectogene region : with a geomorphological map 1:5,000,000*. International Institute for Aerospace Survey and Earth Sciences. Retrieved from https://catalog.hathitrust.org/Record/004250614.
[38] Wang G Xia J and Chen J 2009 Quantification of effects of climate variations and human activities on runoff by a monthly water balance model: A case study of the Chaobai River basin in northern China. *Water Resources Research*, *45*, W00A11. http://doi.org/10.1029/2007WR006768.
[39] Zucca C Canu A and Previtali F 2010 Soil degradation by land use change in an agropastoral area in Sardinia (Italy). *CATENA*, *83*(1), 46–54. http://doi.org/10.1016/J.CATENA.2010.07.003.

IOP Conf. Series: Earth and Environmental Science **148** (2018) 012018 doi:10.1088/1755-1315/148/1/012018

Sustainability of three modified soil conservation methods in agriculture area

M A Setiawan[1-4], F H Sara[4], N Christanto[1,4], J Sartohadi[2,4], G Samodra[1,4], A Widicahyono[3], N Ardiana[3], C N Widiyati[3], E M Astuti[3], G K Martha[4], R F Malik[4], A P Sambodo[4], R P Rokhmaningtyas[4], G A Swastanto[4] and C Gomez[5]

[1] Faculty of Geography, Universitas Gadjah Mada, Indonesia
[2]Faculty of Agriculture, Universitas Gadjah Mada, Indonesia
[3] Ministry of Environment and Forestry, Indonesia
[4] Transbulent Research Group, Indonesia
[5] Graduate School of Maritime Sciences, Kobe University

anggri@ugm.ac.id

Abstract. Recent innovations in soil conservation methods do not present any breakthrough. Providing more attractive soil conservation methods from the farmer's perspective is however still of critical importance. Contributing to this soil research gap we attempt to evaluate the sustainable use of three modified conservation methods, namely JALAPA (*Jala Sabut Kelapa* - geotextile made of coconut fibres), wood sediment trap, and polybag system compared to traditional tillage without conservation method. This research provides both qualitative and quantitative analysis on the performance of each conservation measures. Therefore, in addition to the total sediment yield value and investment cost – as quantitative analysis, we also evaluate qualitatively the indicator of soil loss, installation, maintenance, and the durability of conservation medium. Those criteria define the sustainability use of each conservation method. The results show that JALAPA is the most effective method for controlling soil loss, but it also requires the most expensive cost for installation. However, our finding confirms that geo-textile is sensitive to sun heating by which the coconut fibre can become dry and shrink. Wood sediment trap is the cheapest and easiest to install; however it is easily damaged by termite. Polybag method results in the highest productivity, but requires more time during the first installation. In terms of the farmer's perspective, soil conservation using polybag system was the most accepted technique due to its high benefits; even if it is less effective at reducing soil loss compared to JALAPA.

1. Introduction

The number of alternative methods for soil conservation is still limited. Soil conservations are commonly conducted in the form of mechanics, vegetative, or chemical methods [9]. In Indonesia, terrace becomes the most accepted mechanic conservations among farmers (Bruijnzeel and Van Eijk, [4] [5], in spite of its ineffectiveness to reduce soil loss [11]. The vegetation-based method is considered to be the most efficient among the others [2]. In some areas, the vegetative method has been recognized to come through local wisdom within the society. Contrastingly, chemical method is less implemented due to high cost investment.

In 2014, the Indonesian government has officially established the Act Number 37 for Soil and Water Conservation. This Act legislated serious fines for anybody who intentionally ignore the implementation of soil conservation on their land. However, soil conservation is still regarded as a non-profitable activity among the farmers. Strict law enforcement of this Act might cause conflicts and difficulties to the society. Moreover, there are numbers of misconception among the farmers related to soil conservation that inhibit the acceptance of soil conservation [5]. Thus, providing methodological and implementation breakthrough and more attractive soil conservation method from the farmer's perspective is of critical importance.

Finding new strategy on soil protection should consequently consider the principal of soil conservation and the local resources [7]. This study aims to evaluate the sustainable use of three soil conservation technologies, namely JALAPA (*Jala Sabut Kelapa*), wood sediment trap, and polybag. The effectiveness of those three methods is compared with the traditional tillage system without any soil conservation method. JALAPA is a geo-textile made of coconut fibres from which the availability of the material is largely found in the study area. Wood sediment trap is simply made from piece of wood placed on certain point that could block the overland flow. Polybag is commonly used by the farmers for seedling phase. Soil erosion responses of those four conditions are monitored using plot erosion situated in Magelang District, Indonesia. We conducted a combination of quantitative and qualitative in evaluating each conservation method.

2. Material and Methods

2.1. Study area

This study was based on experimental study using series of field erosion plots. The plots area is located in Bompon Sub-catchment, Magelang District, and Central Java Province, Indonesia (figure 1). It is in the transition zone between Young - Old Sumbing Volcanoes and Menoreh Mountain. It is situated on 540 amsl. The morphological situation of this area is dominated by hill and valleys complex – with ±60 metres relative height difference. The volcanic ash of Sumbing Volcano has been highly weathered resulting in high content of clay within the soil layer. The volcanic intrusion process during Tertiary period of Menoreh had altered the bedrock into halosite clay mineral. The overlaying weathered volcanic ash on the altered clay mineral generated such a super deep soil (> 2 metres depth). High content of clay created pseudo-sand feature caused by clay oxidation process. The annual rainfall is quite high up to 3000 mm/year.

Combination of erodible material from the sensitive clay formation and high rainfall intensity control the erosion rate of this area. The pseudo-sand micro-relief is also eroded and transported during the erosion process, causing high sediment yield along the stream. Those situations are exacerbated by the tillage practice, mainly from the dry land agriculture. Cassava becomes favourite plantation among the farmers in this area (figure 2). During the beginning of growth (first 4 months), there is no significant canopy cover on cassava plant. The soil surface becomes completely bare without any protection against the rainfall and the overland flow energy.

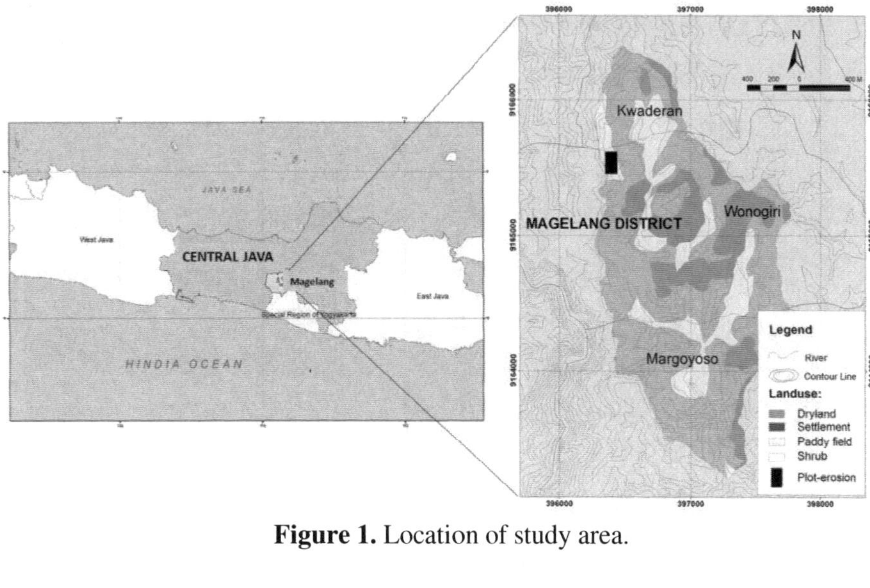

Figure 1. Location of study area.

Figure 2. Intensive tillage activities on cassava plantation.

2.2. Erosion control plot set-up
The main concept behind this research is promoting conservation strategies that consider the ease of implementation, availability of material and resources in surrounding area, and provide the best economic return. In that sense, the local farmers may easily adopt the idea of this research (de Graaff et al., 2008). To fulfil those necessities, four erosion plots were constructed (figure 3).

Plot A implemented the geo-textile technology made of coconut fibre. Coconut becomes one of the valuable vegetation types for community livelihood in the study area. Despite its high availability, coconut fibre has not been utilized for soil conservation by the local farmers. Plot B was designed to apply the principal of sediment trap. A wood branch was placed right on each terrace edge along the plot. Plot C introduced the use of polybag plastic as planting medium. Polybag plastic has been used widely in Indonesia for seedling purpose. However, this planting method may become a solution in initiating the zero tillage concepts. Plot D was prepared for traditional planting method without conservation method, representing the common land management in this area. Each of plots was planted with local Chili (*Capsicum annum L*) that has good market price during that season.

Figure 3. Plots construction in the upper slope with terrace system modifier.

The plots were situated in the upper slope with 20 degrees of inclination. The original slope had been modified with the construction of a terrace. Each plot had 1 m of wide and 10 m of length. Each plot was separated with 60 cm polymeric-rubber sheet. One third of the sheet was placed on the

IOP Conf. Series: Earth and Environmental Science **148** (2018) 012018 doi:10.1088/1755-1315/148/1/012018

ground and supported with bamboo sticks. The plot outlets were directed into a set of drums that could collect up to 400 liters of volume.

2.3. Erosion observation and sediment measurement
The plots had been monitored for 90 days, from October 2016 to January 2017. An automatic weather station was installed nearby these plots to collect rainfall event at 15 minutes intervals. Sediment samples were taken from the drum after daily rainfall event. The bed load and water inside the drums were mixed using a stick; and then immediately sampled by using 300 ml bottles. Those samples were sieved with 42Ø wattman paper. Drying processes were conducted for 8 hours using 105° C in the oven. To get the actual sediment content, each dried sediment sample was weighted and multiplied with the total measured volume in the drum for each sampling period.

3. Results and Discussion
This research provides both qualitative and quantitative analyses on the performance of the tested conservation measure. The total sediment yield is of critical quantitative information in most similar studies [8] [3]. There are only a few studies that thoroughly observe the mechanical process of erosion using similar comparison test setting. Therefore, in addition to the total sediment yield value, we also evaluate the indicator of soil loss, installation and maintenance, investment cost, durability of conservation medium, and productivity. Those six criteria define the sustainable use of each conservation method.

3.1. Soil loss indicator on each soil conservation methods
The complex mechanical processes of soil erosion are imprinted on soil surface. Most of the time, each soil erosion typology can be differentiated during field observation. Specifically, [10] elaborate a method for identifying and measuring indicators of soil erosion. This method is quite common for erosion rapid assessment or for spatial validation of any erosion model simulation. Each of our plots shows different indicator of soil loss.

Figure 4 (a) confirms the evidence of intensive splash erosion on Plot B (wood-sediment trap) and Plot D (traditional tillage). During the splash erosion, the raindrop forces and throws out fine surface material to several millimeters [9]. Those fine materials – mostly silt and clay- is then blocked and sticks on plot borders. Some of the splashes exceed the border's height (±30 cm). In contrast, both Plot A (JALAPA) and Plot C (plastic polybag) do not result in significance splash feature (Figure 4-b). The groundcover created by JALAPA mats protects 50% of the surface area; thus raindrop does not produce significant splash effect. Raindrop gives effect to geo-textile mainly on the space between the mats; and it causes 40 mm of soil loss during observation period (Figure 5). However, soil beneath the geo-textile remains the same depth.

(a) (b)

Figure 4. Indicator of splash erosion on (a) plot 2 and 4 are more evidence rather than on (b) plot 1 and 3.

Figure 5. Splash erosion on JALAPA.

Continuous splash energy from raindrop can cause soil crusting on any uncovered surface (figure 6 (a)). Moreover, soil in the study area is subject to soil crusting due to domination of clay and silt texture. Consequently, among plot B, C, and D experience soil crusting feature. In contrary, the JALAPA on Plot A effectively protects the soil surface from soil crusting effect. Soil crusting can reduce the soil surface infiltration and thus result in higher overland flow rate. A high overland flow rate can commence sheet erosion and even transport the available detached material from splash erosion. Figure 6 (b) confirms the intensive sheet erosion accumulated at the wood-sediment trap (Plot B).

(a) (b)

Figure 6 (a) Soil crusting feature and sheet erosion on uncovered surface soil; (b) eroded material from sheet erosion accumulated on the edge of wood-sediment trap.

The development of adverse erosion features during the observation period is dominated by rill erosion. Rill erosion is generated from concentrated overland flow that can cause soil loss 48 times higher rather than that of sheet erosion rate [9]. At our experiment sites, rill erosion occurs on both plot B and D (figure 7). However, the rills development between Plot B and D are moderately different in terms of rill dimension and morphology (table 1). Rill development on Plot B is obstructed by the wood-sediment trap, situated on the edge of terrace riser. Thus, the total length of rill in Plot B is shorter than that of Plot B. Rill development on Plot D continuously extend from the upper slope until the outlet. We do not find any rill initiation on plot A neither plot B. JALAPA on plot A increases in the surface roughness that prevents excessive overland flow rate and infiltrates more water into the ground. In the same sense, polybag planting system becomes obstacle for overland flow and rill generation, since there are limited open spaces between polybags.

(a) (b)

Figure 7. (a) Rill erosion feature on Plot B; (b) Rill erosion feature on Plot D.

Table 1. Rill development.

Rill development	Plot A	Plot B	Plot C	Plot D
Rill head from upper plot (m)	-	2.3	-	1.2
Average depth (mm)	-	52	-	56
Average wide (mm)	-	130	-	139
length (m)	-	5.7	-	7.6

3.2. Soil erosion rate

The traditional tillage exacerbates soil erosion, as evidenced by the test-plot D (Fig.8). Soil erosion rate in this plot reaches up to 195,9 Kg - equal to 23.2 mm of soil depth loss. The embedded terrace system is not sufficient to protect the direct impacts of splash, sheet, and rill erosion in Plot D. Consequently, an unprotected terrace riser with steep inclination has promoted severe rill development (figure 7 (b)). This result is in line with the work of [9] explaining that uncovered terrace riser may cause serious soil loss problem. Considering this plot observation is only conducted in a period of one third of the rainy season. In the next 10 years soil loss could reach up to 695.0 mm if the land tillage management remains the same. Such a soil loss will become a serious issue to land productivity that will affect the annual income of the farmers.

One solution of this soil loss is the usage of JALAPA, which could effectively reduce the soil loss up to 43% compared to the traditional method of tillage (Figure 8). As a conservation medium, JALAPA has some advantages. It directly protects the ground surface from raindrop effects and intercepts the splashed material. Additionally, JALAPA considerably improves surface roughness on which runoff development could be limited. Its thick fibre absorbs water and slowly infiltrate to the ground. This result is in accordance with [1] who confirms the effectiveness of geotextile as soil and water conservation medium.

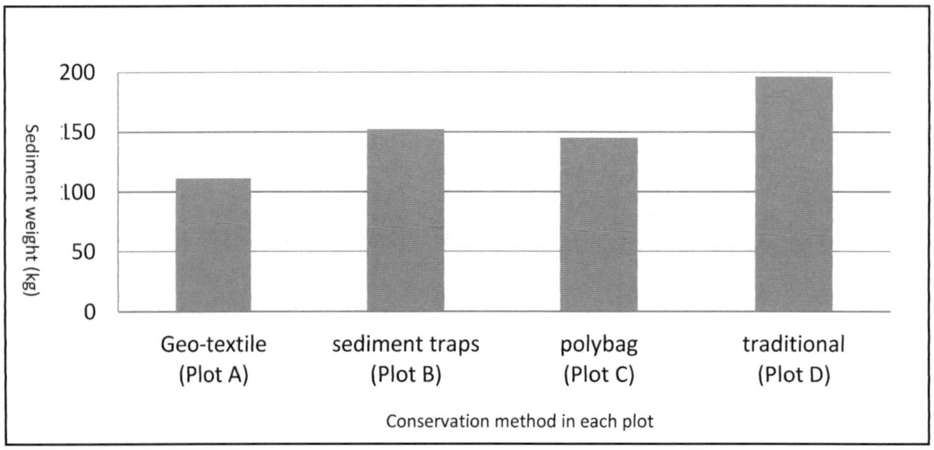

Figure 8. Comparison of sediment yield (in Kg) resulting from each plots erosion.

Both wood sediment trap and polybag medium also could control the soil loss rate up to 22% and 26% less than traditional method, respectively. Wood sediment trap has a maximal capacity to storage the sediment depending on the wood height. When the sediment load already reaches the maximum capacity, sheet erosion will overtop the wood height and transport the material dawn ward to the outlet. It implies that there should be extra maintenance for clearing the sediment trap periodically. In the case of polybag, it is less effective rather than JALAPA, but it is still slightly more effective than the method of sediment trap. The effectiveness can be increased by placing more polybag within the plot, since it will completely protect the soil material. However, we should consider the planting space to maintain an optimal growth of the crop.

3.3. Sustainability of conservation methods

In terms of soil loss protection, the three proposed conservation methods are already proven to reduce the soil loss rate (Figure 4-a). Nevertheless, the sustainability of a soil conservation method is not merely based on how much soil can be protected. Thus, we summarize the qualitative and quantitative performances of each plot that can imply their sustainability use.

Table 2 shows that none of the methods can completely fully the criteria of sustainability. Geo-textile are the best performer for soil loss control, but this method also requires the highest installation cost. Although coconut fibre is barely eaten by termite, our finding shows that geo-textile is sensitive to sun heat by which the coconut fibre can become dry and shrink (figure 4 (a) (b)). Wood sediment trap is the cheapest and easiest in installation process; however it is easily damaged by termite. Polybag method results in the highest productivity, but requires more time and money for first installation. Polybag is not only moderately effective in reducing the soil loss, but it also protects against the pest, i.e. mole cricket (*Gryllotalpa gryllotalpa*). A kilogram of chilli was valued up to 25.000 IDR, thus the farmers received 55.000 IDR from plot C with polybag. In terms of farmer's perspective, soil conservation using polybag system can be more accepted due to its high beneficiary, even less effective in reducing the soil loss compared to JALAPA.

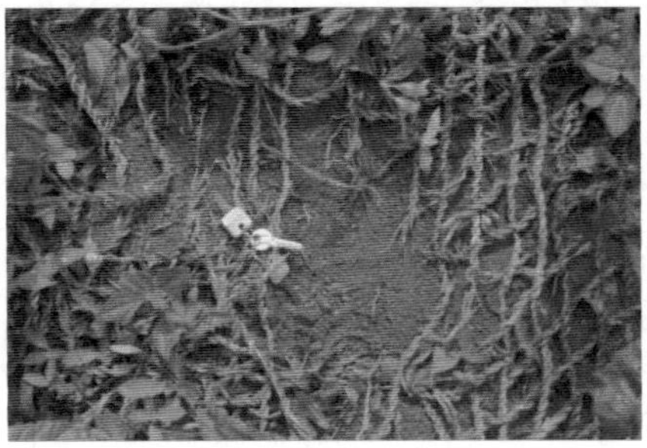

(a)

(b)

Figure 9. (a) Broken geo-textile due to continuous sun heating; (b) habitat of termite beneath JALAPA.

Table 2. Performance of soil conservation method.

Conservation methods	Soil Loss Indicator	Soil loss depth (mm)	Investment Cost	Installation and maintenance	Durability	Productivity
Geo-textile	Low splash erosion, low sheet erosion	13.2	100.000 IDR	• Quick installation (less than 1 hour) • Rolled up during dry and hot weather	• Moderate durability • Sensitive to sun heating	Low productivity (0.1 Kg of Chillies)
Wood sediment traps	High splash erosion, moderate sheet erosion, moderate rill erosion	17.9	Free of charge	• Quick installation (less than 1 hour) • Dredge and put back the sediment	• Low durability • Damaged by termite	Low productivity (0.05 Kg of Chillies)
Polybags	Moderate splash erosion, moderate sheet erosion	17.1	Rp.1000/pc (65.000 IDR)	• Requiring more time (2 days) • No maintenance required	• High durability • Can be used up to 2 years	High productivity (2.25 Kg of Chillies)
Traditional	High splash erosion, High Sheet erosion, High rill erosion	23.2	-	-	-	High productivity (0.08 Kg of Chillies)

4. Conclusions

We have evaluated the sustainable use of three soil conservation methods, namely JALAPA, wood sediment trap and polybag system. Comprehensive empirical analyses on the effectiveness and the efficiency of each method have been carried out to define the sustainability of each proposed conservation methods. Results show that JALAPA geotextile is the most effective method for soil conservation, but it is the most expensive one and it requires detailed maintenance for effective implementation. A negative aspect however is its sensitivity to sun heat and hot climate, making JALAPA a suitable method for areas with cool temperature regime and higher economic farming system (e.g, potato, cabbage, carrot, etc). Meanwhile, despite the easiness and readily available resources, the method based on wood sediment trap has proven to be insufficient to control sheet and rill erosion. Consequently, from the farmer's perspective, soil conservation using polybag conservation method is the best acceptable practice due to its high beneficiary; even it is less effective than JALAPA in reducing soil loss.

5. Acknowledgments
This research was supported with the Research Grant by Faculty of Geography, Universitas Gadjah Mada and BPDAS Serayu Opak Progo, Ministry of Environment and Forestry. We thank to the *Transbulent* Field Laboratory in Bompon that provided full support during our research.

References
[1] Bhattacharyya R Fullen M A Booth C A Kertesz A Toth A Szalai Z Jakab G Kozma K Jankauskas B Jankauskiene G Bühmann C Paterson G Mulibana E Nell J P Van Der Merwe G M E Guerra A J T Mendonca J K S Guerra T T Sathler R Bezerra J F R Peres S M Yi Z Yongmei L Li T Panomtarachichigul M. Peukrai S Thu D C Cuong T H and Toan T T 2011 Effectiveness of biological geotextiles for soil and water conservation in different agro-environments. *Land Degrad. Dev* 22, 495–504.

[2] Blanco H and Lal R 2008 *Principles of soil conservation and management*. Springer: USA.

[3] Boix-Fayos C Martinez-Mena M Arnau-Rosalen E Calvo-Case A Castillo V and Albaladejo J 2006 Measuring soil erosion by field plots: Understanding the sources of variation *Earth-Science Reviews* 78, 267-285.

[4] Bruijnzeel L A and Van Eijk B 1998 Runoff and soil loss from bench terraces in upland West Java, Indonesia and implications for process modelling. Modelling Soil Erosion, Sediment Transport and Closely Related Hydrological Processes, *Proceeding* of a symposium held at Vienna July 1998. IAHS Publ. No. 249.

[5] de Graaff J Amsalu A Bodnar F Kessler A Posthumus H and Tenge A 2008 Factors influencing adoption and continued use of long-term soil and water conservation measures in five developing countries. *Applied Geography*, Vol. 28, 271-280.

[6] de Graaff J Aklilu A Ouessar M Asins-Velis S and Kessler A 2013 The development of soil and water conservation policies and practices in five selected countries from 1960 to 2010. *Land Use Policy 32*, 165-174.

[7] Dumanski J and Peirettin R 2013 Modern conceps of soil conservation. *International Soil and Water Conservation Research Vol 1.*

[8] Maetens W Poesen J and Vanmaercke M 2012 How effective are soil conservation techniques in reducing plot runoff and soil loss in Europe and the Meditterranean *Earth-Science Reviews* 115, 21-36.

[9] Morgan R P C 2005 *Soil Erosion and Conservation 3rd edition*. Blackwell Publishing : United Kingdom.

[10] Stocking M A and Murnaghan N 2001 *A handbook for the field assessment of land degradation.* Earthscan Publication Ltd: London.

[11] Van Dijk A I J M Bruijnzeel L A and Wiegman S E 2003 Measurement of rain splash on bench terraces in a humid tropical steepland environment. *Hydrological Processes* 17, 513-535.

IOP Conf. Series: Earth and Environmental Science **148** (2018) 012019 doi:10.1088/1755-1315/148/1/012019

Achieving sustainable ese of environment: a framework for payment for protected forest ecosystem service

A Widicahyono[1,2], S A Awang[3,4], A Maryudi[3], M A Setiawan[5], A U Rusdimi[6], D Handoko[2] and R A Muhammad[5]

[1]Doctoral Program of Faculty of Forestry, Universitas Gadjah Mada, Yogyakarta, Indonesia
[2]Watershed Management and Protected Forest Agency (BPDAS HL) Musi, Indonesia
[3]Faculty of Forestry, Universitas Gadjah Mada, Yogyakarta, Indonesia
[4]Research and Development Ministry of Forestry Republic of Indonesia, Indonesia
[5]Faculty of Geography, Universitas Gadjah Mada, Yogyakarta, Indonesia
[6]Master of Planning and Management of Coastal and Watershed, Faculty of Geography, Universitas Gadjah Mada, Yogyakarta, Indonesia

anangwidi09@gmail.com

Abstract. Over the last decade, deforestation in Indonesia has reduced the forest area down to more than 6 million hectares. There is conflict that the protected forest ecosystem service is still often perceived as public goods. Many of them went unrecognized in planning process and continue to be undervalued. The challenge lies in maintaining socioeconomic development and ecosystem services sustainability without overlooking the people's opportunities and improving their livelihoods over the long term. An integrated approach is required to understand the comprehensive concept of protected forest ecosystem service. This research aims to formulate a scheme of payment for ecosystem service (PES) in a protected forest. It is a first step towards the attempt for the value of ecosystem services to be reflected in decision-making. Literatures, previous researches and secondary data are reviewed thoroughly to analyze the interrelated components by looking at the environment as a whole and recognize their linkages that have consequences to one another both positive and negative. The framework of implementation of PES schemes outlines the complexity of human-environment interconnecting relationships. It evaluates the contributing actors of different interest i.e. long term use and short term use. The concept of PES accommodates the fulfillment of both conservation and exploitation with an incentive scheme to the contributing parties who are willing to implement conservation and issuance of compensation expense for any exploitation means. The most crucial part in this concept is to have a good and effective communication between every policy makers concerning the forest ecosystem and local communities.

1. Introduction

The increasing deforestation in Indonesia reflects the poor management of forestry. The indirect causes of the deforestation and forest degradation in Indonesia are divided into four: 1) ineffective land use plan, 2) land ownership problems, 3) ineffective and inefficient forest management, and 4) poor law enforcement and corruption. The reformation era and regional policy initiate the pressure on protected forest and conservation forest. Illegal logging, forest fire, and land conversion are few examples of disturbance to protected forest function. Today, the total area of protected forest in Indonesia has reached down to 29.9 million Ha. The condition raises the urge to provide sustainability of its services.

The environmental services from protected forest come as various forms, such as water management, biodiversity, carbon storage and landscape view. Many of the products of forest services are regulated in the Regulation No. 32 year 1999 about Environmental Management and Regulation No. 37 year 2014 about Soil and Water Conservation. Both regulations mention that the payment for environmental services (PES) is an economic instrument to manage the sustainability of the environment. In Java Island, the protected forest is organized under the *Perum Perhutani* as a Forest Management Unit (FMU). The challenges on each FMU vary depending on its local condition and external pressures. One of interesting cases can be seen in FMU of North Kedu, that is located in Serayu Watershed, Central Java Province. The pressure on agriculture within this area is incredibly high over the last decades. Land conversion from protected forest area into agricultural land was initiated in the early 2000 and still continues until present day. The potential of high income from agricultural products drive the local community and involving investors to cultivate land without proper conservation [5].

A conflict of multi-stakeholders comes as the protected forest lies within the Sikunir and Prau Volcanic Complex. The magnificent view from the peak of Sikunir Hill attracts as a new tourism object. An environmental service of the protected forest now comes in form of landscape beauty. Several environmental problems are triggered around the area of tourism area of Sikunir Hill, Sembungan village, such as the alteration of initially natural areas into shops, disturbance to local plants and animals, litter problems and soil compaction. The human activities within the area of FMU of North Kedu, such as agriculture, mining and tourism, tend to exploit the protected forest function for a short-term use. Only few parties consider the long-term sustainability of the forest. In return, the conservation measures are still temporary and depend on the popularity of a policy and resources from the government. The poor law enforcement gives impacts to another damage of the functions of protected forest.

The environmental issue particularly in protected forest consists of complicated elements. Many different perspectives, uncertainties and complexities are difficult to formulate and solve [9]. There are only few previous research regarding "ecosystem services" as the concept was first acknowledged only in late 1980s [2]. The concept of ecosystem services can be set in different scales of space and time. The challenge lies in the need for a comprehensive and practical framework to simplify the complexity of social-ecological problems. This research aims to formulate a scheme of payment for ecosystem service (PES) in a protected forest. This is a first step to be applied in the decision-making process of maintaining the value of ecosystem services.

2. Methods for data collection

Secondary data collection was initiated by literature studies through previous research, regulations and other articles. Information regarding the study area is delivered through basic and thematic maps, statistical data and other spatial information. Primary data was collected through field survey and in-depth interview towards important actors as key respondents, such as visitors of Sikunir Hill, farmers, service owners (for tourism) and the member of Forest Village Community Institution *(*FVCI*)*.

3. Result and Discussion

3.1 Understanding human environment system of Sembungan Village

3.1.1 Environment System. Sembungan Village is situated within Dieng Volcanic Complex in Central Java Province. It is located within the Kejajar Sub-district, Wonosobo Regency, and Central Java Province. Sembungan village lies in the northern part of the Dieng Volcanic Complex where the highest altitude is approximately 2350 metres above sea level and the coldest temperature is around 5^oC at nights. Cebong Lake lies in the midst of the Sembungan village, surrounded by small hills. The northeast part of Sembungan village covers a Plot7 Forest Management Unit (FMU) North Kedu. It is a 513,1 Ha area of protected forest covering the volcanic slope of Sikunir Hill, a well-known area of tourism attraction. Protected forest in Sembungan is classified as community-based forest and managed under the authority of state-run forestry company (PT. Perhutani) since 1961 [9]. The vegetation covering most of the forest is dominated by pine and mahogany trees.

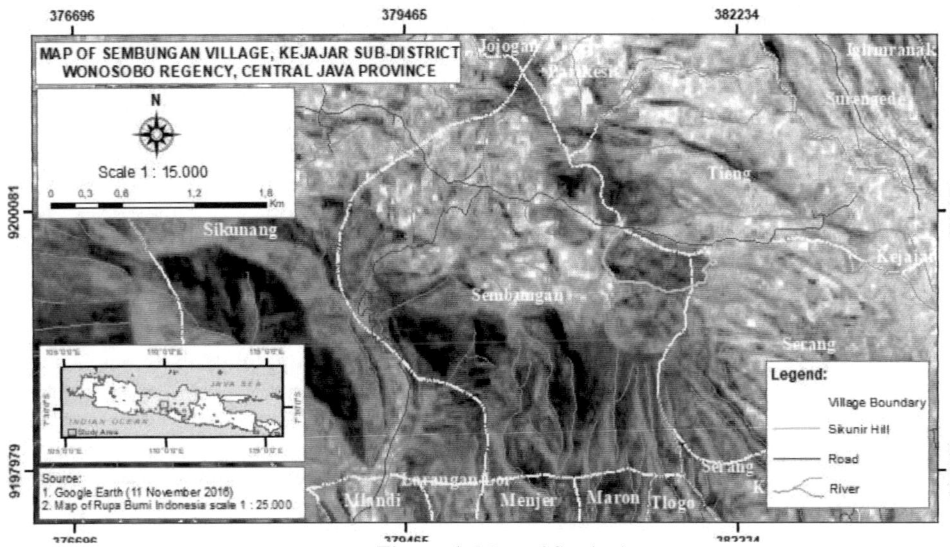

Figure 1. Map of Study Area.

Local people in Sembungan Village mostly rely on their livelihood on agricultural products. The intensive weathering process over porous volcanic material resulted in fertile soil and high productivity. Many farmers cultivate dry land agriculture with crop rotation as one of land management measures. Potatoes (*Solanum tuberosum*) and cabbages (*Brassica oleracea*) are the most cultivated vegetables within the area. The incomes from potato farming have also madeset the living standard of farmers in Sembungan village. Other commodities vary from corns *(Zea mays*), chinese cabbages (*Brassica rapa*), spring onions (*Allium cepa*), tobacco (*Nicotiana tabacum*), coffee (*Coffea arabica*) and rhubab *(Rheum officinale*) [9]. The agricultural land in Sembungan Village is distributed along the moderately slope, strongly sloping up to very steep slope. Monoculture system is used as the cultivation method. Bench terrace system is the most popular cultivation strategy applied in Sembungan Village. Parallel farming on steep slope is cultivated based on the consideration that it would provide good drainage facility [10]. The farmers also use sprinkles to spray water through pipes along the cultivated land as irrigation system. Land use alteration from agricultural land into settlements within the last decades has not been significant due to preference of keeping the lands to be cultivated instead of houses or buildings [9].

Being situated within Dieng Plateu, Sembungan village also suffered from deforestation that put the village at risk of erosion, landslides and floods [7]. Forest clearance and improper potato farming system play significance role to the driving factors of soil erosion and sedimentation. The erosion of Serayu Watershed has led to the increasing sedimentation and pollution in the river. The sedimentation also contaminated some lakes, including Cebong Lake in Sembungan village. It not only reduced the size of the lake, but also caused silting. Recovering the environmental damage in Sembungan village is not easy due to multi-sectoral issues that involve many aspects [6].

3.1.2 Human system. There are many actors involved in the application of forest management in Sembungan Village. Different actors with multi-interest interact in the management process, although not all of them are actively participate. PT. Perhutani holds a central role as an authorized agency that controls any activities in the protected forest within Dieng Volcanic Complex. Towards the initiation of Community-Based Forest Management (CBFM) as PT. Perhutani realized that the utilization of forest function cannot be separated from local community participation. Local farmers are starting to get involved in the maintaining forest sustainability. Together with CSR, financial aids are given to local farmers with intention to increase their welfare. Based on interviews with key persons, it is discovered that the financial aid can range from 4 million to 8 million IDR per ha and is given to any farmer who is willing to cultivate the available land. However, the program goes practically uncontrolled since the farmers are expanding their agricultural land, replacing the trees in the forest area. Therefore, PT. Perhutani take a step to reorganize the forest management program by forming organizations for each forest plot called Forest Village Community Institution (FVCI).

The members of FVCI consist of local community within Dieng Volcanic Complex. FVCI practically not only supervise the agricultural-related activities within protected forest area, but also manage the forest tourism (non-agricultral activities) through the formation of community groups called Tourism Awareness Group (*Pokdarwis*). Pokdarwis is the practical team who works directly on the field regarding the management of forest tourism, initiated by the village government. The interconnection and relationship between those actors describe the governance state in Sembungan Village regarding its contribution to manage the protected forest sustainability towards expanding tourism activities.

3.1.3 Human-Environment system. According to regulations, the utilization of protected forest can be done by not altering its landscape and only limited to the use of forest land and forest non-wood products. Such three kinds of exploitation also require supervision. The case of multi-stakeholder involvement in the protected forest of Sembungan village will be explained using a system thinking to break down the complex element and to show its relationship among them. The system thinking approach recognizes the interconnections and their consequences accross different scales of time and space, depending on the value-laden decisions that we consider [4]. It allows us to see the causal-effect relationships between physical and social elements in the utilization of protected forest of Sembungan village. The positive and negative relationships are also shown in each arrow within the diagram to indicate a positive or negative causal link.

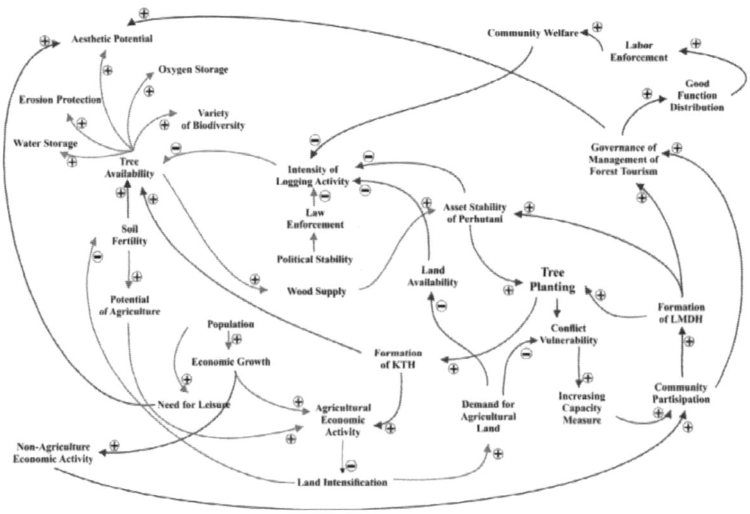

Figure 2. System Thinking of Forest Management in Sembungan Village.

Field observation and interviews are conducted to collect data and information on the stakeholders and their conflict of interest. The system dynamic of multi-interest conflict within Sembungan village is divided into three main stages. The first stage is regarded as the period where the nature of the forest area in Sembungan village is still well managed. A work plan to organize the sustainability of the forest is managed by the PT. Perhutani without community participation. As the soil fertility is still preserved, replanting the forest is done, thus the tree population grows. In this period, the economic needs driven by the dynamic of population is still limited to land cultivation. Although the needs for more space for agricultural purpose tend to increase, the political stability is able to control the law enforcement, thus the intensity of logging is still low and supervised. Therefore, the protected forest functions of conserving water management, recharge area, biodiversity and oxygen storage are also still well preserved. In figure 2, this stage is depicted using the green arrows.

A change in involving actors in the forest area management initiates the second stage. Within this phase, the forest governance grows by involving the local community to step in together with the PT. Perhutani to manage the sustainability of the forest. This phase is also characterized by the formation of Community-Based Forest Management (CBFM) by the PT. Perhutani. An organization of farmers called Forest Farmers Group *(FFG)* is built. An effort by the PT. Perhutani to preserve their asset of wood supply is by replanting the forest together with FFG. However, lack of law enforcement such as legal problems towards forest damage in this period leads to poor forest management. Each member of the organization was offered a financial aid to help them cultivating the land. Farmers tend to use the financial aid to expand their agricultural land. Such situation leads to uncontrolled forest logging and cause a collapsing function of the forest. The red arrows in figure 2 explained this stage.

The last stage is shown in figure 2 using blue arrows. This stage is characterized by the need for leisure as population grows. It initiates the efforts to increase economical need by expanding the non-agricultural measures. Community participation is strongly involved within this phase. They also stand as measure to develop capacities towards conflict vulnerability due to previous replanting program. The main activities that are very depending on agricultural products alter into the exploitation of tourism attraction particularly in Sembungan village. Shops, home stays and restaurants are built and expanded to the areas around Sikunir Hill and Cebong Lake. The *Perhutani* then form an organization involving local community called Forest Village Community Institution (FVCI). The formation is under the concern that the sustainability of forest began to be threatened by such activities. Supervised by *Perhutani,* FVCI works directly in controlling and managing the tourism in forest area within Dieng Volcanic Complex. Several programs encouraged by the FVCI are developing access for tourist, replanting the forest and managing waste. A community group is built to involve most of the local people in Sembungan village in managing the tourism activities, called Tourism Awareness Group (*Pokdarwis*). The governance of tourism in Sembungan village in particular interacts harmoniously as the functions of each involving actor are specific and clear. The community welfare is also well preserved because the activities also require employees from local people in Sembungan village.

3.2 PES Framework for protected forest

Payment for Environmental Service (PES) occurs when the service user offers payment to the service provider whenever the service provider conducted activities that give benefit to the user [3]. The ecosystem of protected forest provide many forms of services such as function of recharge area, landscape beauty, biodiversity protection, oxygen supply and protection from landslide and flash floods. These benefits of services are utilized by local communities. Based on its time scale, human needs can be divided into short term and long term period. Fulfilling the short-term needs tend to exploit the potential of the forest, while fulfilling the long-term needs comprises conservative measures. Conservative measures bring optimalization to the environmental services for communities by preserving the sustainability of the forest. The exploitation to the forest is commonly regarding the economical orientation, where the local communities intend to expand their land by altering the forest area. In the end, these measures lead to collapsing function of the forest which then leads to expensive cost of the environmental services. Figure 3 depicts the scheme of environmental service as a solution to balance the conservative and exploitative human activities within the context of forest area management. The concept of payment of environmental service accommodates both kinds of activities. Conservative measures will be awarded by incentives to the actors, while compensation cost will be given to any exploitative measures. These forms of payment of environmental services will be given back to both service provider and user.

In case of Sembungan Village, exploitation of the forest goods dominates the phase when the FFG is formed and given opportunities to cultivate the land within the forest area. In short term context, the ecosystem services provided by the forest such as wood supply and the benefits of tree availability are being utilized in an exceeding amount. In this period, the relationship between the local community and Perhutani was not in a good term, as there was violation to the cutting of trees by the local people and led to the arrests of the committed people by the Perhutani. This tense situation created non-conducive circumstances in the ongoing forest management. In result, deforestation and forest degradation occurred significantly, leading to secondary environmental impacts. Such impacts comprise roads damage, water scarcity and deeper water table during dry season, erosion due to land clearing and cultivation, and landslides in wet season.

Community-Based Forest Management (CBFM) is formed by PT. Perhutani according to provision No. 682/KPTS/DIR/2009 about CBFM. It is a forest resources management system, carried out by PT. Perhutani and local community together with stakeholders. Community participation incorporates the FFG and FVCI work with the agenda of planning, conducting, utilization, monitoring and evaluation of the forest resource management. The formation of CBFM poses as a resolution to the multi-interest conflicts in using the ecosystem services provided by the Sikunir forest. Its effectiveness can be seen by the active participation of the community and good communication with the PT. Perhutani. While the utilization of forest goods by the local community keeps on going, the presence of CBFM helps with the cooperation between the local communities an PT. Perhutani. Thus the forest resources and products are well managed and supervised. The performance of CBFM can be seen as a long-term conservation measure in preserving forest sustainability in Sembungan village. With the well-supervised sustainable forest management, the benefit of forest service comes back to the local community for their livelihood.

The emergence of other economic interests beyond the agricultural cultivation drives the local community to start exploiting the forest for tourism activities. New buildings, infrastructures and facilities are built and started to spread, thus the sustainability of forest nature is threatened. Concerning this situation, FVCI together with *Pokdarwis* and Perhutani work on managing the forest tourism in Sikunir village, ensuring the activities are well-organized. Under the supervision of the village government, division of tasks and profit sharing of the tourism activities are arranged. The entrance ticket fees to the tourism objects are delivered as income for the village. It is fairly shared between the Perhutani, *Pokdarwis*, FVCI, regional development, social activities, cleaning service and group interest. Supported by the funding, the local communities incorporated in FVCIand *Pokdarwis* are active in succeeding a sustainable management of the forest ecosystem in Sikunir village. This mechanism poses as a design of payment of environmental services (PES) in Sikunir village. The practice of PES in Sembungan village is implemented through communication and negotiation between the affected parties. The process and relationships of these components are explained in figure 3.

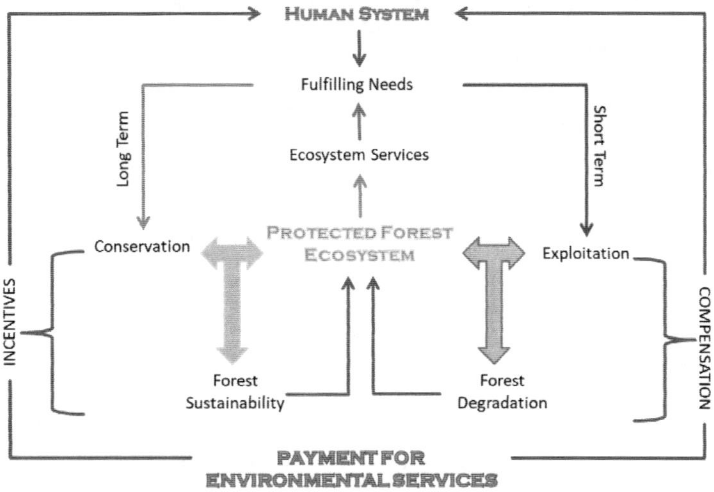

Figure 3. PES scheme of forest management in Sembungan Village.

4. Conclussion

The payment for environmental service in Sembungan village is implemented through management measures toward protected forest sustainability. Compensation expense is issued for any exploitation means. Good communication between stakeholders and involving actors is crucial in implementing conservation means. Consideration towards sustainability of forest ecosystem and local community participation is the key to accommodate the long-term conservation of protected forest. The biggest challenge in this concept is the difficulties in communicating the involving actor that determine the policy of the forest area and local communities.

References

[1] Arifandy I 2015 *Efektifitas Pengelolaan Hutan Bersama Masyarakat sebagai Resolusi Konflik Sumberdaya Hutan* (Bogor: Institut Pertanian Bogor).
[2] Dick J M Smith R I and Scott M 2011 *Environmetrics*. 22 598.
[3] Engel S Pagiola S and Wunder S 2008 *Ecological Economics*. 65 663.
[4] Kmon M K 2016 *System Thinking in the Forest Service: A Framework to Guide Practical Application for Social-Ecological Management in the Enterprise Program* (Portland: Portland State University).
[5] Lavigne F and Gunnel Y 2006 *Regional Environmental Change* 6 86.
[6] Pradana A Sara F H and Wahdaningrum W 2015 *International Journal of Environmental Science and Development* 6 861.
[7] Respatiadi H 2016 *Implementing Community Forestry in Indonesia The Tale of Two Villages* (Jakarta: Center for Indonesian Policy Studies).
[8] Rittel H W J and Webber M M 1973 *Policy Sciences* 4 155.
[9] Setiawan M A 2012 Integrated Soil Erosion Risk Management in the Upper Serayu Watershed, Wonosobo District, Central Java Province, Indonesia *Disertation* Institute of Geography, University of Innsbruck).
[9] Shresta N Rusdimi A U Arias A G Rosidi M Knuver M Ghimire R M Shresta S A and Setiawan M A 2013 Soil Erosion Reduction A Case of Kejajar Sub-District, Dieng Complex *Proc. Joint Scientific Program and One Day Seminar On: Ecosystem Based Disaster Risk Reduction (Yogyakarta, Indonesia, 8 December 2012)* ed Sartohadi J Pokharel J R Nehren U

M Fernandez G and Sandholz S *(Universitas Gadjah Mada and Center for Natural Resources and Development)* pp 21-31.
[10] Situmorang A W Nababan A Kartodihardjo H Khatarina J Santosa M A Safitri M Soeprihanto P Effendi S and Sunaryo 2013 *Indeks Tata Kelola Hutan, Lahan dan REDD+ 2012 di Indonesia.*

IOP Conf. Series: Earth and Environmental Science **148** (2018) 012020 doi:10.1088/1755-1315/148/1/012020

Estimating the age of oil palm trees using remote sensing technique

A C Fitrianto[1,2], A Darmawan[1,3], K Tokimatsu[1] and M Sufwandika[2]

[1]Department of Transdisciplinary Science and Engineering, Tokyo Institute of Technology, 2-12-1 Okayama - Meguro-ku -Tokyo, Japan
[2]Geospatial Information Agency (BIG), Jl. Raya Jakarta Bogor KM 46 Cibinong, Bogor, Indonesia
[3]Agency for the Assessment and Application of Technology (BPPT), Indonesia

aangtimor@gmail.com

Abstract. One of renewable energy that can be converted into electricity is biomass. Biomass energy or bio energy is the largest source of domestic renewable energy in Indonesia. Since palm oil development is rapidly increasing, Empty Fruit Bunch (EFB) and Mesocarp Fiber (MF) are becoming the highest contributor of oil palm waste. Understanding biomass waste potential is very important for further utilization. Remote sensing technique can be used to detect oil palm trees age based on the canopy density and to estimate the amount of EFB in further analysis. In this research, the percentage of canopy density of oil palm trees/stands depends on their ages and the age is divided into four classes; seeds (<3 years old; <10%), young (3-8 years old; 10-40 %), teenage (9-14 years old; 41-80 %), and mature (15-25 years old; >80 %).

1. Introduction

Currently, most countries depend on fossil fuels such as petroleum, natural gas and coal as the main source to generate electricity. Depletion of fossil fuels has been causing the increasing price of crude petroleum and demanding an alternative energy resource which is renewable and environment-friendly. Renewable energy such as biodiesel has the potential to replace petroleum-derived transportation fuel in the future. Biodiesel is defined as the mono-alkyl esters of long-chain fatty acids derived from vegetable oils such palm oil, rapeseed and soybean [1]. The energy output of oil palm is almost three times higher compared to soybean and rapeseed oil [2]. Moreover, other products are also produced from oil palm such as empty fruit bunch (EFB) fibers, shells, fronds, and trunk.

Palm oil is the largest agricultural industry in Indonesia. The total of harvested oil palm area grew from 4.1 million ha in 2006 to an estimated 8.9 million ha in 2015 [3]. According to [4], Indonesia is estimated to have produced 570 million tons of oil palm biomass, among which 299 million tones is OPF (oil palm fronds), 134 million tones is OPT (oil palm trunk), and 28 million tones is EFB (empty fruit bunches). It means that the potential source for renewable energy especially from oil palm was abundant. In addition, one of the important factors influencing fruit bunches production is the age of oil palm trees. According to [5] the data of oil palm trees age can be used in precision farming. Information about oil palm trees age and their spatial distribution is very useful in organization or management of oil palm plantation and one of important variables affecting profit.

On the other hand, lack of information about the age of oil palm trees and changes; also their spatial distribution is mainly problem for energy planning. According to [5], collecting age data of oil palm trees especially on a large area or regional scale is time consuming and costly. Remote sensing provides an effective and efficient approach for oil palm monitoring including the age of oil palm trees. Remote sensing has been widely proven to be essential in monitoring and mapping land use and land cover (LULC), including forest and plantation area [6]. Geospatial data derived from remote sensing can be easily integrated in geographic information system (GIS) for further analysis.

Our study, which is aimed at utilizing optical remote sensing data for oil palm plantation management including mapping, estimation of fruit bunch production and energy potential for electricity, has just been started and is in the preliminary study. In this paper, we focus on investigating the age of oil palms based on Landsat 8 OLI. The study area is the oil palm plantations areas in Cimulang, Bogor Indonesia.

2. Material and Methodology

The methodology consist of collecting Landsat 8 OLI and inventory data on the study area; digital processing of Landsat 8 OLI including of advance vegetation index (AVI), bare soil index (BSI), shadow index (SI), and thermal index (TI); principle component analysis; and performing relationships analysis between canopy density (FCD) and age of oil palm trees. Generally, the methodology on this study can be seen on figure 1.

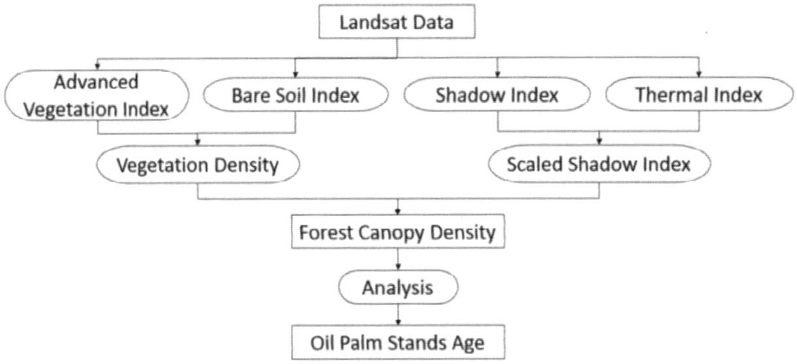

Figure 1. Methodology of estimation of oil palm trees age.

2.1. Data Collection

In this research, we used Landsat 8 OLI path/row 122/065 (*Figure* 2) that was acquired on August 17, 2016. Landsat 8 OLI consist of 11 bands with four visible bands, four infrared bands, one panchromatic, and two thermal bands. The authors also collected field data about age, canopy density, and yield of fruit fresh bunches (FFB) in the study area, which was based on correlation between canopy density class and age class. Composite band 543 of Landsat 8 was employed to delineate oil palm plantation area. This band utilizes near infrared band (NIR) to detect vegetation especially for oil palm trees.

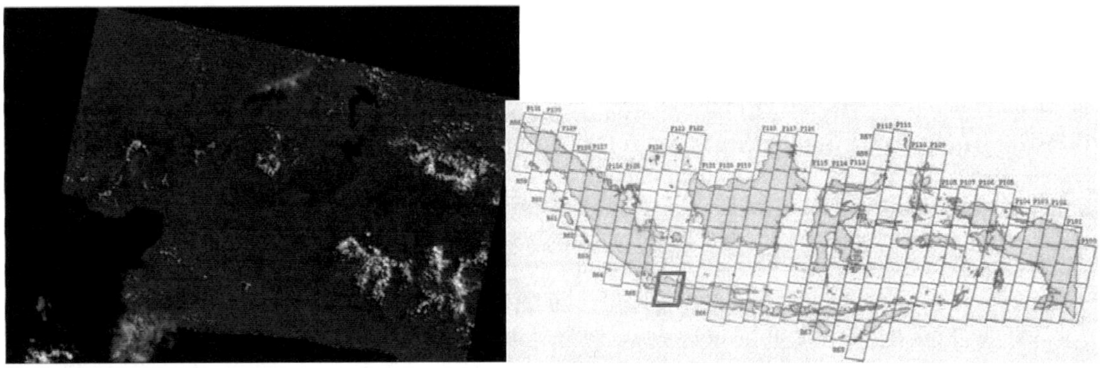

Figure 2. Landsat 8 OLI path/row 122/065.

2.2. Image Processing

The age of oil palm trees was predicted using remote sensing technique based on canopy density of oil palm trees. Forest canopy density (FCD) was utilized as essential method to estimate stand age. This model involved bio-spectral phenomenon modelling and analysis utilizing data derived from four indices as follows:

- Advanced Vegetation Index (AVI)
- Bare Soil Index (BSI)
- Shadow Index (SI)
- Thermal Index (TI)

The result of four indices integration in this modelling is percentage of canopy density in each pixel. Zero percent is no vegetation and 100 percent is very high density.

2.2.1. Advanced Vegetation Index (AVI). This indices was used to measure green vegetation. In remote sensing field, healthy vegetation is characterized by high absorption and low reflectance in visible region. On the other hand, they will have low absorption and high reflectance in near infrared wavelength. AVI has been calculated using *Equation 1.*

$$AVI = (B5 * (1 - B4) * (B5 - B4))^{1/3} \tag{1}$$

The value of AVI ranged from minus one (-1) to plus one (+1) for each pixel which is no vegetation at zero, highest possibility of green leaves if close to +1, and non-vegetation surface or water for negative value.

2.2.2. Bare Soil Index (BI). Similar to AVI, BI is normalized indices that used to separate vegetation and their background. This indices have opposite result with AVI because this is used to detect soil, it means near infrared wavelength have low reflectance due to absorption by soil moisture. BSI has been calculated using Equation 2.

$$BSI = \frac{(B6 + B4) - (B5 + B2)}{(B6 + B4) + (B5 + B2)} \tag{2}$$

The value of BSI will have ranged from zero to 200 for each pixel which is highest possibility of green leaves or canopy density if close to zero and non-vegetation surface or bare soil if the value close to 200.

2.2.3. Shadow Index (SI). Canopy density can also be detected by shadow characteristics derived from spectral information and thermal information of the forest shadow. In remote sensing, shadow characteristics are defined by shadow index (SI) as spectral information because crown arrangement of oil palm trees will have shadow pattern which affects spectral responses. The young age will have low value of SI compare to the old oil palm trees. The shadow index is derived from the low radiance of visible bands of Landsat 8 OLI image [7]. SI has been calculated using Equation *3.*

$$SI = ((1 - B2) * (1 - B3) * (1 - B4))^{1/3} \tag{3}$$

2.2.4. Thermal Index (TI). This index was used to check if the black or shadow is real shadow not black soil. Land surface that is close to shadow will have low temperature because leaf surface blocks and absorbs energy from the sun. The source for this index comes from thermal information infrared band of Landsat 8 OLI data (band 10 and band 11). TI has been calculated using Equation 4 and Equation 5.

$$L_I = M_I * Q_{cal} + A_I \tag{4}$$

where:
LI = Top of Atmosphere (TOA) radiance in (Watts/m2*srad*um)
MI =Band specific multiplicative rescaling factor from the metadata (RADIANCE_MULTI_BAND_x, where x is the band number)
Qcal = Quantized and calibrated standard product pixel values (DN)
AI = Band specific additive rescaling factor from the metadata (RADIANCE_ADD_BAND_x, where x is the band number)

$$T = K_2 \ / \ ln(\frac{K_1}{L_I} + 1) \tag{5}$$

where:
T = at-satellite brightness temperature (0K)
K2 = band specific thermal conversion constant from metadata (K2_CONSTANT_BAND_x; where x is band number 10 or 11)
K2 = band specific thermal conversion constant from metadata (K1_CONSTANT_BAND_x; where x is band number 10 or 11)
LI = product of the radiance formula

2.2.5. Forest Canopy Density (FCD). FCD value was shown as percentage of canopy density. Integration of four indices was used to calculate this model. AVI and BSI were integrated to produce vegetation density (VD). Processing method employed principal component analysis. This method showed that AVI and BSI have high negative correlation. Higher AVI value has low value in BSI because canopy density or vegetation cover is high and no bare soil or open surface at there. After that, VD will be set in the percentage scaling from zero percent to a hundred percent point.

Another calculation is scaled shadow index (SSI) that was derived from integration of SI and TI. In areas where the SSI value is zero (low), this corresponds to oil palm plantation that have lowest shadow value or minimum canopy density, opposite with areas where the SSI value is 100 meaning have the high canopy density. VD and SSI were integrated to achieve FCD value with Equation 6.

$$FCD = (VD * SSI + 1)^{1/2} - 1 \tag{6}$$

For accuracy assessment and collected ground truth, the distance between classes were changed to the form below (Table 1):

Table 1. Distance between classes for accuracy assessment.

Class	Category	Density (%)
1	Water and Cloud (W&C)	-
2	No Forest (NF)	0-5
3	Low Forest (LF)	5-40
4	Middle Forest (MF)	41-70
5	Dense Forest (DF)	71-100

2.3. The Age of Oil Palm Tree

Forest canopy density (FCD) was used to estimate oil palm trees age with the assumption that older trees will have higher percentage of canopy density. The class number of stands age divided into 4 classes, according to oil palm classification from Plantation Education Agency of Indonesia. Correlation between oil palm trees age and canopy density was shown in Table 2.

Table 2. Correlation between oil palm tree age and canopy density.

Nr.	Class	Stands Age	% Canopy Density
1	Seed	0 - 3	< 10
2	Young	3 - 8	10 - 40
3	Teen	9-14	41 - 80
4	Mature	15 - 25	> 80

In this research, we have assumption if some variable was equal for all area such as soil type, planting pattern, type of fertilizer, and another plantation management, because this entire variable has big correlation with canopy. Other plantation management have big correlation with canopy density is pruning tree branches. Many plantations have this policy to keep the amount of sunlight in that plantation and it will make the dense of oil palm trees canopy will different compare with natural oil palm trees. In the future, field survey will be conducted to verify the model result.

3. Result and Discussion

3.1. Forest Canopy Density Model

Advanced Vegetation Index (AVI), Bare Soil Index (BSI), Shadow Index (SI) and Thermal Index (TI) were calculated using GIS software (QGIS 2.8.2 and Arc GIS 10.2) to obtain a canopy density percentage (%). This indicator was adopted to analyze remote sensing measurements to observe contains live green vegetation that was used to find canopy cover density. Oil palm canopy cover, also known as canopy coverage or crown cover, is defined as the proportion of the forest floor covered by the vertical projection of the tree crowns [9] that was useful for distinguishing seed, young, teen, and mature old oil palm stands.

First step of image analysis was to obtain the value of AVI ranging from minus one (-1) to plus one (+1) for each pixel which was no vegetation at zero. The result of AVI in Cimulang area showed index range from 0.21 – 0.55 for each pixel with mean value at 0.44. This result showed that all of the research area was covered by vegetation.

The next step was BSI that it has opposite result with AVI because this indices was used to detect soil. The value of BSI ranged from zero to 200 for each pixel which had the highest possibility of green leaves or canopy density if close to zero and non-vegetation surface or bare soil if the value close to 200. The result of BI in Cimulang area showed index 58.79 – 97.86 for each pixel with mean value at 66.25. This result had same implication to AVI meaning that all of the research area was covered by vegetation or no bare soil.

The next step is vegetation density (VD) which is the procedure to synthesize AVI and BSI, because essentially, AVI and BSI have high negative correlation. Processing method employed principal component analysis and set the scaling of zero percent point and a hundred percent point. The result of VD in Cimulang area showed index 58.786 – 97.856 for each pixel with mean value at 66.25. Comparison of AVI and BSI, and VD was shown in figure 3. High value of AVI will have low value of BSI. The high value of AVI had high percentage in VD and this results show that AVI and BSI have negative correlation.

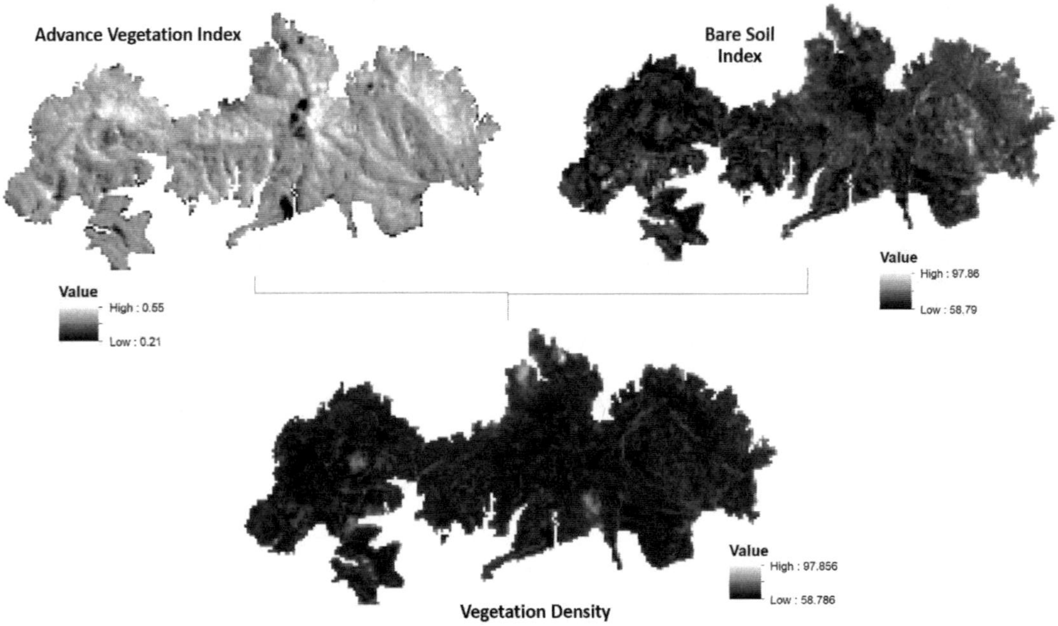

Figure 3. Procedure of VD mapping.

Canopy density can also be detected by shadow characteristics that were derived from spectral information and thermal information of the forest shadow. The result of shadow index (SI) shows the young stands have low value of canopy shadow index compared to the old stands age shadow. The result of SI in Cimulang area shows index 0.79 – 0.97 for each pixel with mean value at 0.96. This result shows that the research area was covered by vegetation and this result will be compared to thermal index.

Thermal index is used to check whether the shadow is real shadow or black soil. Land surface closed to shadow will have low temperature because leaf surface will block and absorb energy from the sun. The result of TI in Cimulang area shows ranges from 293.24 K – 295.56 K for each pixel with mean value at 294.15 K. This result shows the research area was covered by vegetation and showed by low temperature at that area.

Another calculation is scaled shadow index (SSI) that was derived from integration of SI and TI. In areas where the SSI value is zero low, this corresponds with oil palm plantation that have the lowest shadow value or minimum canopy density, opposite with areas where the SSI value is 100 that have high canopy density. The result of SSI in Cimulang area shows ranges from 79.73 – 97.42 for each pixel with mean value at 96.12. Comparison of SI, TI, and SSI will be shown in figure 4.

Figure 4 shows high value of SI and low value of TI. The high value of SI also have high percentage of SSI and this results show that the area with high value of SI was covered by vegetation and have low temperature.

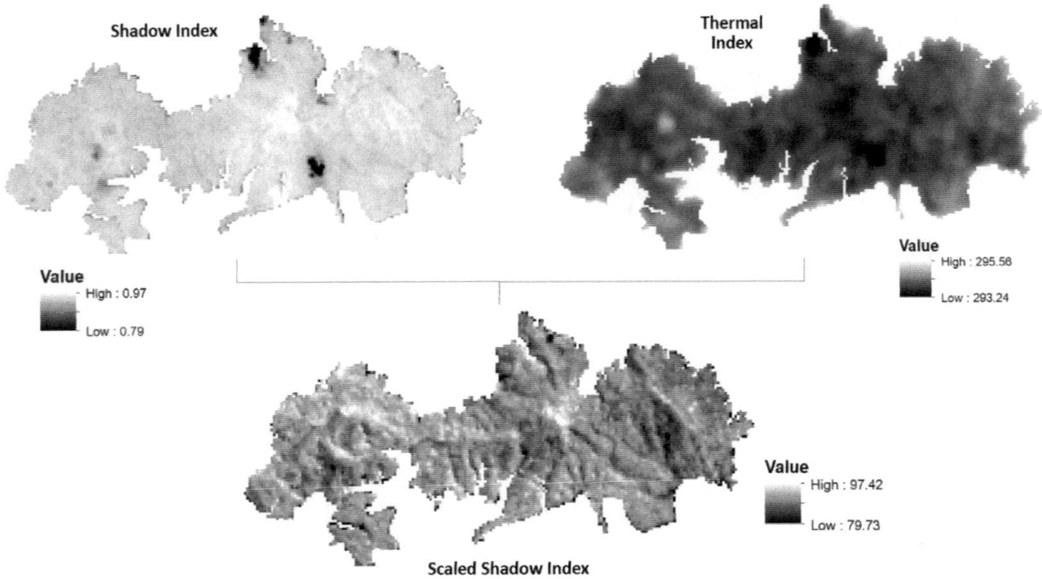

Figure 4. Procedure of SSI mapping.

The last step from this model is FCD that was obtained from the integration of VD and SSI. FCD percentage value have ranges from 0% to 100% meaning that 0% indicates this area is lack of coverage of the canopy and dominated by bare land. On the other hand, the area with greater percentage of the FCD value indicates more covered by canopy.

Figure 5. Forest canopy density.

The result of FCD in Cimulang area shows ranges from 74.66 – 94.79 for each pixel with mean value at 78.77 (*Figure 5*). Forest canopy density indicates that the area covered by vegetation with class of canopy density was Dense Forest (DF). The result from satellite image analysis shows that there is 100% dense forest cover in Cimulang area (figure 6).

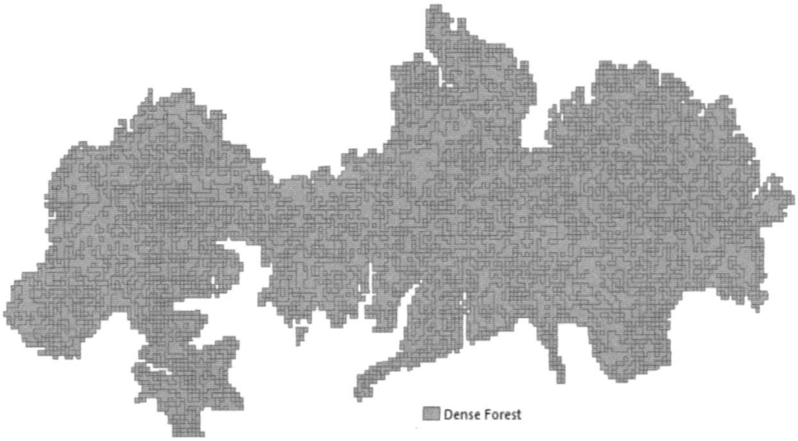

Figure 6. Class of forest canopy density.

3.2. Oil Palm Trees Age

Forest canopy density was used to estimate the age of oil palm trees with assumption older trees will have higher percentage of canopy density. Satellite image analysis shows the results for Cimulang area covers with FCD value ranges from 74.66 – 94.79 for each pixel. These results were used to estimate oil palm trees age (Table 2). This area consist of two classes of oil palm trees age i.e. teen with canopy density 41–80% and mature with canopy density more than 80%.

The result from satellite image analysis shows in Cimulang plantation teen old oil palm trees covers around 81.5% of that area, followed by mature oil palm trees with 18.5% or corresponding to 100 hectares. In this area we cannot find seed or young oil palm trees and it means no new plantation in this area.

The class number of stands age divided into 4 classes i.e. seed, young, teen, and mature. Table 2 shows that almost all of this class has small range of age difference such as seed class from 0 – 3 years old, young class from 4-8 years old, and teen from 9 – 14 years old. But, this condition is different for mature class where it has large range of age class from 15 – 25 years old. This condition was happened because, in this class, it was difficult to distinguish the mature (15-20 years) and old stands age (more than 20 years old). Characteristic from oil palm trees after 20 years old is different, even the length of canopy or steam leaves increases is equal with growing phases, but the direction of steam leaves will grow with smaller angle and tend to lead downwards.

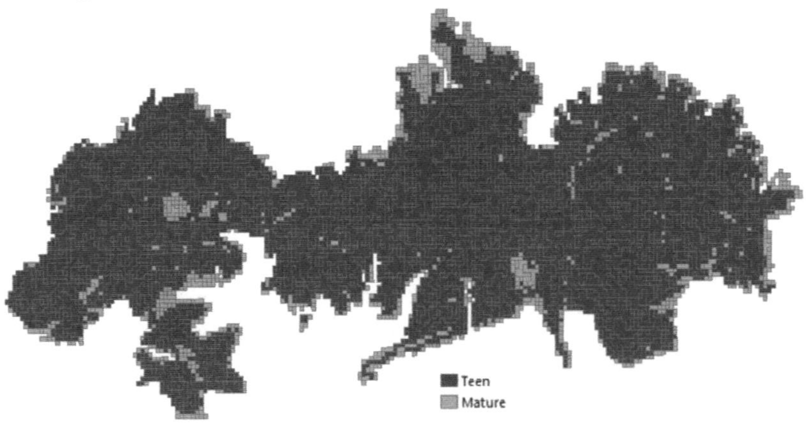

Figure 7. Class of oil palm trees age.

4. Conclusion and Future Work

Forest canopy density can be used to estimate the age of oil palm trees with the assumption that older trees will have higher percentage of canopy density. Result from this research is Cimulang plantation area consists of teen old oil palm trees covering around 81.5% of the area, followed by mature oil palm trees with 18.5% (100 hectares).

Remote sensing technique, especially using Landsat can be used to estimate oil palm trees age even though this method can only divide 4 classes i.e. seed, young, teen, and mature. There is a difficulty to distinguish the mature (15-20 years) and old stands age (more than 20 years old). This problem was happened because oil palm trees have steam leaves growing with smaller angle and tend to lead downwards after 20 years old.

Multi-temporal satellite image analysis of oil palm plantation area will also be conducted in the future. This step is very useful for monitoring and detailing information of oil palm trees age particularly to distinguish the mature (15-20 years) and old stands age (more than 20 years old).

References

[1] Tan K T Lee K T Mohamed A R and Bhatia S 2009 Palm oil: Addressing issues and towards sustainable development *Renew. Sustain. Energy Rev. 13*, vol. 13, May 2007, pp. 420–427.

[2] Sumathi S Chai S P and Mohamed A R A 2008 Utilization of oil palm as a source of renewable energy in Malaysia vol. 12, pp. 2404–2421.

[3] Petrenko C Paltseva J and Searle S 2016 Ecological impacts of palm oil expansion in indonesia," *Int. Counc. Clean Transp.*

[4] The C B S 2016 Availability, use, and removal of oil palm biomass in Indonesia," *Rep. Prep. Int. Counc. Clean Transp.*, pp. 1–39.

[5] Tan K P Kanniah K D and Cracknell A P 2015 Use of UK-DMC 2 and ALOS PALSAR for studying the age of oil palm trees in southern peninsular Malaysia *International Journal of Remote Sensing* February 2015, pp. 37–41.

[6] Darmawan M N S S Takeuchi W Abidin H and Rasid M N 2016 An investigation of age and yield of fresh fruit bunches of oil palms based on ALOS PALSAR 2 *8th IGRSM Int. Conf. Exhib. Geospatial Remote Sens.*, vol. 5, pp. 57–65.

[7] Banerjee K Panda S Bandyopadhyay J and Jain M K 2014 Forest Canopy Density Mapping Using Advance Geospatial Technique *vol. 1, no. 7*, pp. 358–363.

[8] Abkar A A Detection C and Timber I T Forest canopy density monitoring using satellite images.

[9] Azizi Z 2008 Forest canopy density estimating , using satellite images *Int. Arch. Photogramm. Remote Sens. Spat. Inf. Sci. Vol. XXXVII. Part B8. Beijing 2008*, pp. 1127–1130.

[10] Vadivelu S Ahmad A Choo Y H 2014 Remote sensing techniques for oil palm age classification using Landsat 5 TM Satellite. 2014;26 (4):1547–51.

[11] Danoedoro P 2012 Pengantar Penginderaan Jauh Digital, Penerbit ANDI Yogyakarta.

[12] LPP 2013 Buku pintar mandor (BPM), Seri Budidaya Tanaman Kelapa Sawit Lembaga Pendidikan Perkebunan.

IOP Conf. Series: Earth and Environmental Science **148** (2018) 012021 doi:10.1088/1755-1315/148/1/012021

Scenarios reducing greenhouse gas emission from motor vehicles in State University of Malang

I W Agustin[1] and C Meidiana[1]

[1]Department of Urban and Regional Planning, Engineering Faculty, Brawijaya University, Malang City, Indonesia.

immasaitama@ub.ac.id

Abstract. State University of Malang (UM) is one of the universities in Malang city. It has the second largest number of student after Brawijaya University (UB) with the growing number of students each year, resulting in increase the amount of motorized vehicle usage on campus. The State University of Malang condition shows the number of motorcycles in the provided parking area exceeded the capacity, causing the emergence of the improperly parking area. The condition causes the increase of mileage for vehicles that do not get a parking space. They must find and move to another parking area were still empty. The movement to another parking area resulted in the increase of exhaust emissions from motorized vehicles into the air. The main purpose of the research was to create alternative scenario of greenhouse gas emissions reduction in the State University of Malang. Alternative emission reduction based on strategies of Avoid-Shift-Improve (A-S-I), and the importance level of alternative determined with Multi Criteria Analysis (MCA). The result showed that selected alternative in emission reduction with the highest score of 40 per cent was centralized parking management.

1. Introduction

The convention resulted in the consent of The Kyoto Protocol by reducing the amount of greenhouse gas emissions generated from various sources so as not to exceed the limits set. In The Kyoto Protocol, there are sections Annex A and Annex B. The Annex 'A' section mention types of Greenhouse Gases (CO_2, CH_4, N_2O, HFCs, PFCs, and SF_6) along with the source of the producer. While in Annex 'B' there is a list of member countries of The Kyoto Protocol which amounts to 39 countries [1]. Indonesia was not a member of The Kyoto Protocol, so it has no obligation to reduce its emissions.

The Government of Indonesia participates in efforts to reduce the effects of Greenhouse Gases. In an effort to fulfil the greenhouse gas emission reduction commitment at the G-20 meeting in Pittsburg, the Government of Indonesia has prepared steps to be taken in [2] on the National Action Plan for Greenhouse Gas Emission Reduction (RAN-GRK). Greenhouse gas emissions reductions in Indonesia for 2020 are targeted at 26% and 41% if there is international assistance. One of the sectors included in the greenhouse gas reduction section is the energy sector. Transportation sector is part of the energy sector in the calculation of greenhouse gas emissions, because in the use of transportation means energy is always needed as a driving force [3, 13]. CO_2 emission reductions are targeted at 0.008 Giga tons (26%) and 0.011 Giga tons (41%) for energy and transportation. Some policies related to the transport sector include reducing the need for travel (avoid), shifting the model of private vehicle use into the model of low carbon transport such as non-motorized transport and public transport (shifts), and increasing energy efficiency and reducing carbon expenditure on motor vehicles (improve).

RAN-GRK is a guideline for the making process of greenhouse gas reduction plans. East Java Province has issued East Java Governor Regulation No. 67 of 2012 on Regional Action Plan for Greenhouse Gas Emission Reduction East Java Province. Malang city has a number of college institutions that became the goal of many prospective students continue their studies. Some of the existing universities are Brawijaya University, State University of Malang, State Islamic University of

Malang, National Institute of Technology, University of Islam Malang, State Polytechnic of Malang, and several other universities [4]. Due to the high number of students whose can be accommodated by the existing universities in Malang resulting in an increase number of motorists (student). The increase in the number of motor vehicles is always followed by the rising level of fuel oil consumption as the driving force of motor vehicles that will eventually cause an increase in the amount of exhaust gas in the air. This also happened to the State University of Malang which has 24,989 undergraduate students in 2015, State University of Malang is in second place after Brawijaya University with the number of undergraduate students 50.015 students [5].

Some problems arise such as increasing the number of students at the State University of Malang each year. The number of students in the State University of Malang increased by 8,287 people from 2013 to 2014 [6]. The increasing number of students resulted in an increase in the number of motor vehicles within the campus. The use of motor vehicles, the amount of emissions discharged into the atmosphere also increased. Impacts that occur due to emissions of motor vehicle exhaust are the emergence of air pollution and the number of motorcycles parked in the parking area provided by the State University of Malang exceeds the existing parking capacity (Field Survey, 2017). Due to this reason, the path that originally for pedestrians closed and replaced into a parking area. Moreover, when the number of vehicles is excessive, it is anticipated by the impromptu parking area at the location that was originally not destined for parking vehicles (Parking Section in the State University of Malang). These conditions cause the increase of vehicles mileage that does not get parking spaces. They had to find and move to another empty parking area. Moving to another parking area leads to an increase in the exhaust from the vehicle to the air. This study was conducted to determine how much greenhouse gas emissions mainly generated from motor vehicles in the campus area and provide an alternative reference for the reduction of greenhouse gas emissions in the State University of Malang.

2. Methods

The research location is at the State University of Malang with the boundary of Ambarawa Road in the West, Veteran Road in the North, Senior High School Laboratory of State University of Malang (UM) and Jakarta Road in the East, and Terusan Surabaya Road in the South. State University of Malang has 4 main gates as access to and exit of the Veterans Gate (entrance and exit), Semarang gate (entrance and exit), Gate of the Surabaya Canal (entrance point), and Ambarawa Gate (special entrance and exit of motorcycle / MC). The map of The State University of Malang is shown in figure 1.

2.1. Origin destination matrix

Origin destination matrix is a two-dimensional matrix that contains information about the amount of movement between locations (zones) within a certain area. The line represents the origin zone and the column represents the destination zone, so its matrix cell represents the amount of current from the origin zone to the destination zone [7]. Examples of matrix form of origin of destination can be seen in table 1.

Table 1. Origin Destination Matrix.

Zone	**Zone**				**O_i**	**A_i**
	1	**2**	**3**	**n**		
1	T_{11}	T_{12}	T_{13}	…	O_1	A_1
2	T_{21}	T_{22}	T_{23}	…	O_2	A_2
3	T_{31}	T_{32}	T_{33}	…	O_3	A_3
n	…	…	…	T_{nn}	O_n	A_n
D_d	D_1	D_2	D_3	D_n	Σ	
B_d	B_1	B_2	B_3	B_n		

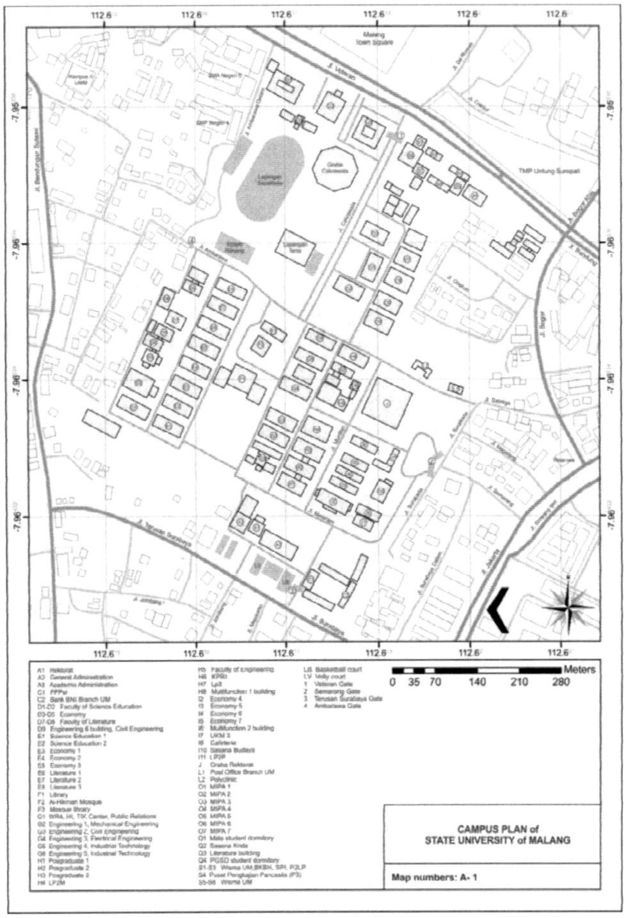

Figure 1. Research location.

It is necessary to set up some data such as data on the number of vehicles, the data capacity of parking, and the data mileage to create an origin destination.

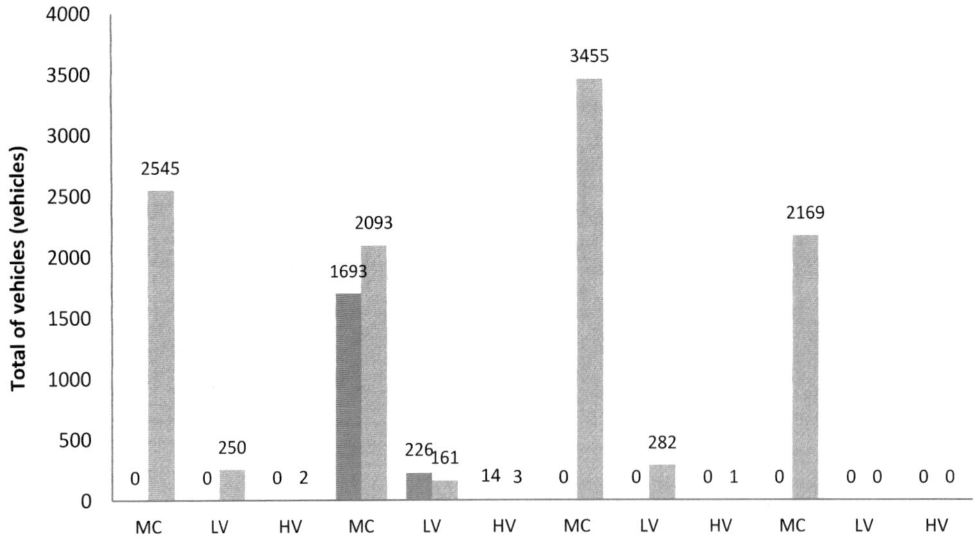

Figure 2. Total vehicles recapitulation

IOP Conf. Series: Earth and Environmental Science **148** (2018) 012021 doi:10.1088/1755-1315/148/1/012021

The number of vehicles data obtained through the primary survey is the recording of the number of vehicles entering the State University of Malang through the main gates. Recording is done on Wednesdays and Saturdays with three shifts: morning shift (06.15 am - 07.45 am), afternoon shift (11.45 am - 13.15 pm), and evening shift (15.00 pm -16.30 pm). The selection of days and hours obtained from the results of the initial survey. The results of the survey of vehicle numbers are shown in figure 2.

Parking capacity is shown in table 2. The parking area is divided into 3 types: student parking area (P code), parking area of lecturer / employee (Q code), and other parking area (R code).

Table 2. Parking capacity (Vehicles).

Code	Location	Capacity		
		MC	LV	HV
P1	Gombong Road	537	-	-
P2	MIPA I	426	-	-
P3	MIPA II	725	-	-
P4	MIPA III	-	33	-
P5	MIPA IV	-	40	-
P6	Center (near monument)	1066	-	-
P7	Block D (near pool)	1052	-	-
P8	Block D (north of the mosque)	336	-	-
P9	Block D (car)	-	96	-
P10	Ambarawa Gate	1087	-	-
P11	Back of The Economy	940	-	-
P12	Economy (car)	-	33	-
P13	C Block (economy)	440	-	-
P14	D2 (car)	-	9	-
P15	VIP (motocycle)	645	-	-
P16	VIP (car)	-	75	-
P17	G3 – G4 car	-	37	-
P18	G2 Mechanical Engineering	282	-	-
P19	A Block (car)	-	79	-
P20	A Block (motorcycle)	619	-	-
P21	H Block (canteen)	627	-	-
P22	H Block (fence)	268	10	-
P23	P Block (Psychology)	370	-	-
Q1	Civil Engineering side	-	7	-
Q2	South of the Mosque I	36	4	-
Q3	Across the economy	90	6	-
Q4	H6 (KPRI)	-	15	-
Q5	I5	21	32	-
Q6	Parking leader	-	14	-
R4	SLB	89	6	-
	Total	**9656**	**482**	**-**

Distance data is the length of the road from each gate to each parking area at the State University of Malang. The value obtained initially has a unit of meter (m), so it needs to be converted to units of kilometer (Km). There are only two suitable places for parking that are R2 (outside parking of Graha Cakrawala) and E7-E8 (Building of Literature Faculty) for heavy vehicles. The full results are shown in table 3.

Table 3. The gate distance to the parking area (km).

MC				
Parking	**Veteran Gate**	**Semarang Gate**	**Terusan Surabaya Gate**	**Ambarawa Gate**
P1	0,042	1,013	0,904	0,531
P2	0,159	0,956	0,845	0,475
P3	0,188	0,987	0,878	0,508
P6	0,342	0,689	0,578	0,246
P7	0,396	0,75	0,541	0,169
P8	0,412	0,768	0,659	0,179
P10	0,56	0,963	0,853	0,026
P11	0,774	0,749	0,637	0,345
P13	0,709	0,683	0,573	0,281
P15	0,83	0,576	0,464	0,401
P18	0,603	0,482	0,37	0,572
P20	0,75	0,262	0,152	0,652
P21	0,901	0,101	0,24	0,8
P22	0,87	0,072	0,268	0,772
P23	0,903	0,337	0,009	0,803
Q2	0,508	0,865	0,754	0,274
Q3	0,655	0,584	0,473	0,349
Q5	0,634	0,244	0,272	0,53
R4	0,854	0,287	0,062	0,754
LV				
Parking	**Veteran Gate**	**Semarang Gate**	**Terusan Surabaya Gate**	
P4	0,21	1,01	0,899	
P5	0,183	0,854	0,742	
P9	0,413	0,768	0,658	
P12	0,655	0,73	0,617	
P14	0,752	0,641	0,53	
P16	0,723	0,52	0,398	
P17	0,947	0,553	0,438	
P19	0,742	0,28	0,169	
P22	0,87	0,072	0,268	
Q1	0,529	0,86	0,748	
Q2	0,508	0,865	0,754	
Q3	0,655	0,584	0,473	
Q4	0,509	0,273	0,405	
Q5	0,634	0,244	0,272	
Q6	0,793	0,329	0,218	
R4	0,854	0,287	0,062	
HV				
Parking	**Veteran Gate**	**Semarang Gate**	**Terusan Surabaya Gate**	
R2	0,226	0,827	0,722	
E7-E8	0,565	0,85	0,738	

2.2. CO_2, CH_4, and N_2O Emission

Greenhouse gases are the atmospheric gases and cause the greenhouse effect that is where the infrared radiation of the reflected sun cannot escape from the atmosphere causing the earth's surface temperature to rise [2]. This study selected CO_2, CH_4, and N_2O gases because it is a type of gas from fossil fuel combustion from motor vehicles that is directly the cause of greenhouse gas [8]. Fuel consumption is the amount of fuel used by motor vehicle to travel a certain distance. Fuel consumption values for motors (MC), light vehicles (LV), and heavy vehicles (HV) are obtained through sales brochures, manufacturing sites, and other sites that test fuel consumption. The fuel consumption value of motor vehicles is shown in table 4.

Table 4. Fuel Consumption of Vehicles Recapitulation.

Average of Fuel Consumption		
Type		**Fuel consumption lt/km**
Motorcycle	MC	0,0251
Light Vehicles	LV	0,0866
Hight Vehicles	HV	0,1667

In this study, the types of fuel used are Pertamax (Gasoline) and Solar (Diesel Fuel). The CO_2, CH_4 and N_2O emission factors for both types of fuel are shown in table 5.

Table 5. Emission Factor of CO_2, CH_4, and N_2O.

Fuel	CO_2 (ton/GJ)	CH_4 (g/GJ)	N_2O (g/GJ)
Pertamax	0,069	5,00	0,60
Solar	0,074	5,00	0,60

The result of the origin destination matrix shows the number of distribution vehicles to each parking area. Based on [8], the number of vehicles acquired and then entered (Equation 1) to calculate the fuel used by all motor vehicles entering the State University of Malang.

$$Fuel = \sum_{i,j}(V_{i,j} \, . \, Di_{i,j} \, . \, C_{i,j}) \qquad (1)$$

Notes:
Fuel = Total fuel (Lt)
$V_{i,j}$ = Total vehicles (vehicles)
$Di_{i,j}$ = Vehicles mileage (Km)
$C_{i,j}$ = Fuel consumption (Lt/Km)
i = Type
j = Fuel type

Since the emission factor has a Kg/GJ unit, the amount of fuel having a unit Liter (lt) is converted to Giga Joule (GJ) units using a conversion factor of 1 Liter of fuel to Joule (table 6). The amount of fuel obtained, put into Equation 2 to calculate the amount of CO_2, CH_4, and N_2O emissions.

Table 6. Conversion Factor of Fuel.

Fuel Type	Conversion 1 Lt of Fuel (GJ)
Gasoline	0,0322
Diesel Fuel	0,0359

$$Emission = \sum_a(Fuel_a \, . \, EF_a) \qquad (2)$$

Notes:
Emission = Total emission (Kg)
Fuel = Fuel used (GJ)
EF = Emission factor (Kg/GJ)
a = Fuel type

2.3. Multi Criteria Analysis

Data and information from various sources, alternative choices, and assessments are required to make decisions. The difficulty is the existence of various criteria in choosing or determining alternatives, opposing goals, and improper decision makers. The Multi Criteria Analysis method emerges as an aid to decision makers in establishing alternatives that have multiple criteria [9,10].

In this study the decision makers (expert) are the people who participate in the determination of regulations and understand about the condition of transportation at the State University of Malang. Some of the resource persons encountered by researchers are:

- Head of Household Subdivision
- Head of Law and Administration Subdivision
- Head of Sub Administration.
- Head of Planning Sub-division
- Head of Planning and Information System
- Members of the security guard
- Posko Satpam Admin Staff
- Head of Center for Research and Dedication for Environment and Natural Disaster Management (P3LHMBA) of The State University of Malang (UM).

Table 7. Important Level.

Important Level	Definition
1	Just as important
3	Quite important
5	Important
7	Very important
9	The most important
2,4,6,8	The level of importance between the values above
The opposite value above	The value of importance is seen from the opposite direction

The experts filling out questionnaire about comparison of interest rates between alternatives and their importance, as shown in table 7. The next comparison results will be incorporated into a pair wise comparison matrix to obtain priority values.

3. Result and Discussion

3.1. Origin destination matrix

Table 8 shows the distribution of the result matrix of destination matrix on Wednesday. The total value for each gate is equal to the number of motors (MC) entering the State University of Malang in figure 2. The calculation results of motor distribution (MC) and then made into a Map of Motor Destination (MC). Specified five segments for the number of vehicles that indicated by line thickness. The thicker the line means the number of vehicles going to the parking lot more and more. Next the line is drawn from the entrance gate as the origin of the movement and ends at the destination parking area.

3.2 Emission of CO_2, CH_4 and N_2O

Equation 1 generates the total fuel amount of vehicles entering the State University of Malang on Wednesday; Motor (MC) uses the most fuel of 50.9 Litre, followed by light vehicles (LV) of 20.2 Litre, and heavy vehicles (HV) of 0.6 Litre. While on Saturday the motor (MC) is still the largest fuel user with 28.6 Litre followed by light vehicle (LV) with 11.2 Litre. Heavy vehicles (HV) on Saturday experienced an increase in fuel consumption of 1.9 Litre, due to the number of heavy vehicles entering the State University of Malang more than Wednesday. The amount of fuel used is shown in figure 3.

Table 8. Motor Distribution (MC) on Wednesday.

Parkir	Wednesday			
	Veteran Gate	Semarang Gate	Terusan Surabaya Gate	Ambarawa Gate
P1	537	0	0	0
P2	426	0	0	0
P3	725	0	0	0
P6	857	0	0	109
P7	0	0	0	549
P8	0	0	0	175
P10	0	0	0	567
P11	0	0	0	490
P13	0	0	0	229
P15	0	15	1087	2
P18	0	6	478	0
P20	0	17	1047	0
P21	0	1406	0	0
P22	0	601	0	0
P23	0	0	634	0
Q2	0	0	0	19
Q3	0	1	57	29
Q5	0	47	0	0
R4	0	0	152	0
	2545	2093	3455	2169
Total		10262		

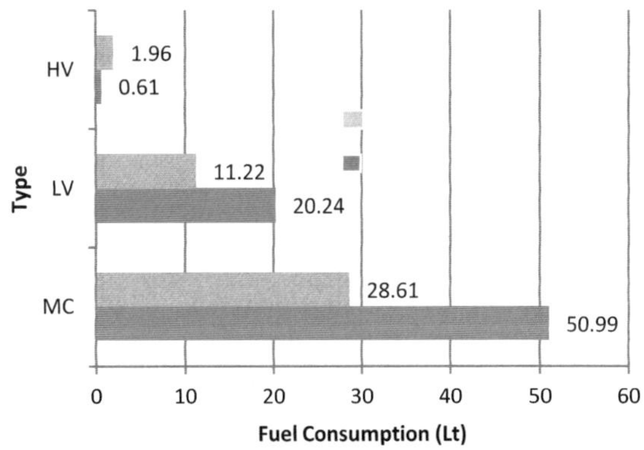

Figure 3. Total fuel consumption.

To get the average emission value per hour, the value divided by 4.5 hours (table 9). The average emission value is then multiplied by the number of working hours at the State University of Malang. Based on [14] on working days and hours at the State University of Malang, working hours are from 07.00 am to 16.00 pm (9 hours).

Table 9. Total Emission per hour.

Type	Total Emission (Kg)					
	CO_2		CH_4		N_20	
	Wednesday	Saturday	Wednesday	Saturday	Wednesday	Saturday
MC	25,18	14,13	0,00182	0,00102	0,00022	0,00012
LV	9,99	5,54	0,00072	0,00040	0,00009	0,00005
HV	0,36	1,16	0,00002	0,00008	0,00000	0,00001
Total	35,53	20,83	0,00257	0,00150	0,00031	0,00018

3.3 Alternative to reduce CO_2, CH_4, N_2O emission

Based on the results of the questionnaire survey on the level of importance, then searched the average score so as to get one value as input in a matrix of pairwise comparisons. The research used the average equation of geometry in Equation 5 to obtain the input value. Table 10 shows stakeholder's opinion for alternative selection.

$$a_{ij} = (z_1, z_2, \dots, z_n)^{1/n} \quad (3)$$

Notes:

a_{ij} = The average value of the opinion of expert

z_n = Expert opinion

n = Total expert

Table 10. Pairwise comparison matrix.

	Alternative 1	**Alternative 2**	**Alternative 3**	**Alternative 4**
Alt. 1	1	0,867	0,454	3,784
Alt. 2	1,153	1	0,837	4,587
Alt. 3	2,201	1,195	1	5,002
Alt. 4	0,264	0,218	0,200	1
Σ	4,619	3,281	2,491	14,373

After pair wise comparison matrices are made, and then the value is summed per column for subsequent normalization. The results of normalization then added per line, to get the Priority (Priority Vector) value of each Alternative (table 11).

Table 11. Alternative Priority Value.

	Alternative 1	**Alternative 2**	**Alternative 3**	**Alternative 4**	**Priority Vector**
Alt. 1	0,217	0,264	0,182	0,263	0,232
Alt. 2	0,250	0,305	0,336	0,319	0,302
Alt. 3	0,477	0,364	0,401	0,348	0,398
Alt. 4	0,057	0,066	0,080	0,070	0,068
Σ	1,000	1,000	1,000	1,000	1,000

Consistency Index gets value $0.01 < 0.1$. While the ratio of consistency with random index value for 4x4 matrix of 0.9, gets the value of Consistency Ratio $0.011 < 0.1$. From the result of consistency check, it can be concluded if the opinion of the informants related to alternative selection is consistent. Furthermore, the priority level of Alternatives can be ranked from the highest percentage to the lowest as follows:

a. Alt. 3 Centralized parking management 40 percent
b. Alt. 2 Increase in non-motorized infrastructure investment 30 percent
c. Alt. 1 Reduction of travel distance 23 percent
d. Alt. 4 Hourly parking fee 7 percent

4. Conclusion

The total emissions for Wednesday are 319.79003 Kg of CO_2, 0.02316 Kg CH_4, and 0.00278 Kg of N_2O. While total emission of CO_2 for Saturdays are 187.42971 Kg, 0.01353 Kg CH_4, and 0.00162 Kg of N_2O. Alternative sequences of CO_2, CH_4 and N_2O emissions reductions from motor vehicles at the State University of Malang are centralized parking management, additional non-motorized infrastructure investment, reduced travel distance and hourly parking fees.

5. Acknowledgements

Special thank to Galih for helping to collect practical data, improve the quality of the work and all of EIIS Laboratory Members for their support and kindly help.

References

[1] United Nations 1998 Kyoto Protocol to the United Nations Framework Convention On Climate Change.

[2] Regulation of the President of the Republic of Indonesia Number 61 Year 2011 on National Action Plan for Green House Gas Emission Reduction (RAN-GRK). Jakarta.

[3] Ministry of Energy and Mineral Resources 2012 Greenhouse Gas Emission Study Transportation Sector. Jakarta: Ministry of Energy and Mineral Resources.

[4] Regional Action Plan for Greenhouse Gas Emission Reduction East Java Province 2012 www.malangkota.go.id

[5] Central Bureau of Statistics 2014 Malang City in Figures 2014 Malang: Statistics Center of Malang City.

[6] Academic Bureau of Student Affairs, Planning, Information, and Cooperation (BAKPIK) State University of Malang 2015 Statistics of State University of Malang Year 2010-2014 Edition 2015. Malang: State University of Malang.

[7] Tamin O Z 2007 *Transport Planning and Modellin.* Second Edition Bandung: Bandung Institute of Technology.

[8] Intergovermental Pannel on Climate Change 2006 IPCC *Guidelines for National Greenhouse Gas Inventories* Volume 2 Energy. Japan: Institute for Global Environment Strategies (IGES).

[9] Mateo J R S C 2012 *Multi Criteria Analysis in the Renewable Energy Industry.* London: Springer-Verlag London.

[10] Department for Communities and Local Government 2009 *Multi Criteria Analysis: A Manual* London: Communities and Local Government Publications.

[11] Saaty T L 1980 *The analytic hierarchy process.* United States of America: McGraw-Hill.

[12] Hofstrand D 2008 *Liquid Fuel Measurements and Conversions File C6-87.* Iowa: Iowa State University.

[13] Indartik 2009 Determinants of Successful Implementation of Reduced Emissions from Deforestation and Forest Degradation: A Case Study of Riau. *Journal of Social Research and Economic Forestry. 6 (2): 83-98.*

[14] Circular Letter State University of Malang Number: 299 / UN32.II / KP / 2014 About Day and Working Hours.

IOP Conf. Series: Earth and Environmental Science **148** (2018) 012022 doi:10.1088/1755-1315/148/1/012022

Effect of urbanization activities towards the formation of urban heat island in Cameron Highlands, Malaysia

M H Ibrahim[1], N A A Latiff[1], K Ismail[1] and N K M Isa[1]

[1]Department of Geography and Environmental, Faculty of Human Science, Sultan Idris Education University (UPSI), 35900 Tanjong Malim, Perak, Malaysia

hairy@fsk.upsi.edu.my

Abstract. This study carried out to study the effect of urbanization activities towards the formation of Urban Heat Islands (UHI) in Cameron Highlands (CH).The aim of this study is to identify the formation of UHI in CH following the urbanization activities. This study also involved two main data that are primary through field survey and secondary data from collection data. In addition, this study was used qualitative and quantitative method. The data was taken two times a day, at a day and night between the hours of 12:00 to 14:00 and 19:00 to 21:00. Data in this study analyzed by using correlation analysis and analysis of Geographic Information Systems (GIS) which known as interpolation. Result found the formation of UHI in CH was concentrated at city centre namely in Pekan Tanah Rata. From the whole average value, city centre was recorded the highest reading of temperatures which is 30.5°C, while reading of temperature for subtown and suburban was recorded 28.6°C and 23.8°C. Average of humidity in CH during the day was recorded as highest reading in the suburban area namely Tringkap Bee Farm (station 11) which is 58.4%. Then, average of humidity in CH at night was recorded highest reading in the suburban area namely Habu Mini Market 91 (station 9) which the value is 83.2%. The reading of wind speed in CH during the day recorded the highest reading day at the suburban namely Tringkap Bee Farm (station 11) which the value is 2.3 m/s. While, following the reading of wind speed in CH at night, suburban namely Habu Mini Market 91(station 9) was recorded the highest reading which is 0.8 m/s. The Intensity of Urban Heat Island in CH during the day was recorded 2.8°C, while at night intensity of UHI was recorded 1.4°C. Overall, the urbanization activities in CH had caused the formation of UHI. Therefore, measures of legislation such as protect forest from development by control the urbanization activities need to be implemented so that the formation of UHI can be reduced and overcome.

1. Introduction

Urban Heat Island (UHI) is one of the most frequent phenomena in the city, especially in urban areas where the region experiences the fastest and most urbanization process. The formation of UHI phenomena also occurs as a result of the rapid urbanization process associated with anthropogenic factors as stated in the study of Wong [18]. Various urban microclimate parameters varied as noted by Botkin and Keller study [2] [12] according to space and time as a result of the rapid urbanization process. Changes in urban microclimate parameters due to the urbanization process are closely related to urban morphology and urban physical properties.

The morphology of the city can be linked to the surface of the city where the city on the earth surface is stretched and concretized, the ability to absorb heat energy is high enough to be filled with rock 'jungle' which able to radiate heat to the environment [2]. The formation of UHI exists in every city in the world which is considered a manifestation of urbanization and urban growth [4] [8]. The formation of UHI is also a result of human activity in urban areas. Hence, urban growth will encourage human activities in urban areas.

Urbanization activities will not only have a negative impact on the environment but will affect the health of the city's ecosystem [3] [10] [16]. Rapid urbanization will further promote the use of more modern and sophisticated technologies. The use of modern and sophisticated technologies not only speeds up urban development but also affects the quality of the environment as the technology produces various types of pollutants that will be released into the city's atmosphere.

Therefore, steps to reduce the formation of UHI should be identified so this phenomenon will not have a negative impact on future generations. Thus, the impact assessment of urban activity on the establishment of UHI was carried out in the study area which is Cameron Highlands. This study involves the collection of daily temperature within five days from 27 January 2017 to 31 January 2017 covering the selected area of study. The data were analyzed to detect the temperature distribution in the study area and its relation to the formation of UHI.

2.Study Area

In this study, several study areas were selected by stations (table 1). The selection of the stations in Cameron Highlands is based on urban activities that include urban, suburban and rural area. The study involved 13 observation stations covering 5 urban areas, 3 suburban areas and 5 rural areas (figure 1). Station 1 is located at Cactus Point, an area that carries out cactus flower business activities. According to the Cameron Highlands District Council [5], although this station is less focused by tourists as well as locals, the location of this station is close to the city center resulting in higher temperatures recorded at this station than the station in the rural area. Next, station 2 is located at Hotel Kea Valley. This station is a tourist attraction and traders and station 3 is Water Cress Valley an area that runs agricultural activities.

Table 1. Location of Observation in Cameron Highlands, Malaysia.

Station	Observation Area	Latitude (N)	Longitude (E)	Area
1.	Cactus Point (Ringlet)	4°30'5.1264	101°24'12.4164	Suburban
2.	Hotel Kea Valley (Ringlet)	4°30'14.5548	101°24'25.74	Suburban
3.	Water Cress Valley (Ringlet)	4°30'28.6308	101°24'59.4036	Suburban
4.	Pasar Malam Brinchang (Brinchang)	4°29'34.8864	101°23'21.3288	Urban
5.	Kok Lim Strawberry Farm (Brinchang)	4°29'49.92	101°23'42.7236	Urban
6.	Bharat Group Tea (Tanah Rata)	4°27'14.4504	101°22'8.346	Urban
7.	Terminal Freesia, Pekan Tanah Rata (Tanah Rata)	4°28'14.2834	101°22'42.0528	Urban
8.	Hotel Casa Adela Rosa (Tanah Rata)	4°29'7.0224	101°22'40.1376	Urban
9.	Habu Mini Market 91 (Habu)	4°26'0.6216	101°23'20.2596	Rural
10.	Habu Bee Farm (Habu)	4°44'6.666	101°38.9509	Rural
11.	Tringkap Bee Farm (Tringkap)	4°30'39.8052	101°25'40.9728	Rural
12.	Cameron Lavender (Tringkap)	4°31'36.7284	101°25'38.154	Rural
13.	Kedai Emas Fong Sheng (Tringkap)	4°30'56.0196	101°25'41.1924	Rural

IOP Conf. Series: Earth and Environmental Science **148** (2018) 012022 doi:10.1088/1755-1315/148/1/012022

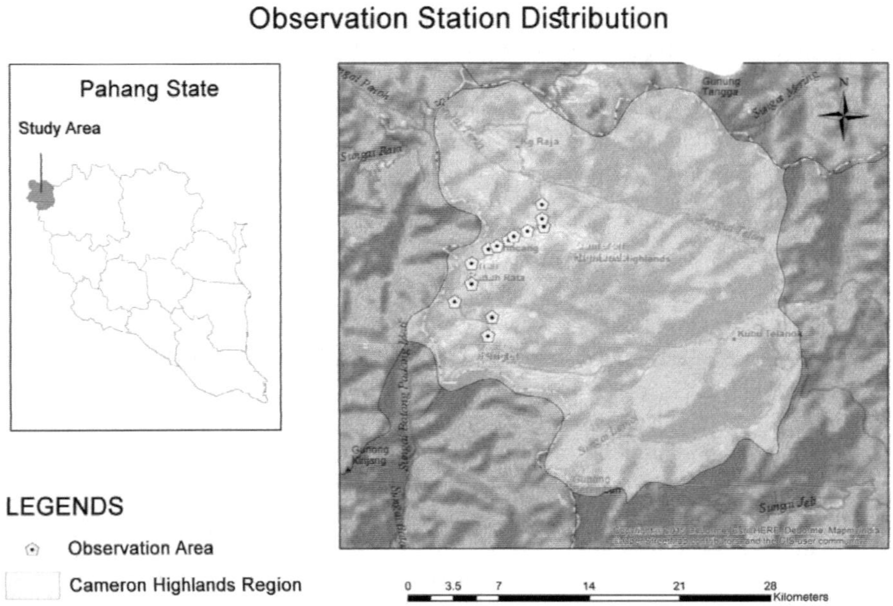

Figure 1. Study area stations.

3. Research Methodology

To carry out this study, the observation process has been conducted to create findings in this study. Subsequently, to ensure that the observation process can be performed well some of the equipment has been used. The selection of equipment to be used for this observation should also be well designed so that the selected equipment is appropriate to the aspect of the study which will be measured [1] [14]. There are two main tools that have been used in the study of UHI formation in CH which is 'Weather Skymaster' and Global Positioning Systems (GPS). The Weather Skymaster is used to record temperature readings, relative humidity and wind velocity. This tool has a reading accuracy of +/- 3 percent for wind speed, +/- 2 ° F for temperature and +/- 3 percent for relative humidity.

To complete the survey, data has been recorded for five days from 27 January 2017 to 31 January 2017. The data has been taken twice daily at 12: 00-14: 00 and 19: 00-21: 00. Weather data readings for each station are taken three times and been totaled to represent the station readings. During the observation, environmental aspects such as human activities, urban activities, number of vehicles and others are also recorded. The data in this study analyzed using correlation analysis to make the relationship between temperature and relative humidity. Then, the average temperature distribution at CH was also analyzed using the Geographical Information Systems (GIS) software which is interpolation in the form of map illustrations via Contouring Inverse Distance Weighted (IDW) method.

The production of the map will involve some of the key processes to produce output that can explain the phenomena of the formation of UHI in CH. Then, this study has also used an average temperature analysis by urban-rural temperature profile to identify the value of IUHI in CH [6] [7] [9] [17]. This temperature-stripping method involves field studies that have involved 13 observation stations which cover urban areas down to rural areas to measure temperature parameters, relative humidity and wind velocity. Hence, IUHI will be evaluated and analyzed by the following equation is derived from equation (1).

2

$$\Delta T\ u\text{-}r \quad (1)$$

Given that, ΔT : Average Temperature Change
U : Average Temperature in Urban Area
r : Average Temperature in Rural Area

4. Research Findings

In this study, the formation of UHI in CH was identified by the average temperature distribution in the area (According Table 2). Previously discussed, Station 4, 5, 6, 7, and 8 are the urban area, stations 1, 2, and 3 are suburban areas while stations 9,10,11,12, and station 13 are rural areas.

In the study on the formation of UHI in CH, the formation of UHI in urban area of CH was identified by the average temperature distribution in the area. As shown in previous chapters, Station 1, 2 and 3 are suburban areas, stations 4,5,6,7, and 8 are central downtown which categorized as urban areas, while for stations 9,10,11,12, and 13 is rural area. Through the analysis, the results showed that the formation of UHI was valid in CH. The formation of UHI at CH during the day was formed in the downtown area compared to suburban and rural areas (see figure 2a).

Table 2. Average temperature, relative humidity and wind velocity in 13 observation area in Cameron Highlands.

Station	Location	Average Temperature (°C)		Average Relative Humidity (%)		Average Wind Velocity (m/s)	
		Day	Night	Day	Night	Day	Night
1	Cactus Point	24.5	19.8	51.7	75.0	0.7	0.2
2	Hotel Kea Valley	26.9	20.5	47.3	65.8	1,1	0.3
3	Water Cress Valley	25.2	18.8	52.3	68.0	0.9	0.4
4	Pasar Malam Brinchang	28.7	22.3	53.3	64.4	1.3	0.2
5	Kok Lim Strawbwrry Farm	27.2	19.8	57.8	70.2	1.9	0.5
6	Bharat Group Tea	28.9	19.5	56.0	65.8	0.6	0.3
7	Pekan Tanah Rata	30.0	22.9	48.6	63.0	0.8	0.1
8	Hotel Casa Adela Rosa	28.6	18.7	57.3	67.8	0.6	0.4
9	Habu Mini Market 91	24.3	16.6	56.4	83.2	1.7	0.8
10	Habu Bee Farm	25.7	17.2	55.0	75.9	2.3	0.7
11	Tringkap Bee Farm	25.0	17.6	58.4	78.8	1.3	0.5
12	Cameron Lavender	24.8	18.3	55.4	82.3	1.2	0.4
13	Kedai Emas Fong Sheng	27.2	21.5	52.7	65.3	0.5	0.2

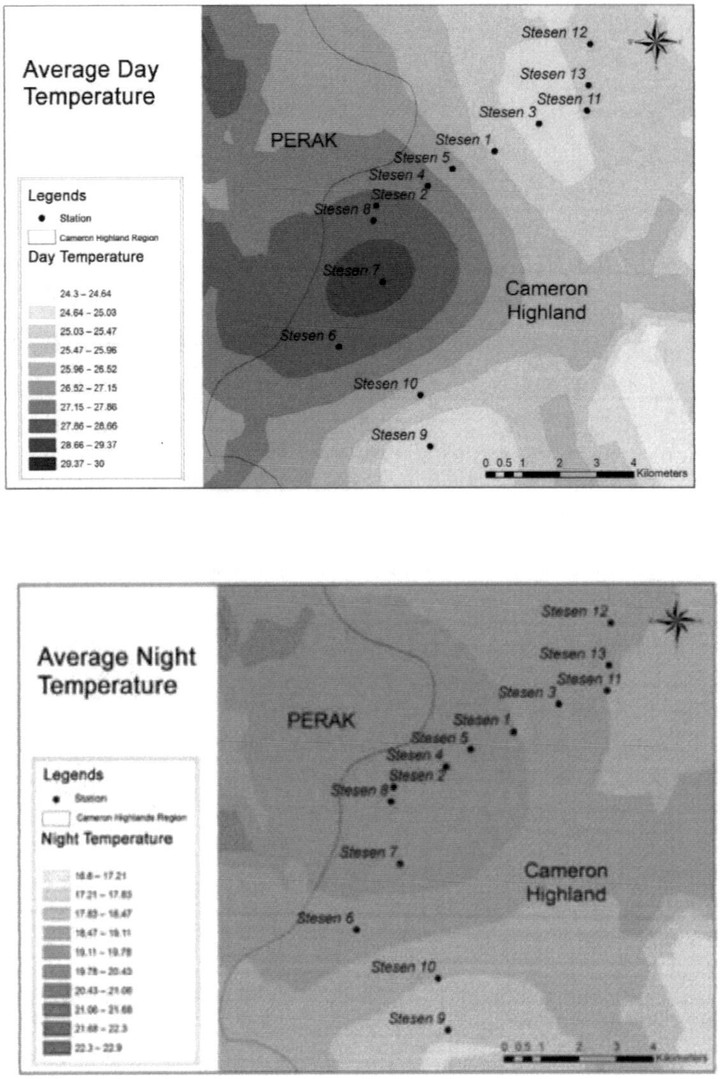

Figure 2. Average temperature distribution in Cameron Highlands using IDW contouring techniques (a) Day time, (b) Night time.

However, the station that shows the highest average temperature distribution in the downtown area is station 7. Station 7 is Tanah Rata Town, Freesia Terminal. This station is the highest area of potential for temperature increase and formation of UHI. The situation clearly demonstrates that UHI phenomena are concentrated in the downtown area at night [13] [15]. Station 7 is also an area that carries out various urban activities and transportation activities. This station is also tourist attraction area until evening. Hence, tourist density and vehicle congestion in the area led to the formation of UHI at night. Next, the average temperature distribution at CH at night shows that the highest reading area in the city center is shown with a ton of dark colors which presented in the temperature distribution map (see figure 2b).

The average temperature distribution analysis in CH shows that the influence of urbanization due to human activity during day and night plays a huge role in establishing a high average temperature distribution especially in downtown areas [6] [27]. Such a situation can be demonstrated through the analysis of the temperature distribution that has been produced and field observations on the dominant urban activity in the downtown area. The downtown area has the highest average temperature compared to suburban and rural areas.

4.1 Cross profile of temperature in Cameron Highlands

The cross profile of the temperature has been developed to determine the formation of UHI in CH. Hence, this temperature cross profile is constructed to make the comparison of the temperature difference between urban, suburban and rural areas (see figure 3). Among the three focused areas, the temperature difference has been detected between the downtown area and its surroundings. This situation was supported by a study conducted by Tan where the temperature difference is about 1-2 ° C between the industrial area and the village area [13]. Hence, the difference in temperature between the surrounding areas proves the existence of UHI formation during the day and night in CH.

The anthropogenic factors primarily from urban activity have led to the formation of UHI in CH [11]. This is due to the rapid urbanization process especially in human activities such as trading, business, transportation and activity which involves the production of large quantities of heat that have caused a high temperature distribution over the surrounding area. For the suburban areas, it is usually free from the influence of human activity and this condition is also characterized by a green area that indicates the 'normal' temperature.

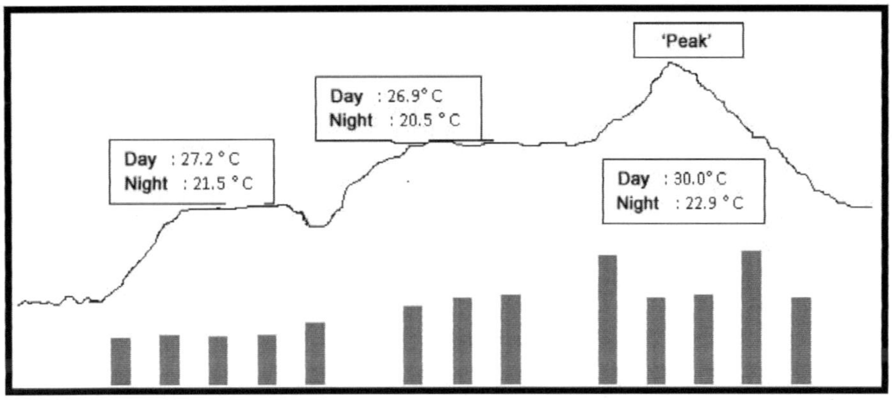

Figure 3. Cross profile of temperature between urban, suburban and rural areas in Cameron Highlands.

Table 3. Station between urban, suburban and rural areas in Cameron Highlands.

S*	Rural	S*	Suburban	S*	Urban
9	Habu Mini Market 91	1	Cactus Point	4	Pasar Malam Brinchang
10	Habu Bee Farm	2	Hotel Kea Valley	5	Kok Lim Strawberry Farm
11	Tringkap Bee Farm	3	Water Cress Valley	6	Bharat Group Tea
12	Cameron Lavender			7	Pekan Tanah Rata
13	Kedai Emas Fong Sheng			8	Hotel Casa Adela Rosa

S = Station*

5. Conclusion

In conclusion, the analysis of the findings shows a significant temperature increase in CH and caused the formation of UHI in CH. This increase in temperature is the result of rapid urbanization activity in CH especially in human activities. The rapidly-evolving urbanization process in CH has led to the production of heat resulting in the formation of UHI. IUHI value for urban and rural area during daytime is 2.8 ° C while at night is 1.4 ° C. IUHI during the day has shown that high temperature increases will affect the comfort of residents and tourists in CH.

For rural areas in CH such as the Habu Bee Farm is relatively less urbanized as it does in the city center of CH. The area is also less saturated with buildings that are capable of blocking the wind. The rural area also has more green areas compared to the downtown area. The area also more spacious and airy and less crowded with residents or tourists. Hence, temperatures in the rural areas are low and do not experience dry climate like in downtown. However, CH's urbanization activities should be controlled and mitigated in order to reduce the impact of climate change on the comfort of residents and tourists in CH.

Acknowledgements

The authors would like to acknowledge the support of the Malaysia Meteorology Department and Universiti Pendidikan Sultan Idris, Malaysia.

References

[1] Ismail A I Ahmad S Hashim N M and Jani Y M 2011 Kesan penyahhutanan ke atas pola suhu dan kelembapan bandingan di Cameron Highlands, Malaysia: Satu analisis awal Malaysia *Journal of Society and Space* 7. Issue 3: :56 - 65.

[2] Botkin D B and Keller E A 2005 *Environmental science: Earth as a Living Planet.* (New Jersey: John Wiley).

[3] Douglas I 1983 The Urban Environment. London: Edward Arnold Environmental Protection.Agency.(2014).http://www.epa.gov/heatisland/mitigation/greenroofs.htm.

[4] Landsberg H E 1981 *The Urban Climate* (New York: National Academic Press).

[5] Majlis Daerah Cameron Highlands 2016 *Rancangan Struktur Tempatan Cameron Highlands 1995- 2020* (Cameron Highlands, Pahang).

[6] Ibrahim M H Ismail K and Nazari S S 2014 Kesan Pembandaran terhadap Pembentukan Pulau Haba Bandar di Kuala Terengganu. *Jurnal Geografi* Vol 2 (2) : 1 - 13 .

[7] Ibrahim M H Zulkifli M R Ismail M I M Isa N K M and Adnan M 2016 Kesan pembandaran terhadap taburan suhu di Malaysia: Kajian kes di Rawang, Selangor *Malaysian Journal of Society and Space 12* Issue 5: (8393) 83.

[8] Hanafiah N A H 2016 *Pengaruh pulau haba bandar terhadap keselesaan terma dalaman bangunan untuk pekerja di Pulau Pinang* (University Science Malaysia).

[9] Oke, T. R. (1987). Urban Heat Storage Derive as Energy Balance residuals. *Boundary–layer. Meteorology*. Vol. 39: 233-245.

[10] Peterson P J Sani S and Nordin M 1999 *Indicators of sustainable development in industrialising countries. Vol. III: Key Indicators for tropical cities* (LESTARI Monographs on Environment and Development. Bangi: LESTARI, UKM).

[11] Maru R and Ahmad S 2011 *Perkembangan Pulau Haba Bandar di Bandaraya Jakarta, Indonesia.* (Kertas kerja yang dibentangkan di Persidangan Kebangsaan Geografi dan Alam Sekitar kali ke-3. Tanjung Malim: Universiti Pendidikan Sultan Idris. [10 Februari 2011]).

[12] Seto K C and Shepherd J M 2009 Global urban land-use trends and climate impacts. *Current Opinion in Environmental Sustainability* 1: 89-95.

[13] Ahmad S Hashim N M and Jani Y M 2009 Fenomena Pulau Haba Bandar dan Isu Alam Sekitar di Bandaraya Kuala Lumpur. Malaysia. *Journal of Soceity and space.* 5(3),57-67.

[14] Sani S 1972 Some aspects of urban microclimate of Kuala Lumpur, West Malaysia. *Akademika* 1:85-94.

[15] Streutker D R 2003 *A Study of the Urban Heat Island of Houston, Texas*. (Ph.D Thesis. Rice University).

[16] United Nations 2008 *World Urbanisation prospects: The 2007 revision population database.* (Departments of Economics & Social Affairs, Population Division).

[17] Voogt J 2014 *How Researchers Measure Urban Heat Islands.* (Available at: http://www.epa.gov/heatislands/resources/pdf/EPA_How_to_measure_a_UHI.pdf (Retrieved on 28 April 2017).

[18] Wong N H Jusuf S K Win A A L Thu H K Negara T S and Xuchao W 2007 Environmental study of the impact of greenery in an institutional campus in the tropics. *Building and Environment* 42: 2949-2970.

Rainfall pattern variability as climate change impact in The Wallacea Region

I Pujiastuti[1] and E Nurjani[1]

[1]Departement of Environmental Geography, Faculty of Geography, Universitas Gadjah Mada

isnapujiastuti@gmail.com

Abstract. The objective of the study is to observe the characteristic variability of rainfall pattern in the city located in every rainfall type, local (Kendari), monsoon (Manado), and equatorial (Palu). The result will be compared to determine which has the most significantly precipitation changing due to climate change impact. Rainfall variability in Indonesia illustrates precipitation variation thus the important variability is the variability of monthly rainfall. Monthly precipitation data for the period of 1961-2010 are collected from Indonesian Agency for Meteorological, Climatological, and Geophysical Agency. This data is calculated with the normal test statistical method to analyze rainfall variability. The result showed the pattern of trend and variability of rainfall in every city with the own characteristic which determines the rainfall type. Moreover, there is comparison of rainfall pattern changing between every rainfall type. This information is useful for climate change mitigation and adaptation strategies especially in water resource management form precipitation as well as the occurrence of meteorological disasters.

1. Introduction

Climate change is one of the important issues nowadays. Climate change resulted in various impact on human life. The most often studied impact regarding the climate change is increasing in global temperature. The temperature is studied very detailed in the term of climate change. In a tropical area like Indonesia, precipitation had a more significant impact on climate change rather than the changing temperature. Rainfall can indicate climate change [6] because precipitation has a direct impact on nature environment that implicate economical sector. The most common precipitation in Indonesia is rainfall. Rainfall plays a core role in every agricultural activity in Indonesia. Moreover, as the basics need for the human with such huge number of population in Indonesia made the detailed rainfall variability and trend study important.

Rainfall variability may cause some problems such as the flood and drought. Wallace city, the cities in Indonesia between Wallace Line and Weber Line, was one of the vulnerable areas with this problem. Manado suffered in may severe flood with periodically in 2000, 2006, 2012, 2013, and 2014, while the biggest at 15 November 2014, 58 households affected their houses were flooded as high as 50-100 cm. Flood often occurred in Kendari, there are 19 floods recorded from 1999 until 2016. The recent flood in Kendari in 2017 caused 25 people evacuated. Hydro-meteorological hazard's priorities in Kendari are a flood and extreme weather. Palu is often flooded, one of them on 25 August 2012 affected two people died, four people washed away, and 32 household or 101 people evacuated. Another loss 9 house washed away, 3 houses heavily damaged, 200 houses lightly damaged, and 25 meters river embankment perforated [4]

The Wallace Region, region between Wallace line and Weber line is chosen because there we can found the complete rainfall type in Indonesia. Rainfall type in Indonesia is: equatorial type with two peaks of rainfall, the monsoonal type that influenced by the Southeast Hindia Monsoon, and local type with very high local influence. Palu station was chosen as the representative of equatorial rainfall type, whereas Manado was chosen as the representative of monsoonal rainfall type, and Kendari was chosen as the representative of local rainfall type. The reason for choosing these station was the station is located in same island, Sulawesi or Celebes Island. The detailed knowledge of rainfall regimes on time series data are important prerequisites for enhancing the water management, planning, and design the buildings and facilities also mitigate the negative impact of drought and floods [1].

Average change of annual and seasonal rainfall data represented the modification of climatic variability in future climate change. Annual rainfall data depicted each year data of rainfall in accumulative. The annual rainfall data were chosen as the fundamental of scientific thought that annual variability is better (than mean value) and one of the most important indicators of reliability of rainfall [5] Monthly rainfall data depicted the availability water in a month that is very important to determine the season and agricultural commodities.

This study aims to analyze the variability of monthly and annual rainfall in 40 years from each rainfall type in Indonesia. The data used as a tool to characterize which rainfall type have most variability due to climate change issue. This study expected as the reference of detailed rainfall study and basic study of climate change impact on the specific rainfall type for further research in rainfall and climate change.

2. The Method

2.1. Study Area

Wallace line is imaginary the faunal boundary in Indonesia separates the Indo-Malaya or Asian-Like ecology and Australian ecology. This line lied in between the Sulawesi Island (figure1). This island is special because we can find the all the rainfall type in one island. The other reason is that the three selected area had suffered from hydro-meteorological disaster, drought, and flood.

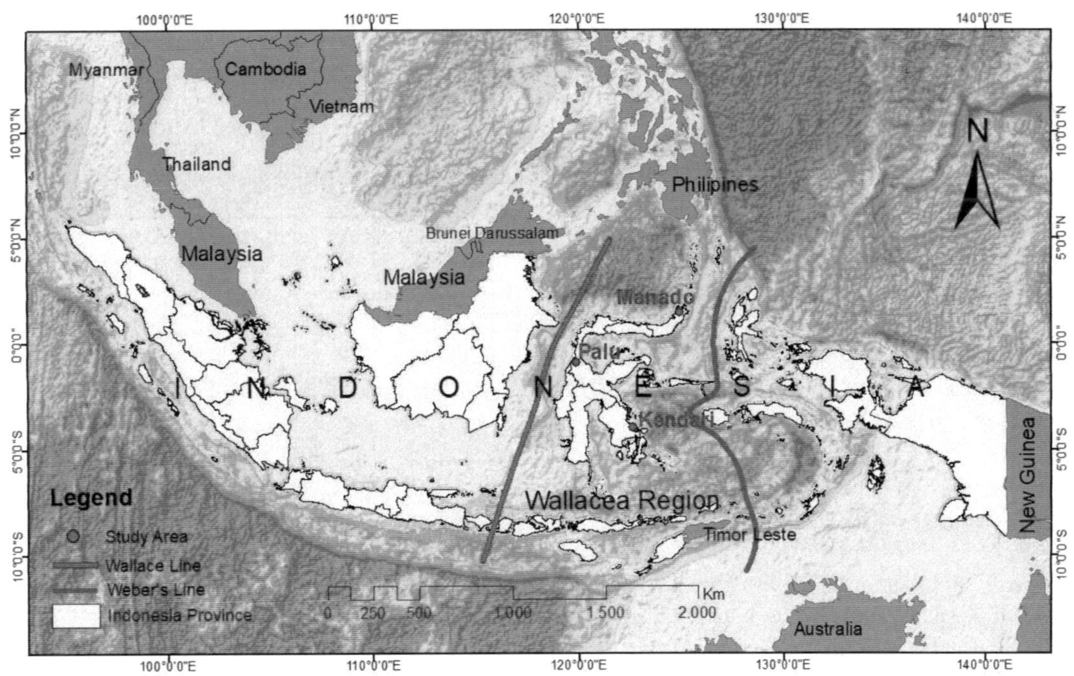

Figure 1. Wallacea Region and Study Area.

Manado is the capital city of North Sulawesi Province which has the monsoonal rainfall type. Manado city is located in 10 30'-10 40' north and 1240 40'-1260 50' west. Palu is the capital city of Central Sulawesi. Palu located in $0^0$36'-$0^0$56' South and $119^0$45'-$121^0$1' East. This city lies in the equator line; the rainfall type is equatorial rainfall type. Kendari Municipality is the capital city of Southeast Sulawesi. Kendari lied in 3°58′2.96″ north 122°35′40.92″ south. Kendari had the local rainfall type.

2.2. Material and Method
Data was collecting from Indonesian Agency for Meteorological, Climatological, and Geophysical Agency [3]. The data used is monthly rainfall data for 41 years (1960-2010) in three cities that have their own type of rainfall. The city was picked up because the Wallace area is the border of the two ecological zone of Indonesia, near the Wallace line. Besides, the city was suffered from hydro-meteorological disaster. Moreover, the area located in common island and the rainfall type is represented all the rainfall type in Indonesia. The rainfall type is equatorial rainfall type, monsoonal rainfall type, and local rainfall type. The island mentioned before is the Sulawesi Island. Manado was chosen as the data representative of monsoonal rainfall. Palu was chosen as the data representative of equatorial rainfall type rather than Gorontalo because the data is more complete. Kendari was chosen as the data representative of local rainfall type.

The data is categorized as the monthly rainfall data and annual data. The annual data was chosen as the variability of annual rainfall while the monthly data depicted how the season and weather changing over the years. The variability for every rainfall type showed which type of rainfall has the most variability during the climate change issue. The data is presented in the monthly rainfall data in a year. The data showed the rainfall pattern in a year.

2.3. Analysis
The rainfall variability calculated with statistical method, to analyze rainfall variability. The basic statistical analysis used in this study is measure of central tendencies (mean and mediation), measure the dispersion (standard deviation), and the degree of symmetry in the distribution (Skewness, SK), the degree to peak of data (Kurtosis/KT), and coefficient of variation (CV). The rainfall trend also applied using the moving average 3, 5, and 10 years and used the simple regression equation. The regression using time as independent variable and monthly rainfall as the dependent variable [8]. The result of the calculation showed with graph and table to describe better the founding.

Table 1. The information about station data length.

Nr.	Station	Coordinate	Rainfall Type	Data lenght
1	Palu	00°41' S - 119°44' E	Equatorial	1960-2010
2	Manado	01°32' N - 124°55' E	Monsoonal	1960-2010
3	Kendari	04°06' S - 122°26' E	Local	1960-2010

3. Result and Discussion

3.1. Rainfall Pattern
The research result showed the pattern of trend and variability of rainfall in every city with the own characteristic which determines the rainfall type. The rainfall type is characterized by the monthly rainfall pattern of peak rainfall. Sun radiation controlled these pattern showed in figure 2.

Palu has the equatorial rainfall type which has two peaks of rainfall that occurred after the equinox. This is manifestation as Palu located close to the equator. The peaks occurred after the sun across the equator line which happened on 21 March and 23 September [2]. However, in this case, Palu has fluctuated pattern. There is only one peak was distinct which form a hill from April until October. The fact that the peak on June-July which the sun in Summer Solstice. The rest monthly rainfall has the fluctuate such as increasing in November, decreasing in December, and the interesting thing is in January the first two decades was increasing but much decreasing in last three decade. In January in the third decade was the less rainfall in these time series (figure 2a).

The fourth decade (1991-2000) showed that the rainfall much more than the other decade in the peak of the season, May, June, and July, also in the low rainfall still the highest in October and December. Other than that, this decade is the most fluctuating monthly rainfall. The only reason was the occurrence of severe El Nino hit Indonesia around 1997. The maximum monthly rainfall recorded in 51 years was 286 mm/month and the minimum was 0 mm/month. The average range is 32-96 mm/month.

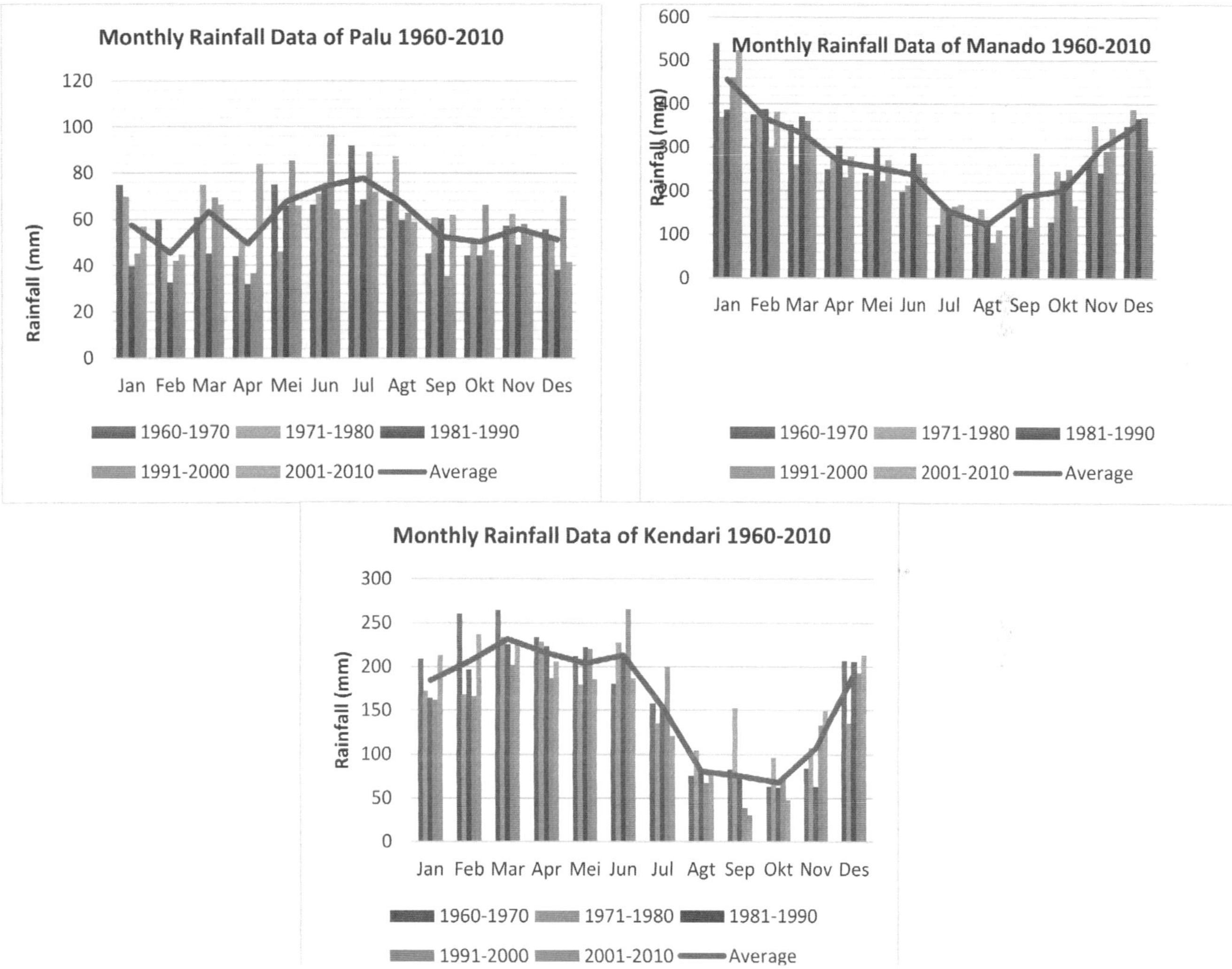

Figure 2. Average Monthly Rainfall in (a) Palu, (b) Manado, (c) Kendari.

Manado monthly rainfall presented in figure 2b. has the monsoonal rainfall type. The monsoonal rainfall type has the V monthly rainfall pattern [2]. The peaks are in January and the minimum in June, July, and August. This data was shifted a little bit to July, August, and September. The second decade showed less rainfall than another decade. The maximum monthly rainfall recorded in 51 years was

1552 mm/month occurred in August 2005 and the minimum was 0. The average range is 123-456 mm/month.

Kendari has local rainfall type which means the local condition control more than other. The pattern is seen as monsoonal rainfall type. But, the difference is the sharp graph resulted in local type. The rainfall drastically decreases in August and increasing in November. The maximum monthly rainfall recorded in 51 years was 755 mm/month and the minimum was 0. The average range is 68-232 mm/month (figure 2c).

The place with highest monthly rainfall is Manado and the lowest is Palu. Monsoon wind gives more influence than the effect of location and condition such as equator and topography (local rainfall type). All of the rainfall types showed that third decade (1991-2000) is the most fluctuating monthly rainfall. Annual rainfall data from 51 years showed the fluctuation data with no certain pattern (figure 3). In general, from the individual year looks like there are increasing and decreasing pattern of annual in one cycle occurred in four years.

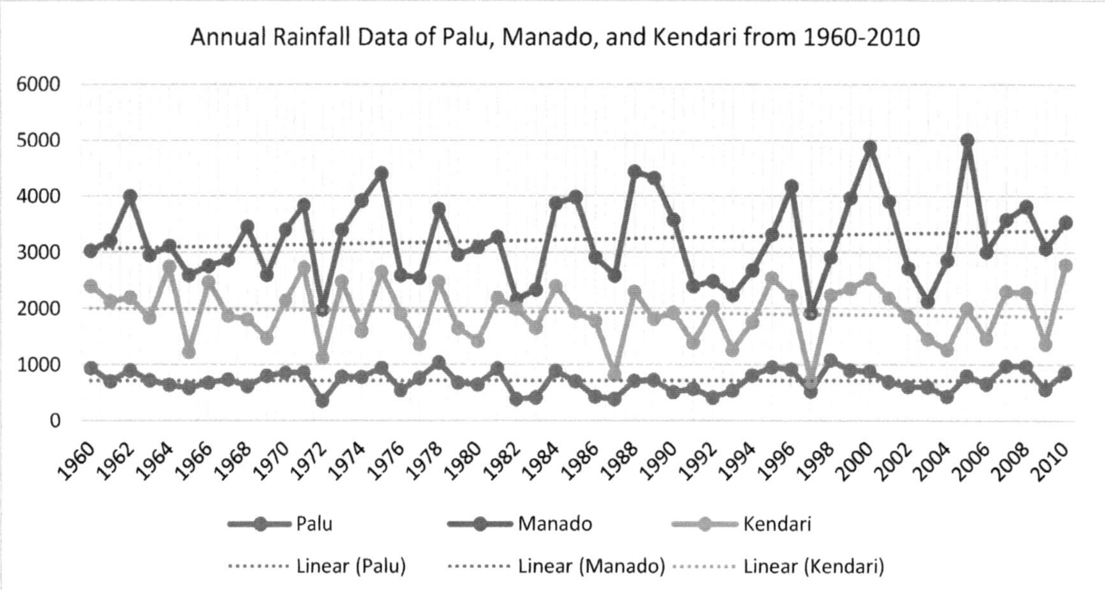

Figure 3. Annual Rainfall Data 1960-2010.

The monsoonal and local rainfall type has the likely specific pattern. The significant decreasing for the first time occurred in 1972, then in 1987, 1997 and 2003. The significant increase occurred in 1975, then 1989, 1995, 2000, and 2005. The pattern showed that the occurrence of extreme peak and base annual rainfall repeated in shorter time. The data showed the extreme peak and base graph extreme cycle first in 28 years and then the extreme reoccurred again only in 16 years. The most extreme are from 1996-2010.

Table 2. Summary Basic Statistic for Annual Rainfall (mm)

No	Location	Min (mm)	Max (mm)	$\bar{x}$	σ	CV	Skewness	Kurtosis
1	Palu	352	1081	713,1	188,0217	0,263683	-0,1368	-0,84545
2	Manado	1920	5010	3228	743,9887	0,230472	0,374267	-0,42651
3	Kendari	709	2778	1931	497,6049	0,257753	-0,36147	-0,47134

3.2. Rainfall Variability

The result of the basic statistical analysis for the parameters annual rainfall presented in table 2. Manado has the highest number of rainfall average and Palu has the lowest number of rainfall average. The standard deviation shows the heterogeneity of data compared to mean. The coefficient of variation showed the similar value from every rainfall type around 0.23-0.26 which small variability. The greater the coefficient value the rainfall variation more fluctuates [7]. Monsoonal rainfall type has the highest variability compared to Palu and Kendari.

The skewness parameter showed the positive and negative value. This means the distribution of the data is positively skewed in Manado means the tendency of rainfall above the average and negatively skewed in Palu and Kendari which mean the tendency of rainfall distribution below the average. The kurtosis showed a negative value for all the rainfall type which means flatter than normal peak distribution. The value ranges between -0.42 to -0.84. Manado and Kendari have the similar value, 0.42; 0.47 and Palu have the -0.84 kurtosis which means the peak flattest than the average.

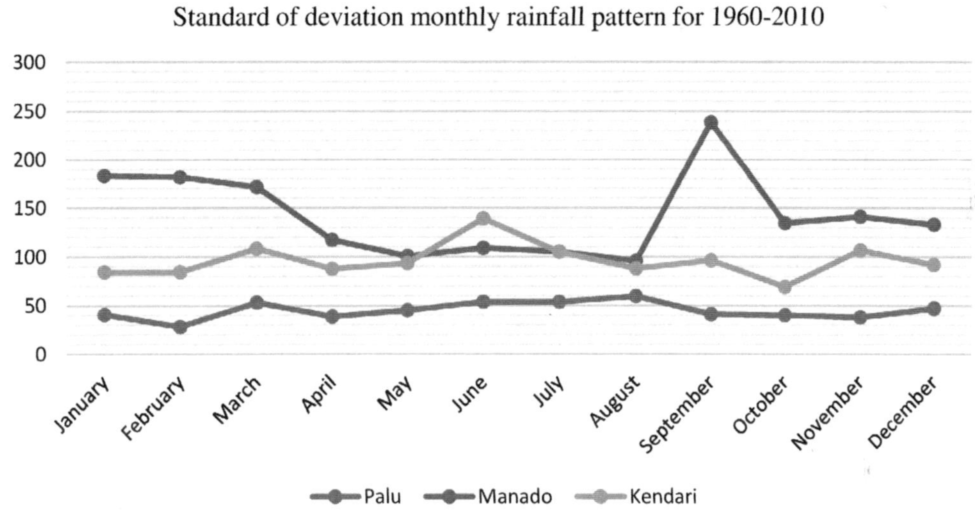

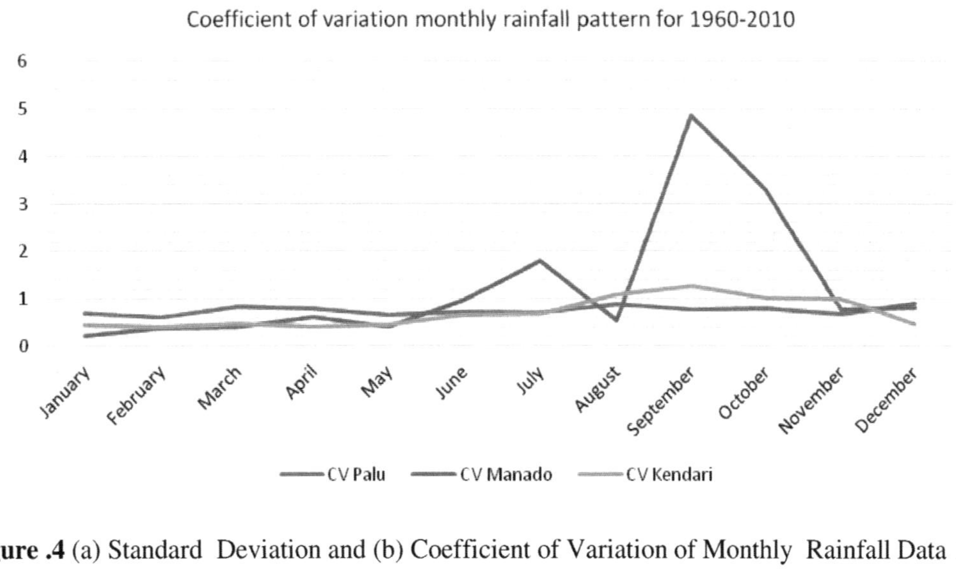

Figure .4 (a) Standard Deviation and (b) Coefficient of Variation of Monthly Rainfall Data 1960-2010.

The monthly rainfall variability is more complicated than the annual rainfall. Monthly rainfall data is extracted by making average every month in the 51 years of data. The standard of deviation monthly rainfall showed the extreme value in the Manado mean that Manado has the most heterogenic data (figure 4a). The most fluctuating data is September. On September occurred the extreme rainfall event in the fifth decade for the exactly in 2005. The data extreme standard of deviation value recorded because September is the originally dry season but in 2005 the number of rainfall as much as the wet season. January also high number but the original month is the wet season so that no many impacts.

Kendari as the local rainfall type representative showed the standard deviation extreme in June and October, where in June is the peak of value and October is the base value. The highest number of rainfall on June occurred in the fourth decade for exactly in 2000. While in October is the lowest rainfall amount. The monsoonal rainfall type has the most distributed data with the highest standard deviation of 238.1078.

The coefficient of variation showed local monsoonal type has the greatest value means the local rainfall type has the most variability (figure 4b). The fluctuation pattern for every rainfall is similar, except for December in Kendari (local rainfall type). The coefficient variation value is less than one except for July with 1.78, September 4.86, and October 3.28 in Monsoonal rainfall type. In monsoonal rainfall type and Augustus, September, and October with value 1.00-1.27. Palu and Manado have the increasing pattern in December while in Kendari is decreasing and reach the lowest coefficient of variation mean rainfall in Kendari on December is less variation. The most variability is monsoonal rainfall type in September and October. The greatest variability caused by the Sun after across the equator and strive to the south (Kendari latitude 40S). This cause greater Sun radiation makes the rainfall number greater.

The skewness and kurtosis data are presented in the graph below (figure 5). The skewness presented the positive skewed means the tendency of rainfall above the mean. The monthly rainfall pretends to increase since the skewness showed the positive value. The most increasing value is in monsoonal rainfall type in September and local rainfall type in November. Because the relative position of the sun, resulted in more precipitation in September. The anomaly of negative skewness found in Manado on December means the data was below the average. The data found in the last decade of the study period with decreasing number of rainfall in 2002 to 2004 with 110mm/month while the average is 353 mm/month. The greatest number found in Manado in September and in Kendari in November. The highest number of rainfall recorded in Kendari in November 2001 with 668 mm/month while the average is 106.7 mm/month.

The kurtosis value ranged between -0.926 to 21,434 which very extreme (Fig.5). The highest range of kurtosis value is Manado ranged between -0.926 - 21.434. Kurtosis parameter represented the peak distribution. Positive kurtosis value indicates the distribution sharper than normal peak and negative kurtosis value indicates the distribution flatter than normal peak distribution. Palu has the positive value more than negative value. The negative kurtosis value only found three months i.e. February, July, and October. These months indicate the average of data flatter than the average data. The positive kurtosis ranged between 0,150 -7,996. The highest value occurred in December means December has the sharpest peak compared the average peak.

Manado has the extreme sharp distribution in September with the highest kurtosis value 21.434. This extreme value supposes of the Sun radiation which across the equator and Manado is close to equator then the rainfall is more simulated. The negative value is dominated this rainfall type means the peak is flatter than the average. Kendari also has the extreme range from -0.471 to 15,041. The highest value indicates the sharpest peak of data occurred in November. This is because there is extreme rainfall in 2001 on October with the amount rainfall 668 mm/month while the average is 107 mm/month.

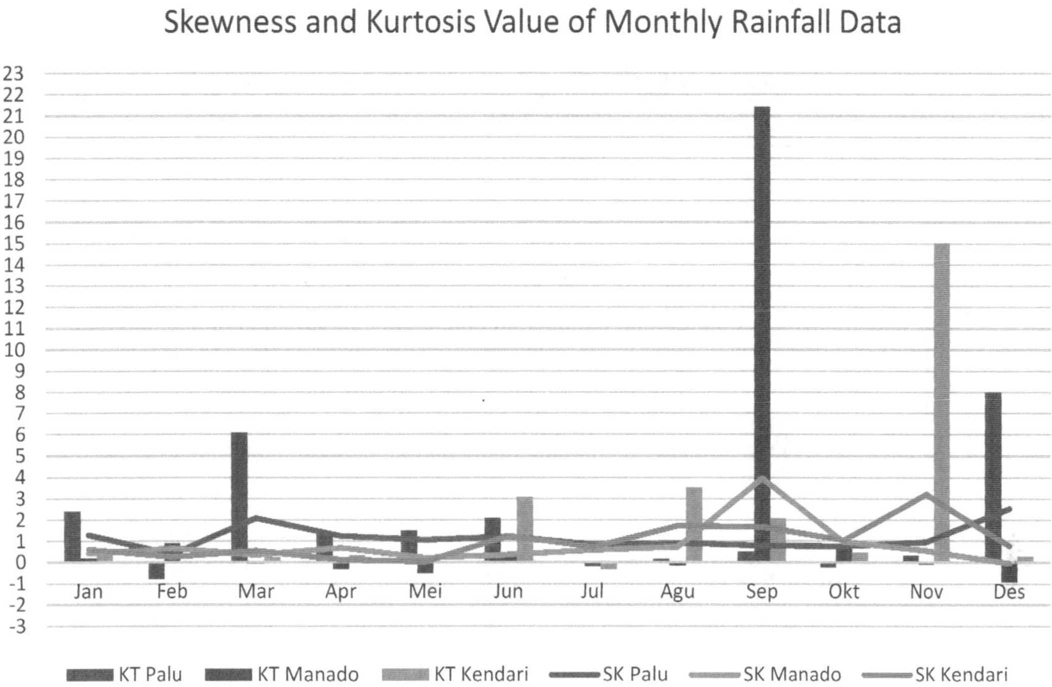

Figure 5. Skewness and Kurtosis Value of Monthly Rainfall Data 1960-2010.

3.3. Rainfall Trend

The rainfall variability can be investigated by the trend. The trend analysis for the data is applicable for the annual rainfall data and presented by 9 years of moving average. The moving average presented for each location in graph figure 6, 7 and 8. The moving average of 9 years used because the three and five years showed the similarity for each data while the ten years of moving average showed the inverse of the trend. This is because of the quality of data which cause the longer trend investigation less reliable than the shorter one. The most extreme trend is Manado which means the most variability data.

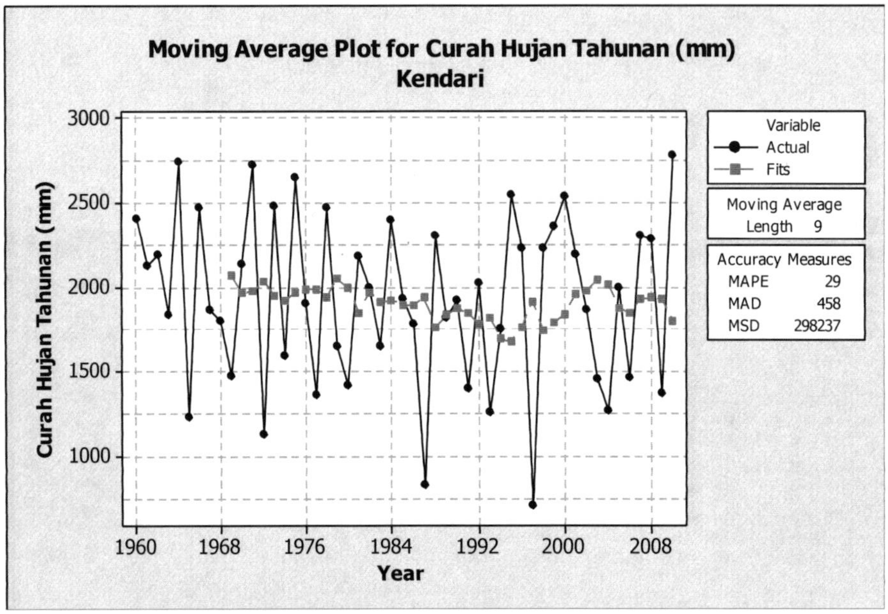

Figure 6. Annual rainfall pattern in Kendari for 1960-2010 period.

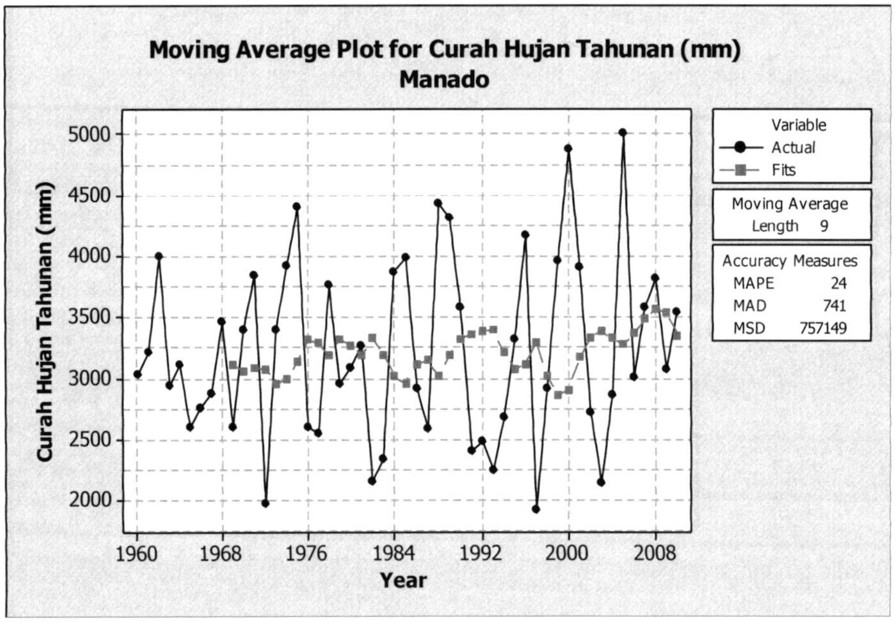

Figure 7. Annual rainfall pattern in Manado for 1960-2010 period.

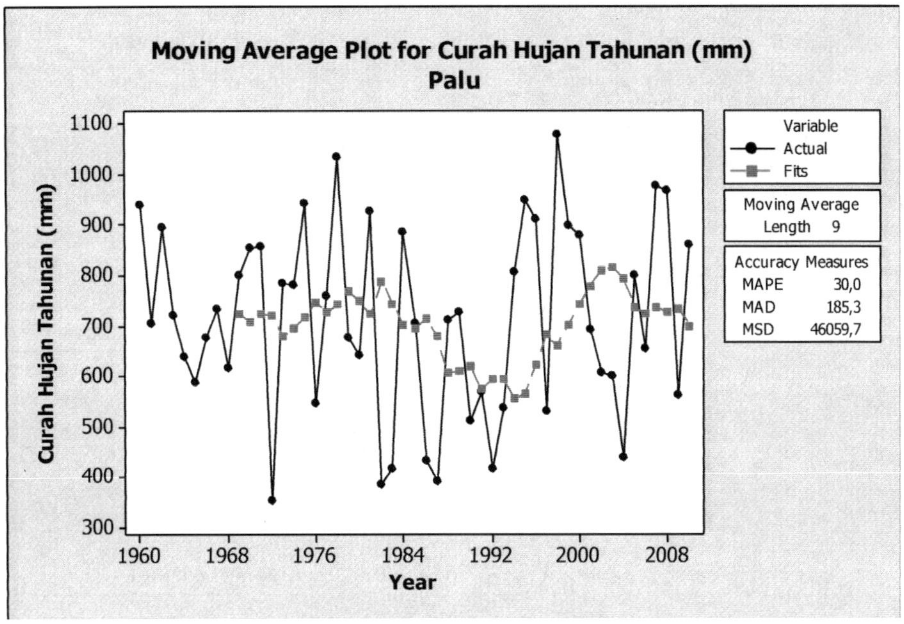

Figure 8. Annual rainfall pattern in Palu for 1960-2010 period.

Kendari showed no significant trend in the first two decades.The decreasing trend showed from 1980 to 1994 and increasing trend from 1995-2005. A full cycle of decreasing and increasing (one wave) is completed from 1975-2005. Manado has the most fluctuating trend. A cycle of increasing and decreasing always completed two decades since 1960. The amplitude of wave around 3000-3500 mm per year except for the lowest trough in 1999 and highest crest more than 3500 mm per year in 2008. These found means that Manado has the most variability. Palu has only one full cycle from 1980-2005 and before these years was no significant fluctuation.

Table 3. The standard deviation and coefficient variation (Palu).

No	Month	Max (mm)	Median (mm)	$\bar{x}$	□	CV	SK	KT
1	January	196	52	57.55	40.56	0.70	1.28	2.42
2	February	110	41	45.53	28.03	0.62	0.38	-0.81
3	March	286	53	63.24	53.06	0.84	2.07	6.09
4	April	161	41	49.41	38.47	0.78	1.24	1.35
5	May	224	57	67.71	44.79	0.66	1.05	1.50
6	June	265	63	74.24	53.48	0.72	1.21	2.10
7	July	204	62	77.73	53.53	0.69	0.84	-0.05
8	August	233	51	67.27	59.32	0.88	0.91	0.15
9	September	181	48	52.65	40.94	0.78	0.78	0.50
10	October	143	45	50.33	39.90	0.79	0.74	-0.23
11	November	153	50	55.94	37.72	0.67	0.94	0.31
12	December	252	37	51.47	46.88	0.91	2.53	8.00

Table 4. The Deviation Standard and Variation Coefficient (Manado).

No	Month	Max (mm)	Median (mm)	$\bar{x}$	□	CV	SK	KT
1	January	904	469	456.49	183.51	0.22	0.41	0.16
2	February	915	364	366.14	181.72	0.38	0.66	0.91
3	March	772	339	332.49	171.67	0.40	0.33	-0.08
4	April	533	240	267.71	117.35	0.60	0.68	-0.32
5	May	469	248	252.41	100.61	0.40	020	-0.52
6	June	558	243	236.92	109.00	0.98	0.36	0.37
7	July	419	143	153.84	105.27	1.78	0.56	-0.18
8	August	378	105	122.69	95.66	0.55	0.72	-0.17
9	September	1552	131	187.36	238.11	4.86	3.95	21.43
10	October	614	176	200.82	134.60	3.28	0.95	0.83
11	November	629	294	298.07	141.21	0.76	0.54	-0.10
12	December	581	337	353.17	132.90	0.82	-0.06	-0.93

Table 5. The Deviation Standard and Variation Coefficient of Kendari.

No	Month	Max (mm)	Median (mm)	$\bar{x}$	□	CV	SK	KT
1	January	430	183	184.24	84.10	0.46	0.61	0.33
2	February	447	213	206.25	84.12	0.41	0.24	-0.03
3	March	514	222	231.53	108.43	0.47	0.53	0.23
4	April	461	222	215.92	87.50	0.41	0.12	0.32
5	May	425	203	203.94	93.07	0.46	0.02	-0.07
6	June	755	201	213.24	139.08	0.65	1.24	3.07
7	July	396	138	154.00	105.03	0.68	0.70	-0.32
8	August	416	55	80.49	88.04	1.09	1.72	3.54
9	September	362	33	75.37	96.16	1.28	1.68	2.07
10	October	271	48	67.82	69.13	1.02	1.04	0.45
11	November	668	88	106.71	106.52	1.00	3.21	15.04
12	December	434	170	191.04	91.79	0.48	0.76	0.29

4. Conclusion

The result showed local rainfall type has the most variability of annual rainfall which the local condition influenced more by the climate change issues. The annual rainfall value showed that Palu as the representative of equatorial rainfall type has the lowest variability rainfall with the highest value of the coefficient of variation and highest (negative) kurtosis values while the skewness is the lowest. While Manado as the representative of monsoonal rainfall type has the increasing trend and highest variability and Kendari as the representative of local rainfall type has the decreasing trend. The monthly rainfall data presented that Manado as the representative of monsoonal rainfall type has the highest variability with the highest extreme trend of standard deviation as well as the coefficient of variation 0,22-4,86, the extreme range of skewness -0,06 – 3,95, and kurtosis -0.93 – 21.43 parameter. Moreover, the moving average test also showed Manado has the most extreme fluctuation trend. The summary of this study is described that the highest variability of monthly and annual rainfall is founded in Manado as the monsoonal rainfall type representative.

IOP Conf. Series: Earth and Environmental Science **148** (2018) 012023 doi:10.1088/1755-1315/148/1/012023

References

[1] Al-Houri Z 2014 Detecting Variability and Trends in Daily Rainfall Characteristics in Amman-Zarqa Basin, Jordan. *International Journal of Applied Science and Technology. Vol. 4, No. 6; November 2014.*

[2] Tjasyono B 2004 *Climatology* Bandung ITB Press.

[3] BMKG 2005-2010 *Rainfall Data*. bmkg.go.id.

[4] BNPB (Indonesia National Disaster Management Agency) 2017 *Indonesia Disaster Data and Information* (in Bahasa Indonesia). Accessed from http://dibi.bnpb.go.id/.

[5] Chakraborty S Pandey R P Chaube U C and Mishra S K 2013 Trend and variability analysis of rainfall series at Seonath River Basin, Chhattisgarh (India). *Int. Journal of Applied Sciences and Engineering Research, Vol. 2, Issue 4, 2013.*

[6] Hanaki K 2008 Global Climate Change and Cities *Urban Environmental Management and Technology*. London: Springer.

[7] Juaeni I 2006 Analisis Variabilitas Curah Hujan Wilayah Indonesia Berdasarkan Pengamatan Tahun 1975-2004. *Journal of Mathematics UNDIP*, 9, N0. 2, 2006.

[8] Setiawan O 2012 Rainfall and Temperature Variability Analysis in Bali. *Jurnal Analisis Kebijakan Kehutanan.* Vol. 9 No. 1, April 2012 : 66 – 79.

An ecohydrological-based management of Lake Beratan in Bedugul, Bali

D M Atmaja[1,3], M S Budiastuti[2], P Setyono[2] and Sunarto[2]

[1] Doctoral student of Environmental Science at Universitas Sebelas Maret, Solo City, Indonesia
[2] Department of Environmental Science, Universitas Sebelas Maret, Solo City, Indonesia
[3]Lecturers at the Geography Education Department, Undiksha, Singaraja City, Indonesia

atmajadewamade@yahoo.com

Abstract. Lake Beratan is one of waterway ecosystems located in the upper land of Bedugul, Bali and has become a tourist object which is visited by many foreign as well as domestic tourists. This is supported by a sufficiently high economic growth which, without the community's being aware of, has caused environmental problems such as the shallowing of the lake, erosion, and water pollution to such an extent that have resulted in the degradation of the function of the lake as the site of catchment. The degradation of the function of the lake can be overcome by ecohydrological-based management. This study was aimed at developing an integrated and long lasting Lake Beratan environment management concept. The study used a descriptive qualitative approach using a survey, by collecting primary and secondary data. On the basis of those data the mapping of the potentials of the lake and problems of the lake which were then integrated to formulate criteria for sustainable use of Lake Beratan waters environment resources. The determination of zonation of the lake was done based on those criteria and the community's existence consideration as well as the exising system of the lake waterway environment use. Based on the study in the field, some recommendations could be made concerning Lake Beratan waterway sustainable and integrated management.

1. Introduction

Ecohydrology studies the interaction between hydrological factors and biota components in an ecosystem and uses the interactional models as the basis for managing the environment, both the hydrological aspects such as water resource conservation, flood control and the biota aspects, especially an effort of conservation and management of the sustainable environmental productivity. Ecohydrological studies for water resource management study ecological factors that consist of vegetation system and land use in the area as the exterminating factors of water system condition in the area [1] [2].

Various ecohydrological studies for water system management environment use hydrological factors as the regulator of ecological integrity which determines the formation of various types' habitat, biodiversity, and productivity of water system environment. [3] uses ecohydrolic approach to develop the dynamic model of the habitat of fish benthic organism as the effect of water system management board action.

Lake Beratan is surface water resource which is very important ecologically, because it plays role as rain water holder in Bedugul high land. Economically, the role of the lake is very vital, since its function as water source for the people around the lake for drink, agriculture, fishery, and tourism object [4]. The presence of Lake Beratan gives an important benefit to the environment, especially as hydrological link, in addition to being the source of revenue for the regency through tourism sector. On the other hand, the increase in economic activity in tourism sector, industrial sector, and agricultural sector cause various problems, such as damage to watershed area, environmental pollution, the siltation of lake, and erosion [5]. The risky activities which cause worsening of lake water resource quality are waste dumping from hotels, households, agricultural waste, and oil waste transported by the water. In addition, there is a need for giving attention to the use of land around the water resource boundaries which cause damage to the environment [6] [7].

Lake Beratan today receive a high pressure, that is high sedimentation, domestic waste pollution and the rapid development of floating net cages fish farming [8] Based on the result of Quickbird Satellite Image analysis in 2005, the surface area of Lake Beratan was 3.827 km^2, while the result of Quickbird Satellite Image Analysis in 2014, showed that the surface area of Lake Beratan was 3.764 km^2. In the nine years time, the decrease in the surface area of the lake was 0.063 km^2 (1.65%) and automatically reduced the lake water volume because of the sedimentation. The sedimentation was caused by high rain fall, young soil structure, and damage in land use system in Lake Beratan rain fall catchment area. A further impact of the problem was damage to the lake water system habitat which caused a decrease in biodiversity [4]. The effort to save the lake water system environment needs to be done, which included the restoration of the lake water system environment as the habitat resources of various water biota and sedimentation rate control. This study was aimed to develop an integrated management concept of sustainable lake water environment

2. Methods

The study was conducted in Lake Beratan of Tabanan Regency, Bali. This study used ecological approach and survey method using instruments such as close-questionnaire. The ecological approach in this study emphasizes the interaction between human actions on the environment [1]. The study used primary and secondary data which consisted of hydro-climatological aspects, water system physical, chemical, and biological aspect, socio-economic aspect, and lake management institution. Based on the data, the mapping of potential and the problems of lake environment were conducted which were followed by the integration to formulate criteria of sustainable lake water system environment resource.

3. Result and Discussion

The sustainable management stressed the important of the efforts of creating the lake's water system resource function to support the life of the people around it (Act No. 7 of 2004). In general, the functions of Lake water system resources can be classified into two big groups which are closely related to each other, that is ecological and socio-economic functions. Ecologically, lake system is an ecosystem in an area in which ecological cycle occurs. The ecological cycle in general is a chained process that occurs among the components in the ecosystem.

The socio-economic function is the function which directly supports the life of community around the lake, such as fishery, transportation, and housing. The socio-economic function depends very much on the lake water system ecological condition. In the sustainable lake water system management concept, the importance of ecological and socio-economic functions is considered fairly and in an integrative way, in which the degree of the lake water system resources used has to be taken into account in order not to exceed the water system environment capacity.

3.1. Biophysical characteristics

Based on the result of a survey in 2015, the maximum depth of Lake Beratan was 20.0 m and the average depth was 12.8 m, while the water surface area was 3.76 km^2 and the catchment area was 13.49 km^2. With those measurements, the lake water volume was 46.22 x 10^6 m^3. The hydro-climatological data shows the relation between water surface height and rainfall in water catchment area, the annual fluctuations of water of lake Beratan in dry months were +0.37 m and -0.27 m where the fluctuation rate was 0.74 m, while in wet months the fluctuations were +0.63 m and -0.67 m or the average fluctuation rate was 1.30 m. The highest inflow which occurs in February was 1.02 m^3/sec and the lowest in September was 0.03 m^3/sec. or the average was 0.37 m^3/sec. The highest outflow occurs in February which was 0.97 m^3/sec. and the lowest in November which was 0.13 m^3/sec. or the average is 0.34 m^3/sec. The highest storage occurs in December, which was 0.50 m^3/sec and the lowest in April, which was 0.09 m^3/sec. or the average is 0.03 m^3/sec. For the detail, see figure 1.

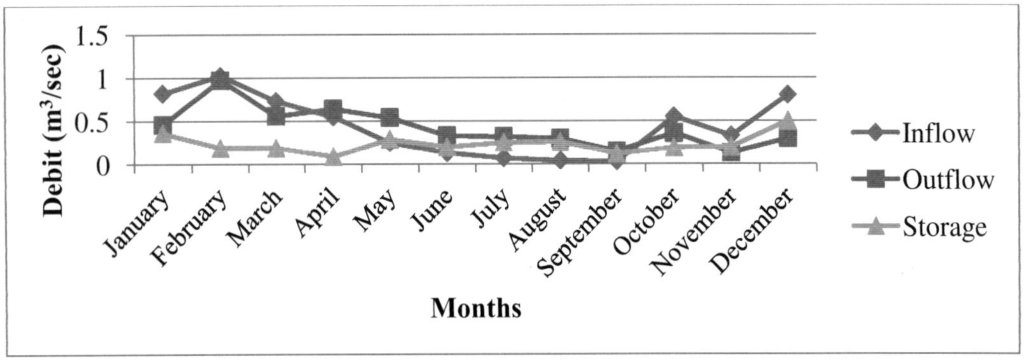

Figure 1. Inflow, outflow and storage in Lake Beratan.

The main problem of Beratan Lake water system environment is the high sedimentation rate which is 320.51 ton/year which causes the very rapid lake siltation process. Based on the 2005 and 2015 Quickbird Satellite Analysis, in ten years time, the decrease in the lake surface area was 0.09 km^2 which automatically reduces the lake volume. In plain view, the sedimentation process produces raised lands which spread in various parts of the lake water system which then stimulates the growth of riparian vegetation and the lake edge that is adapting to dry water condition characterized by the domination of wide leaf decorated grass (*Pennisetum Purpereum)*, coarse grass. While, the water turbidity level with solid suspension value reaches 18 mg/l which also causes the loss of some vegetation plants in the lake water body. Lake Beratan water fertility index shows eutrophic water system status caused by the load of organic sediment pollution from source stream through agricultural areas and residential areas around the lake. The high organic sediment pollution level is indicated from the abundance of benthic biota, especially from Tubicidae class at the lake water system bed, beside the floating net cages fish farming which also has the potential for increasing organic sediment pollution.

Based on the result of Beratan lake water quality analysis in 2015, it can be seen that the water quality is still good enough, almost all of the parameters show that the water quality standard in category A. Only Florida (F) parameter, pH and Coli bacteria show that they exceed the maximum parameters recommended, but the impact from various activities and infrastructure which have been rapidly developed around Lake Beratan have to be supervised. It does not preclude the most serious impact in the future on the existence of the lake itself. This is indicated from the examination that shows that some parameters have exceeded maximum levels. The aquatic flora and fauna of Lake Beratan generally grow as the ecosystem components of the volcanic lake, in general with low level of biodiversity. Based on the survey resulted in 2015, there are 13 species of water plants in the aquatic ecosystem of lake Beratan (*Cyperus ferax, Gygantochloa nerticilat, Panicum stagninum, Hydrocleis nymphoides, Eichomis cracifes, Panicum chamaerthaphoid, Commelina caestitosum, Polygonum caestitosum, Althenanthera sissiles,*

Althenanthera philoxeroi, Cyperus fria, Leersia hexandra, Ludwigia ascendes), 8 nekton species (*Cyprinus carpio, Puntius javanicus, Oreochromis mossambicus, Orechromis nylotucus, Clarias batrachsis, Monoptera albus, Rasbora trillineaota, Osteochilus hasselti*), 18 phytoplankton species (*Goniochloris, Osilatoria, Nostochopis, Thorea ramassisium, Tribonema, Enteromorpha, Rivolaria, Nostoc, Dinobryon, Clorobotrys, Chamaeshipon, Pediastrum, Ulotrik, Diatomae, Basicladia, Amphitrix, Gloetrichia, Goniochloris*), 9 zooplankton species (*Cyelops strenuous, Paramaecium, Euglena, Lepocinelis, Oithona, Dyatilis,Hyperia, Luciper, juvenile*), and 11 species of benthics (*Ephemera donica, Olygoceta, Lompsilis, Actinoneaceae, Anodonta, Pomecea, Balimus, Aeshna, Tryonia, Planorbis, Viviparus*).

The estimated productivity of fishery of Lake Beratan varies depending on the depth of the lake and the level of fertility of its water system. Thus, the control of the area and the depth of lake water body are very important in managing the fishery sector in the lake. Fish production of Beratan Lake in 2015 was 361.6 ton which means 57.7% of the general water system fish production. In addition, in the lake water system has also developed floating net cages fish farming which has not been harvested so that we do not know the production.

3.2. Social economic condition

The residential areas around Lake Beratan are relatively developing with good infrastructure. The lake is surrounded by three villages (Kembang Mertha, Candikuning 1, Candikuning 2). The statistic in 2015 shows the total population of three villages 4.377 with the density rate 257 people/km^2. In general, the people who live around the lake use the water directly for domestic needs through Municipal Waterworks (PDAM). The road infrastructure is generally good up to the residential area at the edge of the lake, so that the people around the lake have a very good access to various basic necessities which can be bought at Candikuning Village. Similarly, electric energy and petrol are distributed well to the residential areas around the lake. The developing infrastructure is an attraction for people who moved to the residential area around the lake. As the result, the rate of population growth will be more rapid and increase the pressure on the lake ecosystem.

The development of the residential area runs parallel to the development of agricultural tourism activities around the lake. People use the protected forest land and raised lands for agricultural field. Based on the survey in 2015, the raised lands have the area of 0.06 km^2 which have been used for growing wide leaf decorated grass as cow feed by the local people while the protected forest land of 6 Ha has been used for growing corn, chilli, string bean, banana, and cassava. The farming is done intensively as shown by the use of pesticide to control pest. The presence of the agricultural field needs to be supervised because of the presence of pesticide residue and nutrient waste which are carried by runoff entering the lake water system.

Socio-cultural survey recorded a relatively good social capital to support the lake water system environment development. This is shown by the high level of trust of the people in the government, and the high level of solidarity among community members. Similarly, there is a law or *adat* village regulation which becomes reference in managing lake water system environment, so that until now there has been no conflict among people.

3.3. Lake Beratan management strategy

The FAO (1976) concept of water resource management suggests the division of river area management into three clusters, that is village area management, watershed sub area management, and watershed area management. When it is implemented in lake water system environment management, the concept can be placed in the context of village area management or watershed sub area, in which people's participation is very important as a subject and at the same time the object of lake water system management. The lake water system management has to cover coordinating efforts to achieve sector objectives optimally by considering the holding capacity of the lake water system environment. Information knowledge about lake water system ecosystem, which includes the component structure and ecological process as well as the community social-economic are needed both for determining the limit of the environmental holding capacity and for determining the value of interest of each sector involved. For this reason, there is a need to develop the lake water system management institution which is participative in nature, in which the government through the related department or offices plays an important role as the facilitators. The actions that need to be done are:

- The empowerment of people through the socialization activities concerning the rules for managing the lake, educating people on the functions of the lake environment.
- Developing a monitoring system and evaluation of the lake environment which is integrated with the lake environment information system.
- Forming a forum for coordinative meeting which involve all the shareholders to write the framework and the structure of institutions.
- Attempting to have a legal agreement in the management which has been determined to become an Act, Government Regulation, and Legally Binding local regulation.
- Implementing the policy and strategy to achieve the objective and writing a master plan for the lake water system area, in writing the master plan, the local people have to involve to gather with the government, the stakeholders and relevant experts.

The lake water system management needs the division of water system area into zones of use which are agreed by all parties involved. In the context of integrated water management (IWRM) World Lake Forum divides the functions of lake water system into four zonings, that is, tourism, residential area, fishery and transportation, the context of sustainable development prioritizes socioeconomic functions, while the consideration from the ecological function perspective gives more stress on water resource conservation and biological resources as the integral part in the creation of the sustainability of socio-economical functioning condition.

3.4. Functions of Lake Beratan

The development of lake functions criteria is intended to optimize the use of human resource potential for the sustainable welfare of the community. In developing lake functions criteria, sector grouping is required to facilitate their application, and has to consider the relations among the sector components to maintain the continuity of the ecological process which is the basis for the sustainability.

3.4.1. Tourism. [11] states that one form of a lake ecosystem management is based on the wet land national strategy, that is, by putting aside a part of the area which has protected endemic and unique bio natural resources to be developed into a conservational area. Hence the tourism criterion consists in the use of various resources and characteristics of the environment. Every lake water system is unique. As an attraction, the lake water system offers some categories of tourism products with different market shares. Lake is not a single destination but is an attraction with other tourism objects.

The activity of tourism in Lake Beratan is associated with natural and cultural tourism object, which is more oriented toward ecotourism. The attractions offered in Lake Beratan such as parasailing, banana boat, ski boat, trekking using a bike, travelling around in boat and fishing. Cultural attractions which can be presented are temples and mosques in addition to the farmers' activity in the field which are also enjoyed by foreign tourists, then, there also scientific tourism in Bedugul Botanical Garden which is managed by Indonesian Institute of Science (LIPI).

3.4.2. Animal Husbandry. Farmers around Lake Beratan besides growing crops in the dry land, also raising cattle such as cows, goats, pigs, ducks, and chicken. Most of the farmers raise cows and goats because they are easier to rise and it is easy to find feed for them. Farmers use sediment land and empty land to grow wide leaf decorated grass (*Pennisetum Purpereum).*

3.4.3. Agriculture. Agricultural sector depends on the availability of fertile land and irrigation system that continues on along the year with enough water supplies while from the environmental aspect, there is a need for cultivation technology that produces the least waste (FAO, 1976). Agricultural activities in catchment area in Lake Beratan are done intensively and the area belongs to a dry agriculture land. Farmers in this area grow crops such as cabbages, mustard greens, green vegetables, broccolis, salad, string beans, eggplants, cucumbers, potatoes, carrots, chillies, paprika, cassavas, corns, strawberries, oranges, and coffee. Thus, various treatments with agricultural activities such as fertilizing and using pesticide have a direct effect on the lake environment.

The effort to overcome the impact of agricultural waste can be done by protecting the lake water system with wet land vegetation which serves to absorb pollution from humus and pesticides before the runoff from the agricultural land enter the lake. In relation to the need for water resources, the lake water body can be considered as water reservoir for dry season.

3.4.5. Residential Area. The criteria of the residential area function are based on three main factors, namely: security, health, and comfort. Lake water system environment area is flooded regularly along with the rainfall in rainy season. This problem can be anticipated by selecting a higher location from maximum water surface and by selecting a house design that is adaptable to the flood.

Health factor is related to the availability of clean water resources for drinking and bathing, washing, and using toilet, sanitation system, both liquid and solid waste, and the risk of disease epidemic. Health factor is closely related to the holding capacity of lake water system. The community around Lake Beratan generally uses lake water directly for households, the potential for the pressure from household waste and also pollution into lake water system should be watched out.

Comfort factor is related to the accessibility of residential location to public facilities such as schools, hospitals, markets, jobs, and transportations. Residential areas around Lake Beratan have generally been equipped with transportation to obtain the basic necessities of the people.

3.4.6. Transportation. The criteria of transportation function needs to consider security, comfort, to smoothness of the traffic, and efficiency factors. The four factors are related to three elements, namely: biophysical condition, water system, and people who use the transportation and the vehicles used. The biophysical aspect of water depth determines the smoothness of transportation traffic, so that the effort to maintain the lake water system at a certain depth level according to the capacity of the means of transportation which will be developed have to be done.

Water flow determines the security and comfort of water transportation; thus, the determination of water transportation route needs to consider the patterns of water movement to select secured route. The effect of climate on water movement pattern needs to be considered since strong wind can cause dangerous water waves for small boats used for transportation. The development of shelters along the transformation route needs to be done to protect people who are trapped in a dangerous condition. Another thing that needs to be considered to develop water transportation is anchoring facility in the public area, like parking lots for cars that are safe and comfortable.

3.4.7. Fishery. The fishery function in the lake water system has to consider fish eco-physiological aspect and the variety of species, water system environment condition, and the condition of the community who get the benefit from fishery resource management effort. The first principle is that fish is viewed as resources since it gives benefit. From the view of the variety of types of fishery development in Lake Beratan is focused on fishery for food, while fishing can be done as part of lake tourism development.

According to [6] [7] water resource living environment cycle can be divided into 4 types of habitat, namely spawning ground, nursery ground, feeding ground, and playing ground. Fishery resources in Lake Beratan water system can be divided into two big groups, namely: lake local fish and introduced fish. Lake local fish is the native fish in the lake water system such as *Osteochilus hasselti, Rasbora trillineaota, Monoptera albus*. The introduced fish is introduced in the lake water system to increase the lake productivity; these include *Cyprinus carpio*, *Puntius javanicus*, *Oreochromis Mossambicus*, *Orechromis nylotucus,* and *Clarias batrachsis*. The development of floating net cages fish farming needs lake water system as water resource for media for fish growth, for this purpose, there is a need for good water quality that is suitable for fish growth, the adequate amount of water and water circulation that keeps oxygen supply in water and removal of feed waste and metabolism.

3.4.8. Zoning. Zoning in the lake is based on the function criteria of the lake to optimize the use of resource potential for the continuous welfare of the people. Zoning also considers the availability of zones, forms of zone proposed and their spread in Lake Beratan water system which can be seen in figure 3. Some forms of zoning are synergistically oriented to increase the productivity and preservation of fishery resources in lake water system which consist of the development of the edge of the lake and riparian vegetation, the development of zone of artificial wetland, and the management of floating net cages fish farming. These zonings are also oriented to control the inflow and distribution of waste in the lake water system. In addition, it is also proposed to determine lake's boundary areas by considering the condition of residential areas and the existing infrastructure.

Lake water fluctuation is a factor which plays the most important role in determining the zoning of the lake, and to accommodate the possibility of using the lake function as water resource conservation area, the development of zoning is done in some alternative lake water surface heights and the optimal water surface level suggested is 0.74 m from the zero surface level of Lake Beratan water. The effort to manage the lake water system environment needs to implement environment engineering technology, especially to adapt the water system condition to the changes of the environment which have been occurring. The aim of implementing environment engineering is to use as maximally as possible the lake water system resources in a sustainable way for the welfare of the community.

4. Conclusion

Based on the result of the study that has been done, briefly it can be concluded that: The lake management should be done in an integrated way between the central government and the local government by involving the community to participate actively. The management is done to maintain lake water potential sustainability. The actions that need to be done are: The empowerment of people through the socialization activities concerning the rules for managing the lake, educating people on the functions of the lake environment. Monitoring system and evaluation of the lake environment which is integrated with the lake environment information system could be developed. Forum for coordinative meeting which involve all the shareholders to write the framework and the structure of institutions should be arranged. Legal agreement in the management which has been determined to become an Act, Government Regulation, and Legally Binding local regulation could then be implemented. Implementing the policy and strategy to achieve the objective and writing a master plan for the lake water system area, in writing the master plan, the local people have to involve together with the government, stakeholders and relevant experts.

5. Acknowledgement
The researchers would like to express their deep gratitude to the Ministry of Research, Technology, and Higher Education for the doctoral program scholarship; The Department of Environmental Science, Sebelas Maret University; Promotor and co-promotor and Sampling Teams.

References

[1] Eamus D Hatton T Cook P and Colvin C 2006 *Ecohydrology: Vegetation, Water, and Resource Management* (Australia: CSIRO Publishing) p 348.

[2] Van Loon A F and Laaha G 2015 Hydrological drought severity explained by climate and catchment characteristics. *Journal of Hydrology* Vol 526 July 2015 pp 3 – 14.

[3] Schneider M 2005 *Water Quantity and Flow as Key Parameters for Benthos and Fish – An Investigation Using Ecohydraulic Approaches*. In: Bloesch, J., D. Gtknetcht, and V. Lordache. Hydrology and Limnology – Another Boundary in the Danube River Basin. (Technical Document in Hydrology. No 75) (Paris: UNESCO) pp 33 – 37.

[4] Atmaja D M Budiastuti S Prabang S and Sunarto 2017 The Understanding of The Farmer Community Around Lake Beratan of Water Resource Conservation *Journal IJAES* Vol 12 (Delhi: Research India Publications) pp 1191-1201.

[5] Arsyad S 2012 *Konservasi Tanah dan Air dalam Penyelamatan Sumber Daya Air* Jakarta: Yayasan Pustaka Obor Indonesia).

[6] Mustapha M K 2011 Perspectives in The Limnology of Shallow Tropical African Reservoirs in Relation to Their Fish and Fisheries *The Journal of Transdisciplinary Environmental Studie.* Vol 10 pp 1-23.

[7] Omondi R Kembenya E Nyamweya C Ouma H Machua S K and Ogari Z 2014 Recent Limnological Changes and Their Implication on Fisheries in Lake Baringo Kenya *Journal of Ecology and The Natural Environment* Vol 6 pp154-163.

[8] Hehanussa P E 2005 *Prosiding Daya Dukung Sumber Daya Air (Denpasar)* Vol 1 (Denpasar: UPT Balai Konservasi Tumbuhan Kebun Raya Eka Karya Bali-LIPI bekerja sama dengan Badan Pengendalian Dampak Lingkungan Daerah Pemerintah Daerah Provinsi Bali) hal 79.

[9] Atmaja D M BudiastutiS Prabang S and Sunarto 2017 *Proc. Int. Conf. on Geography and Education (Malang)* vol 79 (American: Atlantis Press) pp 76-80.

[10] FAO1976 *Frame Work For Land Evolution*.FAO Soil Buletin Soils Resources Management and Conservation Service Land and Water Development Division.

[11] Haeruman H 1999 *Prosiding Semiloka Nasional Pengelolaan dan Pemanfaatan Danau dan Waduk* (Bogor: PPLH-IPB, Ditjen Pengairan PU, Kantor Men. LH) hal 128.

[12] Anonimous 2005 Managing Lakes and Their Basins for Sustainable Use: *A Report for Lake Basin Managers and Stakeholders* (Outsu, Japan: International Lake Environment Committee Foundation) p 146.

IOP Conf. Series: Earth and Environmental Science **148** (2018) 012025 doi:10.1088/1755-1315/148/1/012025

Vertical Subsurface Flow (VSSF) constructed wetland for domestic wastewater treatment

M C Perdana[1], H B Sutanto [2] and G Prihatmo[2]

[1] Environmental Science Program, Graduate School of Gadjah Mada University, Jl. Teknika Utara, Pogung, Sleman, Yogyakarta, Indonesia 55281
[2] Faculty of Biotechnology, Duta Wacana Christian University (DWCU), Jl. Dr. Wahidin Sudiro Husodo No. 5 – 25 Yogyakarta, Indonesia 55224

mayyoung780@gmail.com

Abstract. Vertical Subsurface Flow Constructed Wetland (VSSF) is appraised to become an alternative solution for treating domestic wastewater effectively and efficiently. The system which imitates the natural wetland concept is able to reduce organic material and nutrients in wastewater; therefore, it will be more feasible to be discharged to the environment. This study aimed to compare which species is more recommended to be applied for reducing organic material and nutrients in domestic wastewater. This experimental study applied four treatments, i.e 1) control (unplanted), 2) single species *Iris pseudacorus,* 3) single species *Echinodorus palaefolius,* and 4) combination (*Iris pseudacorus* and *Echinodorus palaefolius*) with three days of retention time. The application of those plants aims for holding the role in increasing wastewater quality and adding aesthetic impression at once. The plants were planted on VSSF media, in relatively same of weight and size to compare their effectiveness in decreasing organic and inorganic load. The parameters measured pervade TDS, pH, BOD_5, COD, Nitrate, and Phosphate. The plants' condition was also observed during and after the system worked. The result showed that the best average value of effectiveness for each of parameters: COD by combination treatment (50.76%), BOD_5 by single *I. pseudacorus* (30.15%), Nitrate by single *E. palaefolius* (58.06%), Phosphate by single *E. palaefolius* (99.5%), and TDS by *E.palaefolius* (3.25%). The result showed that there was a significant difference of Nitrate and Phosphate reduction between control and three other treatments, while pH parameter showed non-significant change among them. In term of performance, *I.pseudacorus* seemed showed a preferable achievement.

1. Introduction

As a developing country, Indonesia is inevitably encountered by the rate of population growth, especially in moderate to big city. Based on "The Indonesian Central Bureau of Statistics" [2], the population growth rate of Indonesia achieved 1.49% per year, with total population of 237,641,326 inhabitants in 2010. This trend definitely drives the increase of domestic activity and will be parallel to greater productivity of domestic wastewater which ends up to high rate of its volume. [31] stated that the domestic activity was the prime contributor pollutant in Indonesia. As usually happen in most of the settlements, the wastewater is directly discharged to water body without any treatment. It is clearly revealed by [33], this practice occurred since many cities in developing countries still have obstacle in constructing and managing sewage treatment facilities. Domestic wastewater which is not treated right will evoke many impacts that threaten ecological function. Eutrophication is the example of ecological damage caused by nutrient content of wastewater [5]

[28]. Eutrophication constitutes the phenomenon of blooming algae raised by nutrient enrichment of a water body coming from the input of organic material or by nitrate and phosphate. It can also deliver to the increase of aquatic plant's growth and reduce light intensity to the water body [12].

Problems such as high cost (construction, operation, maintenance) and aesthetics take important concern in finding out the right treatment. Besides, a tendency of a certain system might not be applicable to some developing countries due to high rate of energy and mechanization [24]. The aspects such as cost-effective and aesthetics presumably deliver to an inevitable consideration in innovating treatment which is applicable for community, anywhere and anytime. As those reasons, the effective and efficient wastewater treatment is considerably required.

One of the efforts to find out the alternative treatment is by using CW (Constructed Wetland). CW is system which is designed to implicate natural process such as wetland vegetation, soils, and the associated microbes for treating wastewaters [30]. CW constitutes low cost and low maintenance technology, has a big prospective to be applied in developing countries [13], fits for either small communities or as a final stage treatment in large municipal system [20], and allows to be applied in various wastewater types [19]. CW is also effectively proven in removing pollutants in municipal wastewater [11] [14]. According to US Environmental Protection Agency [27], CW is an artificial natural wetland involving diverse kind of natural process to treat wastewater, which has planted usually by aquatic vegetation. Plants are holding important role to conduct purification in CW [8]. The objective of the study is for comparing which species is more recommended to be applied for reducing organic material and nutrients.

[19] mentioned that, according to the flow, CW was categorized into two i.e. surface and subsurface CW. Subsurface type is divided into horizontal and vertical flow. Vertical Subsurface Constructed Wetland permits the continuous or intermittent loading with upstream or downstream characteristic [21]. The subsurface type of wetland is considered to have several advantages over the surface type. If the water surface is maintained below the media surface there is little risk of odors, exposure, or insect vectors [26]. Besides, the wastewater was directly connected with the plant's *rhizosphere*, which plays important roles in degrading organic materials and reducing nutrients. Contaminant removal is supported by root system of plants and media (soil and stone), compiling bio-film which maintains the degradation [28].

The plants are used since those two are commonly found in Indonesia as ornamental plants. In consideration, they will bring up the fineness in application. *Iris pseudacorus* (also called "Yellow flag iris") is a perennial herb in the Iris family. *I. pseudacorus* considered as a tough one, since it withstands high levels of soluble organics, reduces the organic load by 25% over one year, and may survive in unfavorable conditions [7]. It also tolerates the low levels of oxygen [25]. [32] observed that *I. pseudacorus* had high nutrient absorption and storage capacity for treating wastewater and concluded as a preferred species for CW in northern China. *E. palaefolius* is also well known or its capability in reducing nutrients of domestic wastewater through some studies in Indonesia. According to Prayitno [16], the efficiency of *E. palaefolius* pervaded 68.75% for COD and 61.79% for BOD in tannery wastewater. *E.palaefolius* could reduce 446.890 mg/L, 172.1748 ppm, and 38.748 mg/L of COD, phosphate, BOD_5 respectively, in laundry wastewater [15].

2. Methods

2.1 Site description of inlet source

The inlet contained domestic wastewater which was collected from primary treatment of Municipality Treatment Plant (MTP) / IPAL Sewon, Bantul. This MTP is located in Sewon District, Bantul, Yogyakarta, positioning at 7°51'39.2"S as latitude and 110°20'06.4"E as longitude. It carries out waste management with centralized system namely the existence of pipes which drain wastewater coming from water disposal/waste households that in the region of Sleman, Bantul, and the city Yogyakarta to Sewon [17].

2.2 The implementation of study

The study was conducted in the Laboratory of Ecology of Duta Wacana Christian University (DWCU) for three months from March to May 2015.

2.3 Media and plant selection

The reactors' media consisted of soil and gravels obtaining from around DWCU's yard. *I. pseudacorus* plants were also taken from DWCU's yard, while *E.palaefolius* plants were obtained from plant market in Bantul, Yogyakarta. The plants were selected in relatively same of weight and size, concerning their good condition. Plant age was negligible. No additional fertilizer was applied before and during the system worked.

2.4 Experimental design

Four different treatments (reactors) i.e. 1) control (unplanted); 2) single species *Iris pseudacorus*; 3) single species *Echinodorus palaefolius*; and 4) combination (*Iris pseudacorus* and *Echinodorus palaefolius*) flowed continuously from 300 L inlet basin. Each of reactors had the dimensions of 33 cm (L) x 24 cm (W) x 40 cm (D). Every treatment was made of two replications. The system was run for 3 days retention time. The flow rate was controlled every day to retain the flow stability. Gravels by diameter ± 1 cm, ± 4 cm, and ± 8 cm were arranged below the soil from up to bottom, respectively. The scheme of experimental design (reactor) is showed in Figure 2. Every reactor was filled with tap water up to 30 cm depth for the first three days, and then flowed by domestic wastewater gradually until the water tap was all replaced by wastewater.

2.5 Analyses of inlet and outlet sample

Dissolved Oxygen (DO) and Chemical Oxygen Demand (COD) were measured at the beginning to observe the readiness of the reactor until reaching the stable value. The performance of media and plants were also monitored when the initial wastewater started to flow. Measurements of Total Dissolved Solid (TDS), pH, and Biological Oxygen Demand (BOD) were conducted in Laboratory of Ecology of Duta Wacana Christian University. The TDS measurement was using TDS meter, while the pH measurement was using pH meter. BOD_5 measurement was using "Micro Winkler" method. Measurements such as Phosphate, (COD), and Nitrate were conducted in Yogyakarta Health Laboratory Service.

2.6 Data and Statistical Analyses

Each of treatments produces six data for each of parameters to calculate the average of removal percentage of organic material and nutrients. The removal percentage was calculated by formula:

$$\text{Removal Percentage} = \frac{\text{Inlet Concentration - Outlet Concentration}}{\text{Inlet Concentration}} \times 100\% \quad (1)$$

One-Way ANOVA was used for verifying significant difference related to the average of removal percentage for both physical and chemical parameters at 5% level of significance using SPSS Statistics 17.0.

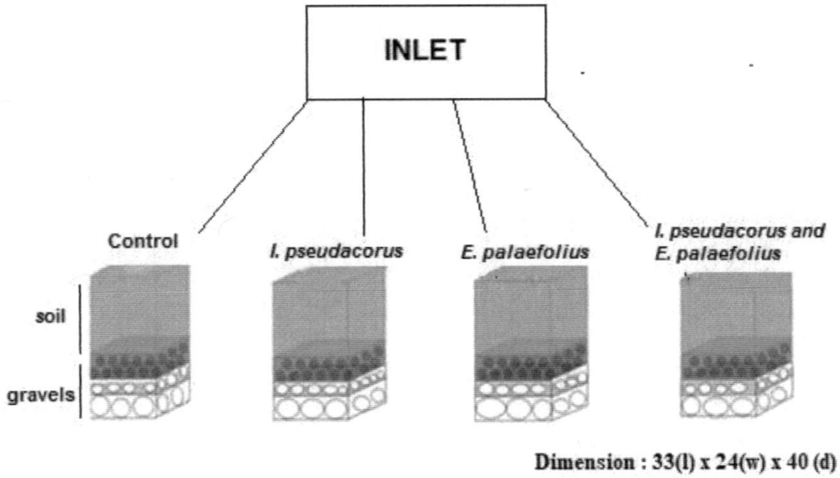

Figure 1. Scheme of experimental design (reactor).

3. Result and Discussion

3.1 Reduction of BOD and COD parameter

In general, result showed none of the significant difference among four treatments in BOD_5 and COD parameter. It implied that each of treatments more or less had the same capability in reducing BOD_5 and COD parameter value. However, combination treatment was appraised having the best performance for reducing COD value (50.76 %). It is assumed that the diversity of microorganism in *rhizosphere* area was high since the mixture of both plants. Averages of effectiveness among the treatments are shown in Table 1.

Table 1. The averages of removal percentage/effectiveness (%) among treatments. The values which were followed by same letter indicate non-significant difference, while those followed by different letter indicate significant difference.

Parameter	Treatments			
	Control	*I.pseudacorus*	*E.palaefolius*	Combination
COD	32.73^{a}	18.91^{a}	27.8^{a}	50.76^{a}
BOD_5	-5.05^{a}	30.15^{a}	-40.15^{a}	6.25^{a}
Nitrate	38.88^{a}	51.36^{b}	58.06^{b}	46.13^{b}
Phosphate	23.2^{a}	87^{b}	99.5^{b}	95.3^{b}
TDS	-21.25^{a}	-41.21^{a}	3.25^{a}	-45.8^{a}
pH	3.56^{a}	1.78^{a}	0.96^{a}	0.88^{a}
	Average value: 6.9	Average value: 7.2	Average value: 7.2	Average value: 7.4

[19] explained that BOD_5 and COD removal is supported by the mutuality interaction between microbial and physical mechanism by involving the dissolved oxygen. BOD_5 removal involves sedimentation and microbial degradation conducted by aerobic bacteria attached to plant roots, while COD removal prefers to sedimentation and filtration rather than biological process. The single *E.palaefolius* seemed to show unfavorable work for these parameters. The authors assumed for some reasons. Along the system run, this plant grew so nice. Many leaf buds appeared and some mature leaves developed wider. However, some of their base stalks were frequently withered, and then they got rotten inside the soil gradually. This condition is assumed contributing organic load and increasing the COD value. The negative value of efficiency was also found in *I.pseudacorus* presumably met the same condition, even though the rotten parts of *I. pseudacorus* plants were not frequently found as *E.palaefolius*. The control is less favorable rather than combination since there was no additional oxygen supply from *rhizosphere*, as it was

unplanted. Nevertheless, the existing aerobic microorganisms that live in soil apparently could still utilize the available oxygen inside the soil to degrade the organic matter. It was often found that soil animal such as earthworm existed in control treatment, thus giving good aeration due to their tracks. It assumed that the control showed a better result than two single planted treatments. This also signified that those plants presumably were better in reducing nitrate and phosphate which are inorganic, rather than the organic ones.

A study [14] stated that microorganisms can multiply themselves and conduct the degradation of organics since the media such as sand contributes a good environmental condition. Therefore, even though the control is unplanted, the COD value is subsided. [23] stated that soil microorganism such as aerobic bacteria, fungi, and *actinomycetes* are commonly found in *rhizosphere* than in non-*rhizosphere* area.

3.2 Reduction of Nitrate (NO_3)

By comparing all the treatments, there was a significant difference result in reducing nitrate. The control had significant result than three others. According to Table 1, the control showed increment of nitrate concentration namely -38.88% removal for the average. This is assumed that NO_3 was formed due to nitrification. Nitrification can be conducted by some existing *heterotroph* bacteria such as *Nitrobacter, Nitrospina*, and *Nitrospira*, converting NO_2 to NO_3 in a system [9]. Through this significant difference between unplanted and planted treatment, it could be concluded that plants played big role in uptaking NO_3. The highest average nitrate removal (58.06%) conducted by single *E.palaefoilus*. As authors stated in previous explanation of how they grew along the system ran (leaf buds appeared, mature leaves developed wider, the base stalks were frequently decayed) became the reason why the nitrate was well removed. The plants do not only uptake the nitrate as nutrient, but also serve the carbon and energy to provide denitrification which was originated from decaying biomass or root release [1]. It is assumed that *E.palaefolius* held those two roles by uptaking the nitrate to establish their growth (indicated by their new buds and wider leaves), and serving the base matter of denitrification namely carbon and energy (from decaying base stalks) to reduce nitrate. [18] conveyed that denitrification results nitrogen gas (N_2), nitrous oxide (N_2O), or nitric oxide (NO) through the facultative bacteria. Some of them namely *Proteus, Aerobacter, Flavobacterium,* were only able to reduce NO_3 to NO_2.

Single *I pseudacorus* was also achieved a good work in reducing NO_3 (51.36%). The processes might similar, nevertheless, since they have different root size and shape, they may have different soil microenvironment due to the diversity of the microorganism. [1] stated that plants support NO_3 removal but the effect is various depend on their species. The combination treatment did well in reducing NO_3, even though single treatments were better. The microbial diversity in this treatment might be the highest, since the combination of those two plants. However, the authors assumed that in combination treatment, the NO_3 was concentrated to be used for plants' growth, rather than converted. This was observed by the performance of the plants in this treatment. During the system worked, those two species grew so well when they are combined, especially *I.pseudacorus,* signed by the leaves that grew higher, and the flowers were formed. Camacho et al [4] in their study observed that *I.pseudacorus* experienced good growth and had 47.8% efficiency in reducing NO_3 in domestic wastewater.

3.3 Reduction of Phosphate

Phosphate is one of the prominent parameters in domestic wastewater since detergent is widely used in every domestic activity, especially for washing clothes and dishes. The form of phosphate was measured as orthophosphate in wastewater [5]. Excessive phosphate in wastewater may raise eutrophication and leads to the blooming algae [12].

In this study, the phosphate generally could be removed by all treatments. Nevertheless, the result showed that there was significant difference between control and three other treatments in reducing phosphate. Therefore, using single and combination for both plants in reducing phosphate were not giving big difference in this study. Single *E.palaefolius* revealed the most effective was by 99.5% phosphate reduction. It indicates that this species is very potential to uptake the containing phosphate. [6] stated phosphate removal relies on plant uptake; hence, observing of accumulation on plant's tissue is important. It delivers to an assumption that planted system took bigger role compared to unplanted. A study [22] observed that root is the key of up taking nutrients due to symbiotic microorganism exists in the roots. The following uptake was conducted by stem and leaf respectively.

Besides plant uptake, the phosphate removal can also occur through ion exchange and adsorption from substrate material [10]. [3] stated that combination of materials such as sand or gravel can enhance P-sorption in subsurface constructed wetland.

3.4 Reduction of TDS

The negative value of TDS removal efficiency was revealed by most of treatments, except single *E.palaefolius,* although the value was so small. There was no any significant difference among them in reducing TDS. It was presumed that after the bio-films were growing in the media especially gravels in large amount, some of the bio-films fell out and got dissolved, and therefore it could enhance TDS value through the outlet. Total Dissolved Solids (TDS) is defined as solid which dissolved in water, comprising inorganic salts (principally calcium, magnesium, potassium, sodium, bicarbonates, chlorides and sulfates) and some small amounts of organic matter [12]. A study conducted by Almeida et al [1] found out the slight increment in TDS in effluent compared to influent and effluent of each trial, however, no significant difference was found. This study indicated that these results were generated by the absence of retention of salts into the vertical flow constructed wetland.

3.5 pH parameter

The capability of maintaining pH was almost the same for each of treatments. It indicated that those four treatments generated pH stability. Wastewater pH is such a prominent aspect promoting the anaerobic organic degradation [18]. pH of wetland should be above 6 in order to support the denitrification process, delivering to N atmospheric [29].

3.6 Condition of plants

During one month of observation, *I.pseudacorus* showed more tenacious than *E.palaefolius.* It had shown by their endurance in tolerating wastewater, proved by their appearance. They were rarely withered during the system ran, only a few of their base stalks rotted occasionally. The leaves grew taller and the flowers bloomed. In combination with *E. palaefolius, I.pseudacorus* is disposed to be more stable than *E.palaefolius.* While *E.palaefolius* was excellent in the rapid rate of shoot and flower growth. The mature leaves were also wider fast. These conditions were compatible with the good capability of this species in absorbing nutrients, as this species showed the best result for reducing some parameters. Figure 2 shows the condition of plants.

Figure 2. Treatment from left to right: control, single *I.pseudacorus*, single *E.palaefolius*, and combination treatment.

4. Conclusion

Vertical Subsurface Flow (VSSF) constructed wetland denoted effective and efficient system of removing nutrient load from domestic wastewater. Chemical, physical, and biological mechanisms are occurred within the constructed wetland for reducing nutrient load. The role of media (sand, gravel), plant, and microorganism within the system support the good environmental condition in enhancing wastewater quality. This study found out that *I. pseudacorus* as vegetation was better in maintaining endurance and tolerance towards wastewater. Nevertheless, *E.palaefolius* was appraised more efficient in reducing some parameters such as Nitrate, Phosphate, and TDS. However, since there were no any significant differences in each of parameters between *I.pseudacorus* and *E.palaefolius*, the authors concluded that *I.pseudacorus* species was more recommended than *E.palaefolius* since *I.pseudacorus* showed better endurance. There was no significant difference between single species and combination treatment in upgrading the quality of wastewater. However, there were significant differences for reducing Nitrate and Phosphate parameter between control (unplanted) and planted system. In this study, those two plants tend to be better in removing inorganic matter, rather than the organic one. VSSF constructed wetland has great potential to be applied in concerning about domestic wastewater. It can also be used as means of education to recognize people about low cost and low maintenance wastewater treatment technology.

Reference

[1] Almeida A Carvalho F Imaginário M J Castanheira I Prazeres A R and Ribeiro C 2017 Nitrate removal in vertical flow constructed wetland planted with *Vetiveria zizanioides*: Effect of hydraulic load *Ecological Engineering* 99 535-542.

[2] Bureau of Statistics of Indonesia *https://www.bps.go.id/Subjek/view/id/12#subjekViewTab3\accordion-daftar-subjek1* Accessed in August, 2017.

[3] Brix H Arias C A and del Bubba M 2001 Media selection for sustainable phosphorus removal in subsurface flow constructed wetlands *Water and Science Technology* 44 47-54.

[4] Camacho J V De Lucas M A Gómez R and Mena S J 2007 A comparative study of five horizontal subsurface flow constructed wetland using different plant species for domestic wastewater treatment *Environmental Technology* 28 1333-13343.

[5] Haritash A K Sharma A and Bahel A 2015 The Potential of *Canna lily* for Wastewater Treatment Under Indian Condition *International Journal of Phytoremediation* 17 999-1004.

[6] Haritash A K Dutta S and Sharma A 2017 Phosphate uptake and translocation in a tropical *Canna*-based constructed wetland *Ecological Processes* 6 12 2-7.

[7] Jacobs J Graves M and Mangold J 2010 Plant guide for paleyellow iris (*Iris pseudacorus)* (Montana: USDA-Natural Resources Conservation Service, Montana State Office).
[8] Kong L Bin W Y Na Z L and He C Z 2009 Enzyme and root activities in surface-flow constructed wetland *Chemosphere* 76 601-608.
[9] Kurniadie D 2011 *Teknologi Pengolahan Limbah Secara Biologis* (Bandung:Widya Padjadjaran).
[10] Mojiri A Ahmad Z Tajuddin R M Arshad M F and Gholami A 2017 Ammonia, phosphate phenol, and copper (II) removal from aqueous solution by subsurface and surface flow constructed wetland *Environmental Monitoring and Assessment* 189:337 1-12.
[11] Mthembu M S Odinga C A Swalaha F M and Bux F 2013 Constructed wetland: A future alternative wastewater treatment technology *African Journal of Biotechnology* 12(29) 4542-4553.
[12] Muigai P Shiundu P M Mwaura F B and Kamau G N 2010 Correlation between Dissolved Oxygen and Total Dissolved Solids and their Role in the Eutrophication of Nairobi Dam, Kenya *International Journal of BioChemiPhysics* 18 :38-46.
[13] Mustafa A 2013 Constructed Wetland for Wastewater Treatment and Reuse: A Case Study of Developing Country, *International Journal of Environmental Science and Development* 4 20-24.
[14] Otieno A O Karuku G N Raude J M and Koech O 2017 Effectiveness of the Horizontal, Vertical and Hybrid Subsurface Flow Constructed Wetland Systems in Polishing Municipal Wastewater *Environmental Management and Sustainable Development* 6 158-173.
[15] Padmaningrum R Tien A and Yuliati 2014 Pengaruh Biomasa Melati Air (*Echinodorus palaefolius*) dan Teratai (*Nyphaeae firecrest*) terhadap Kadar Fosfat, BOD, COD, TSS, dan Derajat Keasaman Limbah Cair Laundry *Jurnal Penelitian Saintek* 19 64-74.
[16] Prayitno 2013 Pengurangan COD dan BOD Limbah Cair Terolah Industri Penyamakan Kulit Menggunakan Taman Tanaman Air dengan Tanaman *Majalah Kulit, Karet dan Plastik* 29 37-42.
[17] Purwanta J 2013 Upaya Pengelolaan Lingkungan Hidup di IPAL Sewon Kabupaten Bantul Melalui Kajian Biaya Pemantauan dan Pengendalian Kualitas Air dan Lingkungan Sistem Jaringan Limbah *Industrial Engineering Conference (2013)* (Yogyakarta: FTI UPN Veteran) p 1-16.
[18] Saeed T and Sun G 2012 A review on nitrogen and organics removal mechanisms in subsurface flow constructed wetlands: Dependency on environmental parameters, operating conditions and supporting media *Journal of Environmental Management* 112 429-448.
[19] Sehar S Sumera S Naeem I Perveen N and Ahmed S 2015 A comparative study of macrophytes influences on wastewater treatment through subsurface flow hybrid constructed wetland *Ecological Engineering* 81 62-29.
[20] Shutes R B E 2001 Artificial wetlands and water quality improvement *Environment International* 26: 441-447.
[21] Stottmeister U Weiβner A Kuschk P Kappelmeyer U Kästner M Bederski O Müller R A and Moormann H 2003 Effect of plants and microorganisms in constructed wetlands for wastewater treatment *Biotechnology Advances* 22 93-117.
[22] Sudarsan J S Annadurai R Mukhopadadhyay M and Chakraborty P 2016 Domestic wastewater treatment using constructed wetland: an efficient and alternative way, Sustainable Water Resources Management 1 89-96.
[23] Supradata 2005 Pengolahan limbah domestik menggunakan tanaman hias *Cyperus alternifolius*,

L. dalam sistem lahan basah buatan aliran bawah permukaan (SSF-Wetland) (Semarang: Universitas Diponegoro).
[24] Sutanto H 2013 Penerapan Lahan Basah Buatan sebagai Alternatif Pengolahan Limbah Skala Rumah Tangga *Kalpataru* (Yogyakarta: Badan Lingkungan Hidup).
[25] Sutherland W J 1990 Biological Flora of the British Isles: *Iris pseudacorus* L. *Journal of Ecology* 78: 833-848.
[26] US Environmental Protection Agency 1993 *Subsurface Flow Constructed Wetlands For Waste Water Treatment A Technology Assessment* (US:Office of Water).
[27] US Environmental Protection Agency 2000 *Manual for Constructed Wetlands Treatment of Municipal Wastewaters* United States Environmental Protection Agency (Cincinnati : Office of Research and Development).
[28] Vasudevan P Griffin P Warren A Thapliyal A and Tandon M 2011 Localized domestic wastewater treatment:part I – constructed wetlands (an overview) *Journal of Scientific & Industrial Reasearch* 70, 583-594.
[29] Verhoeven J T A and Meuleman A F M 1999 Wetlands for wastewater treatment: Opportunities and limitations *Ecological Engineering* 12:5-12.
[30] Vymazal J 2010 Constructed Wetlands for Wastewater Treatment *Water* 2: 530-549.
[31] Wirawan W A Wirosoedarmo R Susanawati L D 2014 Pengolahan Limbah Cair Doe:menggunakan Tanaman Kayu Api (*Pistia stratiotes L*) dengan Teknik Tanam Hidroponik Sistem DFT (Deep Flow Technique) *Jurnal Sumberdaya Alam dan Lingkungan* 63-70.
[32] Haiming W Jian Z Peizhi L Jinyong Z Huijun X Zhang B 2011 Nutrient. removal in constructed microcosm wetlands for treating polluted river water in northern China *Ecological Engineering* 37:560-568.
[33] Qing Z Jinasada K B S N Richard M Gersberg Yu L Jern W N and Keat T S 2014 Application of constructed wetlands for wastewater treatments in developing countries-A Review of recent developments (2000-2013) *Journal of Environmental Management* 141:116-131.

IOP Conf. Series: Earth and Environmental Science **148** (2018) 012026 doi:10.1088/1755-1315/148/1/012026

Minimization of municipal solid waste transportation route in West Jakarta using Tabu Search method

M Chaerul[1] and A M Mulananda[1]

[1]Department of Environmental Engineering, Faculty of Civil and Environmental Engineering, Institut Teknologi Bandung, Indonesia

chaerul_2000@yahoo.com

Abstract. Indonesia still adopts the concept of collect-haul-dispose for municipal solid waste handling and it leads to the queue of the waste trucks at final disposal site (*TPA*). The study aims to minimize the total distance of waste transportation system by applying a Transshipment model. In this case, analogous of transshipment point is a compaction facility (*SPA*). Small capacity of trucks collects the waste from waste temporary collection points (*TPS*) to the compaction facility which located near the waste generator. After compacted, the waste is transported using big capacity of trucks to the final disposal site which is located far away from city. Problem related with the waste transportation can be solved using Vehicle Routing Problem (VRP). In this study, the shortest distance of route from truck pool to *TPS*, *TPS* to *SPA*, and *SPA* to *TPA* was determined by using meta-heuristic methods, namely Tabu Search 2 Phases. TPS studied is the container type with total 43 units throughout the West Jakarta City with 38 units of Armroll truck with capacity of 10 m^3 each. The result determines the assignment of each truck from the pool to the selected *TPS*, *SPA* and *TPA* with the total minimum distance of 2,675.3 KM. The minimum distance causing the total cost for waste transportation to be spent by the government also becomes minimal.

1. Introduction

The rapid development of Jakarta City causes changes including the alteration of environmental quality. These developments include the growth of population, social, economic and cultural. With population of more than 10 Million inhabitants, Jakarta generates 6,595 tons of waste daily [1]. These development implicates for the increasing amount of municipal solid waste (MSW) that should be transported to final disposal site (*TPA*). Indonesia still adopts the concept of collect-haul-dispose for MSW handling system. The concept has consequences for the high operational costs of waste management. Cost for the waste transportation contributes dominantly to total cost of MSW management [2] [3]. The high cost of management will require an increase in fees to be paid by the generator. Higher fees will encourage generator to reduce waste generation rate [4].

Transportation is a sub-system in a MSW Management system to collects and transports the waste from the generator to temporary collection point, to waste treatment facilities, or directly to the final disposal site/landfill (*TPA*). By optimizing the assignment of trucks, the sub-system is expected to transporting waste easier, faster, and lower the cost. Another objective of the waste transportation is to minimize of vehicles used, the distance, and travel time [5]. These waste transportation route must be effective and efficient and so that it results to the optimum route of waste transportation and the minimal adverse effects on the environment [6].

Jakarta has a landfill that located outside the city, namely *TPA Bantar Gebang* in Bekasi City. The location of the landfill outside the city would be required the most cost if all trucks directly entering to *TPA Bantar Gebang*. Moreover, it will cause the queue at the entrance of the landfill site. The study aims to find the shortest route as the proposed route of the existing route by utilizing the compaction facility as a transshipment point as part of Vehicle Routing Problem (VRP).

2. Research Methodology

Transshipment Model is a transportation mode that enables the delivery of goods (commodities) indirectly, where goods from one source can be transported to another source or another destination before reaching its final destination. In this study, transshipment point is a compaction facility (*SPA*) that located in *Sunter*, Jakarta. The waste coming from temporary collection points may or may not transit to *SPA* before transported to TPA *Bantar Gebang*. Other constraints that should be achieved in this study include:

- All truck coming from a pool should go to a *TPS* to transport the waste in these *TPS*;
- All waste generated and stored in all TPS should be transported using a particular truck to *SPA* or *TPA*;
- All waste accepted at *SPA* and *TPA* cannot exceed the respective capacity. *SPA Sunter* will only accept waste from West Jakarta at a maximum 200 m^3. While, there is no limitation for *TPA* capacity;
- A truck may transport waste generated in *TPS* in a maximum 3 times of return a day.

The algorithm to find the solution is divided into two, namely the construction algorithm and improved algorithm. Construction algorithm performed by sequential insertion, while the improvement using the Tabu Search Algorithm. Sequential insertion algorithm is one of the algorithms for generating the initial solution. The algorithm is easily extended to accommodate hard constraints such as time windows, heterogeneous fleet, and pick up delivery [7]. The principle of sequential insertion is to insert the component solutions (in the VRP problem is the customer, in this study refers to *TPS*) one by one according to certain criteria.

Table 1. Algorithm developed for the research.

ALGORITHM 1
TOUR ALGORITHM

Step 1: Construct a giant tour T, where T = $\{v_1, v_2, ..., v_n\}$
Step 2: **for** each arc $(i,j) \in G$, **do**
(a) Calculate c_{ij} and di the travel cost and demand associated to the route.
(b) Calculate F_k, the fixed cost of vehicle k.
end do
Step 3: Calculate the shortest path from node v_i to node v_n of graph G.
Step 4: Repeat steps 2-3 for other possible tours.

ALGORITHM 2
TABU SEARCH ALGORITHM

Step 1: (Initialization)
Calculate an inital solution using the Tour Algorithm
Step 2: (Search Procedure)
(a) Begin with $\lambda = 1$ and define λ_{max}
(b) Select best non-tabu feasible solution $\varepsilon' \in N(\epsilon)$
(c) Keep tabu elements of the movement
(d) Keep the new solution cost f(ε') in array
(e) Do ε - ε' and $\lambda = \lambda + 1$
Step 3: (Swapping Procedure)
(a) Calculate ε' from ε applying the swap procedure at each route ε
(b) Update ε - ε' and $\lambda = \lambda + 1$
(c) If f(ε') < f(ε^{BEST}) then $\varepsilon^{BEST} = \varepsilon'$, $\lambda_{BEST} = \lambda$ and $over = 0$
else
$over = over + 1$
Step 4: (Diversification/Intensification Mechanism)
(a) Split each route of ε
(b) Allocate the cheapest vehicles to satisfy demand
(c) Define the new solution cost f(ε')
Step 5: (Stopping Rule)
if $\lambda > \lambda_{max}$ then stop

Tabu Search is a meta-heuristic algorithm that is designed to escape from a local optimum [8]. Tabu Search has taboo restriction to limit the search space and aspiration criteria to be separated from the barrier that had been determined previously using restriction taboo. Tabu Search also has a memory function with different timescales in the form of short-term memory and long-term memory to run the intensification and diversification strategy in the process of finding solutions. Intensification strategy is a strategy centering on local search solutions that are considered "promising". Meanwhile, diversification strategy is the search strategy by directing the search in a new area.

Tabu Search has been applied for several studies related with various areas [9] [10] [11]. In general, there are two algorithms developed for the research, namely Tour and Tabu Search algorithms (See table 1) [12]. In our study, node, travel cost, demand, and the cheapest vehicles in the algorithms below refers to *TPS*, distance of transportation, waste generated at *TPS*, and the shortest distance of transportation, respectively.

3. Result and discussion

West Jakarta is located at 106°22'42'' E - 106° 58'18'' E and 5°19'12'' S - 6°23'54'' S with an area of 126.15 km^2. The area of West Jakarta is divided into 8 districts including 56 urban villages, with a total population of 2,427,414 people [13]. The amount of waste generated from various sources amounted to 1,528.03 tons per day [14].

West Jakarta has 264 locations of *TPS* where 190 locations of *TPS* are handled by West Jakarta Sub-dept., and the rests are handled by the private sector. In this study, Tabu Search methods are used as a tool to formulate the problems of transporting waste in West Jakarta City. There are 43 locations of *TPS* with container type for storing the waste throughout the city (figure 1). Vehicle considered in this study is Armroll truck with total amount of 38 locations with capacity of 10 m^3 each. Total waste generated at each *TPS* can be found in table 2. The waste coming from TPS may or may not transit to *SPA Sunter* located in North Jakarta City, about 60 km from center city of West Jakarta. *SPA Sunter* serves also other 4 cities in Jakarta Province namely East Jakarta, South Jakarta, East Jakarta, and Central Jakarta, while *TPA Bantar Gebang* located in Bekasi City, about 60 km from West Jakarta.

Figure 1. Distribution of *TPS* in West Jakarta.

In total, waste generated in West Jakarta City is 548 m^3/day. It can be found also from table 1 that TPS#27 contributes waste generation dominantly (30 m^3/day), while TPS#12 stores the minimum waste generation (1.5 m^3/day). The difference of amount stored at each TPS is affected significantly by inhabitants living at the surrounding of the TPS site.

The distance from pool of trucks to each *TPS*, each *TPS* to *SPA*, and each *TPS* to *TPA* was generated using Google maps and the range data is presented in Table 3. In addition, the distance between *SPA* and *TPA* is 40.8 km.

Table 2. Amount of waste generated in *TPS*.

Code	Capacity (m3)	Code	Capacity (m3)	Code	Capacity (m3)
TPS1	10	TPS16	20	TPS31	16
TPS2	10	TPS17	16	TPS32	20
TPS3	12	TPS18	10	TPS33	16
TPS4	16	TPS19	8	TPS34	15
TPS5	10	TPS20	11	TPS35	15
TPS6	7	TPS21	11	TPS36	11
TPS7	7	TPS22	11	TPS37	14
TPS8	12	TPS23	12	TPS38	13
TPS9	9	TPS24	11	TPS39	5
TPS10	13	TPS25	10	TPS40	10
TPS11	12	TPS26	26	TPS41	9
TPS12	**1.5**	TPS27	**30**	TPS42	8
TPS13	10	TPS28	16	TPS43	8
TPS14	17	TPS29	20		
TPS15	17	TPS30	14		

Table 3. Vehicle mileage.

TPS	Distance (Km)			TPS	Distance to (Km)		
	Pool of truck	SPA	TPA		Pool of truck	SPA	TPA
TPS1	6.5	22.3	45.6	TPS23	13.6	22.7	48.5
TPS2	7.4	15.9	46.8	TPS24	11.3	**25**	45.7
TPS3	7.6	19	47.1	TPS25	14.4	23.3	44.6
TPS4	5.5	20.1	51.1	TPS26	14.6	12.7	38.7
TPS5	8.3	17.7	44.8	TPS27	13.2	12.4	38.6
TPS6	5.9	20.4	55.1	TPS28	**15.5**	13	40.1
TPS7	5.5	21.4	55.4	TPS29	13.2	12.9	**37.4**
TPS8	6.6	20.2	55.2	TPS30	12.4	11.7	43
TPS9	**1.5**	19.1	**56.2**	TPS31	11.5	10.4	44.2
TPS10	15.2	23.3	40.2	TPS32	12.6	11.8	43.1
TPS11	12.5	15.6	42.7	TPS33	12	12.1	40.2
TPS12	12.8	18.2	44.5	TPS34	14.5	12.5	50.3
TPS13	12.6	15.6	44.9	TPS35	12.2	**10.3**	47.3
TPS14	11.6	14	42.8	TPS36	11.8	13.4	48.5
TPS15	13.1	21.2	41	TPS37	13.1	13.5	46.9
TPS16	12.5	15.6	44.9	TPS38	15.2	15.6	47.8
TPS17	10.4	16.5	44.9	TPS39	14.5	14.3	51.3
TPS18	12.5	15.6	42.7	TPS40	13.7	16.5	48.5
TPS19	13.2	21.5	46.5	TPS41	12.8	12.1	49.1
TPS20	11.5	24.6	48.3	TPS42	12.8	12.1	49.1
TPS21	10.7	24.8	43.6	TPS43	12.2	14.7	50.8
TPS22	10.2	24.3	43.2				

From table 2 it can be seen that TPS#9 is the closest from the pool, which as far as 1.5 km and TPS#28 is the farthest from the pool as far as 15.5 km. For a distance from *TPS* to *SPA*, TPS#35 is the closest (10.3 km) and TPS#24 is farthest (25 km). For a distance from *TPS* to *TPA*, TPS#29 is the closest (37.4 km) and TPS#9 is the farthest (56.2 km).

Before running the model, the study simulates 2 scenarios: (1) all waste from *TPS* transported directly to *TPA*; and (2) all waste from *TPS* transported to *SPA* before transported to *TPA*. From 1st and 2nd scenarios, it results to the total distance of trucks will be 5,238 Km and 2,194 Km, respectively. Result of the model obtained by processing the waste transportation route optimization is presented in Table 4.

Table 4. Assignment and proposed route of trucks.

Truck No.	Route	Truck No.	Route
1	0-1-45-0-45-0-45	20	0-34-44-34-44-35-45
2	0-6-44-0-44-10-45	21	0-35-44-36-44-36-45
3	0-5-44-4-44-7-45	22	0-37-44-37-44-38-45
4	0-8-44-8-44-3-45	23	0-38-44-39-44-40-45
5	0-9-44-10-44-11-45	24	0-41-44-42-44-43-45
6	0-11-44-12-44-13-45	25	*0-0-45-0-45-0-45*
7	0-14-44-14-44-15-45	26	*0-0-45-0-45-0-45*
8	0-15-44-16-44-16-45	27	*0-0-45-0-45-0-45*
9	0-18-44-17-44-17-45	28	*0-0-45-0-45-0-45*
10	0-19-44-20-44-20-45	29	*0-0-45-0-45-0-45*
11	0-21-44-21-44-22-45	30	*0-0-45-0-45-0-45*
12	0-22-44-23-44-23-45	31	*0-0-45-0-45-0-45*
13	0-24-44-24-44-25-45	32	*0-0-45-0-45-0-45*
14	0-26-44-26-44-26-45	33	*0-0-45-0-45-0-45*
15	0-27-44-27-44-27-45	34	0-4-44-0-45-45
16	0-28-44-28-44-29-45	35	*0-0-45-0-45-0-45*
17	0-29-44-30-44-30-45	36	0-3-45-0-44-2-44
18	0-31-44-31-44-32-45	37	*0-0-45-0-45-0-45*
19	0-32-44-33-44-33-45	38	*0-0-45-0-45-0-45*

This algorithm defines that (0) is the notation for pool of truck, (44) for *SPA* and (45) for *TPA*. From these results, it can be seen that Truck#1 transport waste at TPS1 and subsequently go to the TPA. Truck#1 only has to transport waste at TPS#1, while another truck may have different number of rotations. For instance, Truck#36 transport wastes at TPS#3 and TPS#2. Almost trucks transport the waste from 3 locations of *TPS*. It can be seen also that from total 38 trucks owned by the city, it is only 26 trucks necessary used to transport the waste from 43 locations of *TPS*. It means that 12 trucks can be used to transport the waste from other type of *TPS* in the city. Another result of the program using Visual Basic can generate solution by minimizing the distance. Total minimum distance by considering a set of constraints mentioned above is 2,673 Km.

4. Conclusion

Waste management is an issue that is still frequently occurring in environmental management. The present study focuses to transport all waste generated at 43 locations of *TPS* using container to store the waste throughout the city of West Jakarta. The trucks may or may not transit to *SPA* as a transshipment point before finally dispose the waste at *TPA* as a final destination. The Visual Basic model can generate solutions by minimizing the distance that should be traveled by the trucks is 2,673 Km. Second result from the model is that the number of trucks required to transport all waste generated at all *TPS* are 26 trucks. Minimum distance will generate minimum cost for operational of the transportation. Thus, the model helps the municipality having lack of financial to increase the service level of waste management.

IOP Conf. Series: Earth and Environmental Science **148** (2018) 012026 doi:10.1088/1755-1315/148/1/012026

5. References

[1] Wardhani I 2011 *Policy analysis of integrated waste management based on Zero Waste and Environment Management (ZWEM) System* (in Indonesian) Universitas Indonesia.

[2] Tavares G Zsigraiova Z Semiao V and Carvalho M G 2009 Optimization of MSW collection routes for minimum fuel consumption using 3D GIS modelling *Waste Man.* 29 1176-1185.

[3] Elmira S Nadi B and Mahmud AR 2011 *Optimization of municipal solid waste management* Int. Conf. on Biology, Environment and Chemistry. Singapore.

[4] Widyarsana I M W Damanhuri T P and Tomo H S 2014 *Study of household waste handling scheme using dynamic modelling: A case study of Bandung City* (in Indonesian) Institut Teknologi Bandung.

[5] Byung-In K Seongbae K and Surya S 2005 Waste collection vehicle routing problem with time windows *Comput. Oper. Res.* 33 3624-3642.

[6] Clifford T 2008 *Waste collection optimization tools* Proc. on Waste and Climate Change. London.

[7] Widagdo S J 2014 *Determination of Medical Waste Routing at PT. Jasa Medivest using Tabu Search* (in Indonesian) Institut Teknologi Bandung. Bandung, Indonesia.

[8] Glover F 1999 Tabu Search: A Tutorial. *Interfaces* 20 74-94.

[9] Fan W and Machemehl R B 2008 Tabu Search Strategies for the Public Transportation Network Optimizations with Variable Tansit Demand. *Compt. Aided Civil Infras. Eng.* 23 502-520.

[10] Glover F Laguna M and Dowsland K A 1993 *Tabu Search-Modern heuristic techniques for combinatorial problems* Blackwell: London, UK.

[11] Gómez J R Pacheco J and Gonzalo-Orden H 2015 A Tabu Search Method for a Bi-Objective Urban Waste Collection Problem *Compt. Aided Civil Infras. Eng.* 30 36–53.

[12] Cheeneebash J and Jugee S 2012 Applying Tabu Search Algorithm in Public Transport: A Case Study for University Students in Mauritius *Int. J. Transport. Vehicle Eng.* 6 1794-1797.

[13] Statistic of West Jakarta 2013 *City Statistic year of 2014* (in Indonesian) West Jakarta City.

[14] West Jakarta Cleanliness Sub-Department 2013 *Waste generation data: A yearly report* (in Indonesian) West Jakarta City.

IOP Conf. Series: Earth and Environmental Science **148** (2018) 012027 doi:10.1088/1755-1315/148/1/012027

Hydrograph monitoring and analysis for sustainable karst water management in Nyadeng Spring, East Borneo

M Widyastuti[1], H Fatchurohman[1], W A Fathoni[1], A A Hakim[1] and E Haryono[1]

[1] Karst Research Group, Faculty of Geography Universitas Gadjah Mada, Indonesia

m.widyastuti@geo.ugm.ac.id

Abstract. Karst aquifer stores abundant water resources within its matrix, conduits, and inter-granular pores. Karst aquifer plays an important role in providing water supply, especially in the areas nearby that commonly dry and lack of surface water resources. Karst spring hydrograph analysis is very fundamental step to–assess and determines the condition of the catchment area in karst terrain. Recession curve is believed to be the most stable part in single flood hydrograph that represents the aquifer characteristics. Nyadeng is one of the most significant karst springs that located in Merabu Karst Area, East Borneo. Villagers in Merabu highly depend on Nyadeng Spring for fulfilled their freshwater need. Hydrograph monitoring has been initiated for one year in Nyadeng Spring as a preliminary action for karst water management in Merabu. Water level data series obtained using automatic water level data logger and then correlated with manual discharge measurement to generate stage-discharge rating curve. The stage-discharge rating curve formula for Nyadeng Spring calculated as y = 0,0102e5,8547x with r^2 value =0.8759. From the combination of several single flood events, Master Recession Curve (MRC) was generated to determine flow regime as the main consideration for karstification degree calculation. From the MRC result, flow regimes formula determined as Qt= 3.2-0.001t + 1.2(1-0.012t)+1.6(1-0.035t) indicated that one sub-regime with laminar flow and two sub-regimes with turbulent flow existed. From the MRC formula, the degree of karstification in Nyadeng Spring classified at seventh scale (developed karstification of the aquifer) based on Malik's karstification degree (2012). The degree of karstification in Nyadeng Spring indicates that the aquifer formed by large conduit channels, fissures, and macro fissures which are able to provide significant water sources that can be utilized for multi purposes. Therefore, it is concluded that spring hydrograph monitoring provide essential information in order to establish a careful water resources management actions.

1. Introduction

Environmental problems regarding the quality and quantity of water resources is one of the major issues all over the world. Groundwater and surface water are the most water reservoirs utilized by humans because of their accessibility. Groundwater resources are facing management problem including karst aquifers. Karst aquifers formed in a very complex hydrological system. Developed in fracture media and soluble rocks have led karst aquifers into various porosity types and aquifer properties. Karst terrain stores abundant water resources, despite lack of surface water [6] [16] [16] [17].

One of the most dependable surface water resources in karst area are karst springs [2] [8]. Karst springs are considered as one of the most enormous water resources compared to others springs from other aquifer types. Karst springs emerge as the most important surface water sources in karst area [4]. Nyadeng Spring is located in the middle of the traditional forest of Merabu Village Nyadeng is the biggest karst spring located in Merabu Karst, Berau Regency (figure 1). Nyadeng spring utilized by the villagers to meet their daily need of freshwater. Nyadeng play very important role as the main water resources for the villagers. Nyadeng also initiated as a tourism spot that included in the Merabu Forest and Merabu Village tourism schemes.

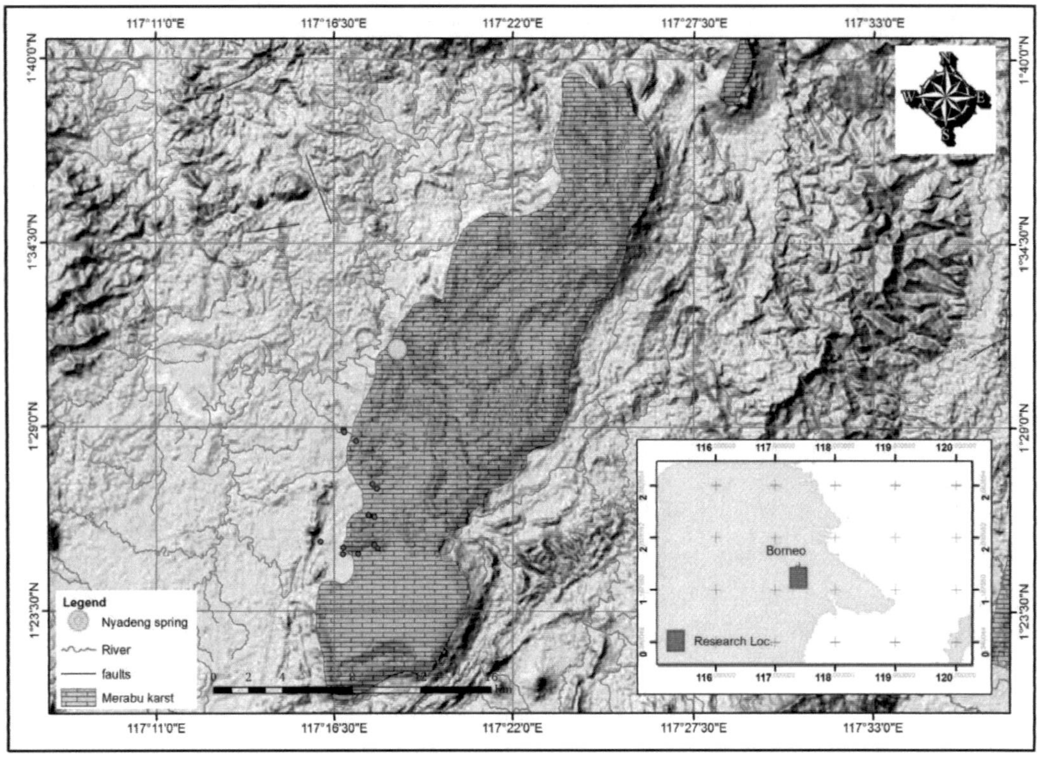

Figure 1. The location of Nyadeng Spring.

Despite intensively utilized and being the most significant water source in Merabu, Nyadeng Spring has not been monitored before. Therefore, a quantitative research based on automatic water level monitoring has been conducted in Kakap Spring to provide hydrograph data. Karst spring hydrograph monitoring is necessary to obtain quantitative parameters of the aquifer. The forms of hydrograph reflect the output discharge and the networking systems of an aquifer. Therefore, karst spring hydrograph is important to determine and characterize the aquifer condition. The spring hydrograph features provide valuable information in order to establish a careful water resources management action [3] [7]. This research aimed to determine the hydrograph fluctuation of Nyadeng Spring. Furthermore, this research also aimed to determine the karstification degree as the main consideration for karst water resource management.

2. Result and discussion

The primary material for karstification degree determination is hydrograph series data that obtained from automatic water level logger installation. Onset HOBO Water level data loggers were installed for annual water level observation. The HOBO water level data logger sensor measures the water pressure and compensated with barometric pressure to generate water level data. Water level measurement accuracy up to 1,5 cm with very minimum error. Manual discharge measurements have been done with velocity area method. From the manual discharge measurements, discharge data and water level data combined to generate exponential regression.

Derived from hydrograph data, several single recession curves were selected and respectively a Master Recession Curve assembled. Master recession curve is a graph that shows the average recession curve shape that assembled from several recession periods in a particular site. Master

Recession Curve (MRC) commonly used to describe and analyze the average recession from a set of hydrograph recessions [12] [13].

The determination of karstification degree is based on the identification of karst discharge regime. The discharge regime defined as a pattern of changing conditions in the groundwater depletion graph. Later Study described that more than two flow components presence in karst spring hydrograph [3] [10] [14]. Other description of recession curve that describes each flow component is also necessary to be considered.

Malik and Vojtkova introduced a quantitative method for karstification degree classification. The development of karstic area classified using a 10-degree scale, from the 1st degree for the lowest karstification, and 10th degree for most developed (Table 1) [10]. Typical recessional equations used to describe recessional types that characterized by the existence of different flow sub-regimes. There are more than two flow components presence in karst spring hydrograph. Exponential function represents diffuse aquifer type with laminar flow characteristics (Eq 1.) [9], while linear function represents conduit channels with turbulent flow characteristics [5].

$$Q_t = \sum_{i=1}^{n} Q_0 e^{-\alpha t} \tag{1}$$

$$Q_t = \sum_{i=1}^{n} Q_0 e^{-\alpha t} + \sum_{j=1}^{m} \left(\frac{1}{2} + \frac{|1-\beta jt|}{2(1-\beta it)}\right) Q_{0j} = Q_0(1 - \beta\, \mathrm{jt}) \tag{2}$$

Water level data monitoring in Nyadeng Spring have been conducting from April 2016-presents. However, the water level data series that available and ready to be analyzed was only until March 2017. A stage-discharge rating curve then constructed based on six manual discharge measurement (figure 2). The stage-discharge rating curve equation in Nyadeng Spring expressed in the Eq. 3:

$$\boldsymbol{y = 0{,}0102e^{5{,}8547x}} \tag{3}$$

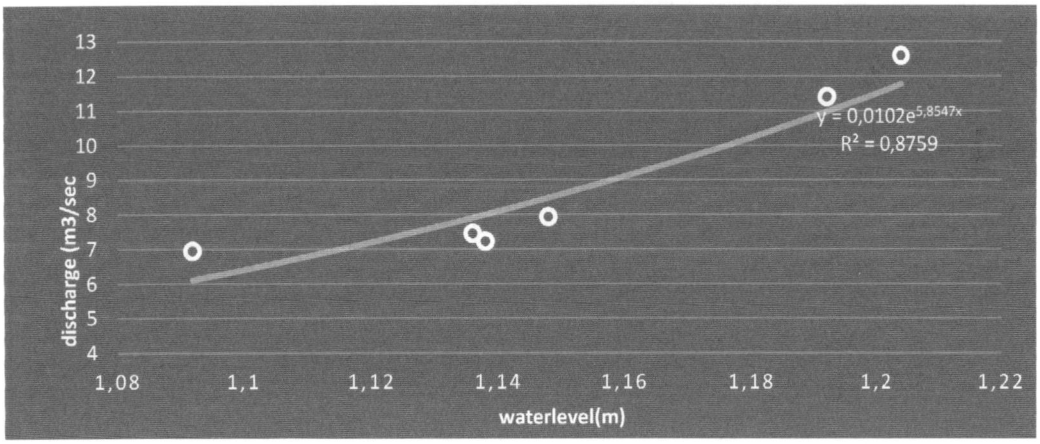

Figure 2. Stage discharge rating curve of Nyadeng Spring.

Table 1. Karstification degree in recharge areas of springs according to recession curves parameters.

Karstification degree	Groundwater flow sub-regimes type	Characteristic recession curve equation	Characteristics of recession-curve parameters
0.5	**Single laminar** sub-regime, lower values of $\alpha 1$	$Q_t = Q_{01}e^{-\alpha 1t}$	$\alpha_1 < 0.001$
1.0			$\alpha_1 = 0.001\text{-}0.0025$
2.0	**Single laminar** sub-regime, higher values of α_1		$\alpha_1 = 0.0025\text{-}0.007$
2.3			$\alpha_1 > 0.007$
2.5	Combination of **two or more sub-regimes** with merely **laminar** flow characterized by	$Q_t = Q_{01}e^{-\alpha 1t} + Q_{02}e^{-\alpha 2t}$	$\alpha_1 < 0.0024$ and $\alpha_2 < 0.033$
2.7	different discharge coefficients, lower values of α_1 and α_2		$\alpha_1 < 0.0024$ or $\alpha_2 < 0.033$
3.0			$\alpha_1 = 0.0024\text{-}0.0045$ $\alpha_2 = 0.033\text{-}0.067$
3.5	Combination of **two or more sub-regimes** with merely **laminar flow** characterized by different		$\alpha_1 = 0.0024\text{-}0.0043$, and $\alpha_2 = 0.060\text{-}0.16$
3.7	discharge coefficients, lower values of α_1 and α_2		$\alpha_1 > 0.0043$ and$\alpha_2 <$ 0.060
4.0			$\alpha_1 = 0.0041\text{-}0.018$ and $\alpha_2 = 0.055\text{-}0.16$
4.3	Discharge hydrogram is composed of **a sub-regime**	$Q_t = Q_{01}e^{-\alpha 1t} +$	$\alpha_1 > 0.018$ or $\alpha_2 > 0.16$
4.7	**with turbulent flow** and a sub-regime with **laminar**	$Q_{04}\ (1\text{-}\beta_1 t)$	$\alpha_1 > 0.018$ and $\alpha_2 > 0.16$
5.0	**flow**.		low values of β and α
5.5	Complex discharge regime, a combination of **one sub-regime** with **turbulent flow** and **two sub-regimes** with **laminar** groundwater flow.	$Q_t = Q_{01}e^{-\alpha 1t} + Q_{02}e^{-\alpha 2t} + Q_{04}\ (1\text{-}\beta_1 t)$	$\alpha_1 > 0$ and $\alpha_2 > 0$ and $\beta_1 > 0$
6.0	Very complex discharge regime, a combination of **two sub-regimes** with **turbulent flow** and **two sub-regimes** with **laminar** groundwater flow.	$Q_t = Q_{01}e^{-\alpha 1t} + Q_{02}e^{-\alpha 2t} + Q_{04}\ (1\text{-}\beta_1 t) + Q_{05}\ (1\text{-}\beta_2 t)$	β_1, β_2, and α_1, α_2 high values
7.0	Discharge regime is a combination of **one sub-regime** with **laminar** flow with **two to three** sub-regimes with **turbulent flow**. Substantial role in groundwater discharge plays by the sub regime with laminar flow	$Q_t = Q_{01}e^{-\alpha 1t} + Q_{02}e^{-\alpha 2t} + Q_{04}\ (1\text{-}\beta_1 t) + Q_{05}\ (1\text{-}\beta_2 t) + Q_{06}\ (1\text{-}\beta_3 t)$	β_3, β_2, β_1 and α high values, $\beta_1 > \beta_2$
8.0	Discharge regime is a combination of **a sub-regime** with **laminar flow** with **two to three** sub-regimes with **turbulent flow**.		β_3, β_2, β_1 and α thigh values
8.5	Groundwater flow regime is represented merely by turbulent flow, with only **one turbulent flow** sub-regime present, which represents turbulent circulation in channel systems (conduits) without hydraulic connection to the groundwater in adjacent rock blocks. Groundwater circulation is mostly connected to vadose zone	$Q_t = Q_{04}\ (1\text{-}\beta_1 t)$	α_1; $\alpha_2 = 0$ and $\beta_1 > 0$
9.0	Groundwater flow regime is represented merely by turbulent flow, consisting of **two turbulent flow sub-regimes.** These represent turbulent circulation in channel systems (conduits) without hydraulic connection to the groundwater in adjacent rock blocks Groundwater circulation is mostly connected to vadose zone	$Q_t = Q_{04}\ (1\text{-}\beta_1 t) + Q_{05}\ (1\text{-}\beta_2 t)$	β_1 dan β_2 low values
10.0	Complex turbulent flow regime of groundwater discharge, consisting of **three different turbulent flow** sub regimes. Probability of a very complex groundwater circulation by occasional flows in the vadose zone. Documented only in perennial flows.	$Q_t = Q_{04}\ (1\text{-}\beta_1 t) + Q_{05}\ (1\text{-}\beta_2 t) + Q_{06}\ (1\text{-}\beta_3 t)$	β_1, β_2 dan β_3 thigh values

Nyadeng Spring catchment area has not been defined yet, due to the minimum information and data in Merabu Karst Area. The flow characteristic of Nyadeng Spring is dominated with significant diffuse flow input, indicated with slow discharge response to the precipitation events. However, Nyadeng also dominated by strong turbulent flow regime, indicated by extreme flood pulses. The lowest discharge in Nyadeng Spring recorded at 1.47 m^3/second, and the maximum discharge recorded was 6.2 m^3/second (figure 3).

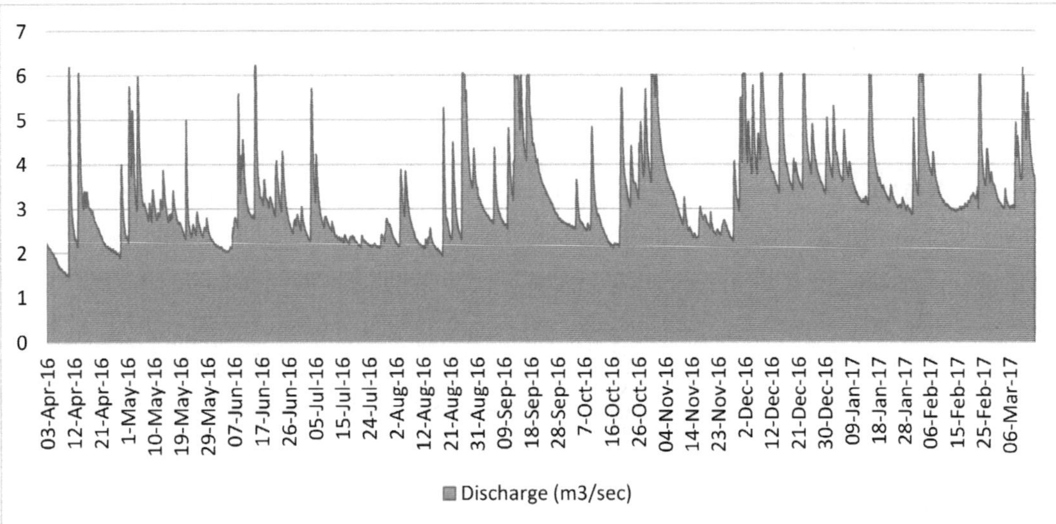

Figure 3. Hydrograph data series of Nyadeng Spring.

Based on automatic recession curves filters, more than thirty (30) flood events were identified. However, only four flood events were ideal for the analysis. The selection of ideal recession curve based on recession length and the inclination of the recession limb. Subsequently, each recession curve was calibrated with the appropriate flow model, based on the shape and inclination of the recession limb (figure 4).

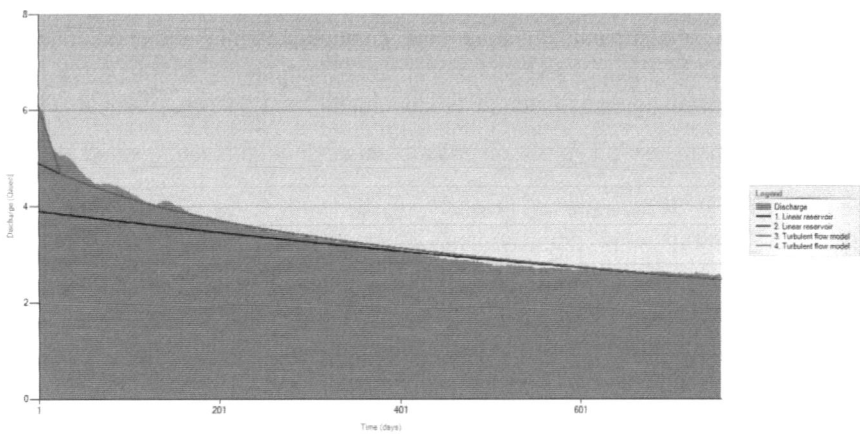

Figure 4. Calibration model of a single recession curve.

The shape of each recession curve is distinctive, depends on several factor such as; the development of conduit, the catchment area size, the aquifer response to the precipitation intensity, etc. After each recession curve has calibrated, all the recession curves subsequently combined to generate Master Recession Curve. The shape of calibrated MRC in Nyadeng Spring shows the combination of two turbulent flow regimes and one laminar flow regime as it presented in Fig. 5. The domination of turbulent flow regimes indicates the reduction of active fissures in the aquifer system (Malik & Vojtkova, 2012). This condition consolidates the hypothesis that open conduit channels dominate the systems. The flow types at Nyadeng Spring can be classified as turbulent fast flow.

Figure 5. Nyadeng Spring located in the middle of Merabu Forest.

However, the diffuse flow at Nyadeng Spring During also play important role during recession periods or the dry season. Diffuse flow provides steady discharge storages with abundant availability. From the hydrograph and MRC can be seen that the diffuse flow with laminar characteristics also exist in every single recession event. This condition indicates that the diffuse flows significantly contribute to the aquifer. The combination of linear and turbulent flow alternately dominate the aquifer, clarify that the aquifer in Nyadeng systems formed in mixed-aquifer type [1].

The determination of karstification degree at Nyadeng Spring derived from four individual recession events. Based on the sub regimes analysis, one laminar sub regime and two turbulent sub regimes were identified. Eq.4 express superimposed equation of flow regimes at Nyadeng Spring.

$$\mathbf{Qt= 3.2^{-0.001t} + 1.2(1\text{-}0.012t)+1.6(1\text{-}0.035t)} \quad (4)$$

Based on the MRC result, Nyadeng Spring dominated by turbulent flow regimes. According to [10] karstification degree classification, Nyadeng Spring classified at the scale of 7 to 8. Discharge regime in Nyadeng formed in a combination of sub-regime with laminar flow with two sub-regimes with turbulent flow. Excessive role in groundwater discharge is played by sub-regimes with turbulent flow. Sub-regime with laminar flow is less significant. This classification expresses a highly developed karstification of the aquifer, characterized by large karst channels/ conduits. The role of phreatic zone is insignificant due to the domination of conduit channels with turbulent flow characteristics.

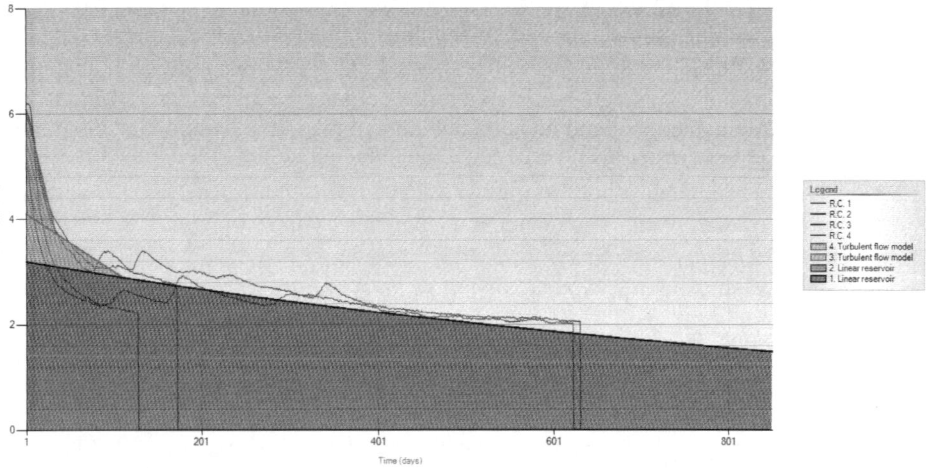

Figure 6. Master recession curve of Nyadeng Spring.

Hydrograph monitoring in Nyadeng Spring is a preliminary research that potentially provides many advantages in the future. This research was facing many technical and non-technical constraints. However, from this minimum result, it can be concluded that Nyadeng Spring play very important role as water resources in Merabu Karst area. This research is expected to provide more scientific data in the future. The determination of karstification degree in Nyadeng Spring shows that Nyadeng Spring formed in developed karstic aquifer with complex discharge regimes. These facts potentially provide a scientific document for karst river basin management.

3. Conclusion

Nyadeng Spring formed in a combination of sub-regime with laminar flow with two sub-regimes with turbulent flow. Excessive role in groundwater discharge is played by the combination of laminar and turbulent flow. The karstification degree in Nyadeng Spring classified at 7th-8th scale. The combination of linear and turbulent flow alternately dominate the aquifer, clarify that the aquifer in Nyadeng systems formed in mixed-aquifer type. These results indicate the development of Nyadeng Spring catchment area. This preliminary research expected to be continued in the future and provides comprehensive scientific considerations for karst water resources management.

4. Acknowledgement

This research is part of the research grant entitled "Karakterisasi dan Revitalisasi Nilai Penting Kawasan Karst Sangkulirang-Mangkalihat". This research was conducted by Karst Research Group, Faculty of Geography, Universitas Gadjah Mada. Authors gratefully acknowledge the research funding and support from Tropical Forest Conservation Act (TFCA) and all parties that involved in the funding scheme.

References

[1] Adji TN 2010 Spatial and Temporal Variation of Hudrogeochemistry and Karst Flow Properties to Characterize Kasrt Dynamic System in Bribin Underground River. *Dissertation*. Yogyakarta: Faculty of Geography, Universitas Gadjah Mada.

[2] Adji TN Haryono E Fatchurohman H and Oktama R 2016 Diffuse flow characteristics and their relation to hydrochemistry conditions in the Petoyan Spring, Gunungsewu Karst, Java, Indonesia. *Geosciences Journal, Vol. 20,* **No. 3**, *p. 381–390, June 2016*

[3] Bonacci O 1993 Karst spring hydrograps as indicators of karst aquifers. *Hydrological Sciences-Journal-des Sciences Hydrologiques* **38,1,2**, 51-62.
[4] Eiche E Hochschild M Haryono E and Neumann T 2016 Characterization of recharge and flow behaviour of different water sources in Gunung Kidul and its impact on water quality based on hydrochemical and physico-chemical monitoring. *Applied Water Science. DOI 10.1007/s13201-016-0426-z.*
[5] Fatchurohman H Cahyadi A and Suprayogi S 2016 Hydrograph Analysis for Sustainable Karst Aquifer Protection and Management. *Geomedia Volume 14 Nomor 1 Mei 2016.*
[6] Fitts C R 2002 *Groundwater Science.* London: Academic Press.
[7] Ford D and Williams P 2007 *Karst Hydrology and Geomorphology.* West Sussex: John Wiley & Sons Ltd.
[8] Haryono E 2000 Some properties of epikarst drainage system in Gunung Kidul Regency, Yogyakarta, Indonesia. *The Indonesian Journal of Geography, vol. 32, No. 79-80.*
[9] Malik P 2007 Assessment of regional karstification degree and groundwater sensitivity to pollution using hydrograph analysis in the Velka Fatra Mountains, Slovakia. *Environ Geol* **vol.51,** 707-711.
[10] Malik P 2015 Evaluating Discharge Regimes of Karst Aquifers. In Z. Stevanovic´, *Karst Aquifers – Characterization and Engineering* (pp. 205-249). -: Springer International Publishing Switzerland.
[11] Malik P and Vojtkova S 2012 Use of recession-curve analysis for estimation of karstification degree and its application in assessing overflow/underflow conditions in closely spaced karstic springs. *Environ Earth Sci (2012)* **65,** 2245–2257.
[12] Posavec K Parlov J and Nakic Z 2010. Fully Automated Objective-Based Method for Master Recession Curve Separation. *Groundwater* **Vol 48**. *No.4*, 598–603.
[13] Rivera-Ramírez H D Warner G and Scatena F 2002 Prediction of master recession curves and baseflow recessions in the luquillo mountains of puerto rico. *American Water Resources Assosiation vol.38 no.3*, 993-704.
[14] Tallaksen L 1995 A review of baseflow recession analysis. *Journal of Hydrology* **165**. 349-370.
[15] Todd D K and Mays LW 2005 *Groundwater Hydrology-Third Edition.* New York: JohnWiley & Sons.
[16] Walton W C 1970 *Groundwater Resource Evaluation.* Tokyo: McGraw-Hill.
[17] Younger P L 2007 *Groundwater in The Environment.* Victoria: Blackwell Publishing.

Trash-polluted irrigation: characteristics and impact on agriculture

D Sulaeman[1], SS Arif[2] and Sudarmadji[3]

[1]Post Graduate School, Universitas Gadjah Mada, Yogyakarta 55281, Indonesia
[2] Faculty of Agricultural Engineering, Universitas Gadjah Mada, Yogyakarta 55281, Indonesia
[3] Faculty of Geography, Universitas Gadjah Mada, Yogyakarta 55281, Indonesia

sigitsupadmoarif@ugm.ac.id

Abstract. Trash pollution has been a problem in sustainable water resources management. Trash pollutes not only rivers, lakes and seas, but also irrigation canals and rice fields. This study aimed to identify the characteristics of solid waste (type, time of occurrence and sources of trash) and its impact on agriculture. The study was conducted in four irrigation areas, namely Gamping, Merdiko, Nglaren and Karangploso in Bantul District, Yogyakarta Special Region. We applied the Irrigation Rapid Trash Assessment (IRTA) as our field survey instrument. The results showed that trash was found throughout irrigation canals and rice fields, and the occurrence was influenced by water flow, time and farmer activities. The irrigation was dominantly polluted by plastic trash (52.2%), biodegradable waste (17.91%) and miscellaneous trash (12.3%). The IRTA score showed that Gamping Irrigation Area was at marginal condition, bearing a high risk of disturbing the operation and maintenance of the irrigation canals as well as farmers' health. Trash in irrigation also generated technical impact of the irrigation operation and maintenance, environmental quality, and social life. This research also offered environmental policy integration approach and water-garbage governance approach as an alternative solution to manage water resources and agriculture in a sustainable manner, under the pressure of increasing amount of trash.

1. Introduction

Trash pollution in water bodies has been widespread and caused serious consequences for the ecosystem and livelihood of people. In terms of water quality control, trash generated from human activities might be identified as litter, debris, rubbish, or refuse [31]. Currently, trash is easily found in rivers, lakes, beaches and seas [16, 17, 33, 35, 37].

Researches on trash in rivers, lakes and seas in diverse areas across the world reveal that no water bodies are free of trash. Like in Manipur, India, where trash was disposed in rivers which were used as the standard water resources [30], trash had also polluted rivers in the urban areas of Ondo State, Nigeria [27], and was found in rivers and seafloor in Los Angeles, USA [31].

Researches on coastal areas and oceans found that trash had polluted coastal and remote areas of oceans [33], up to the Arctic and the Antarctic [35]. According [11] trash produced by 192 coastal countries in 2010 was as much as 275 million metric tons (MT) and 4.8 to 12.7 million MT entered the oceans.

The condition in Indonesia is not much different. The Ministry of Environment and the Ministry of Health recorded that rivers were still used for trash disposal [21, 18, 19]. Cikapundung River in Bandung was polluted by 1,500 m^3 of trash per week [14], Saguling dam was polluted by 250,000 m^3 per year [20], Citarum watershed was polluted by domestic trash of the settlement located around the river [5], rivers and ponds in Jakarta were used for disposing both domestic and industrial trash [9] and

Code River in Yogyakarta was polluted by domestic trash from human settlement [13]. Indonesian seas were estimated to be polluted by plastic trash as much as 0.48–1.29 MMT/year due to mismanaged plastic waste on the land [11].

Not only has trash contaminated natural water bodies, rivers, lakes and seas, but also artificial water bodies such as drainage. Several studies had reported that most urban drainage was polluted by domestic and non-domestic trash [26, 27, 29]. Like in Cikapundung, drainage was polluted by domestic trash up to 44.3 m^3/month [14].

Trash disposed at a site will not only contaminate the site, but also spread to other locations through water flow, wind and sea waves [17]. Waste was disposed or flushed into the rivers [27], pollutant from the municipal landfills/dump sites was carried away by runoffs to rivers [24], trash was dumped near the rivers, and when the rain fell most of these materials were washed away into the rivers [25].

Widespread trash disposal to water bodies has been suspected to be caused by the lack and weak infrastructure, management systems, management institutions and human resources who manage trash [10, 24, 25], and low awareness and community participation [3, 8, 30]. In fact, the impacts of pollution on water bodies have been widely reported, such as causing illness, water dysfunction, endangering aquatic life [10, 24], flood, erosion and infrastructure damage [27].

Irrigation system using river as its water source also has the potential of trash contamination. However, there are not many researches on trash contamination in irrigation that have been carried out. Previous researches only mentioned indications of pollution contamination in irrigation, but the information was limited to information on the trash in irrigation canals and its suspected impacts on irrigation system in the form of inhibition of water flow and farmers' working hours [1, 12, 15]. Therefore, research on the characteristics of trash in irrigation and its impacts on agriculture are needed to ensure that irrigation systems can proceed as planned and decision on food supply can be made.

2. The Methods

The study was conducted in four irrigation areas in Bantul Regency, namely Gamping Irrigation Area, Merdiko, Nglaren and Karangploso in Bantul Regency (figure 1).

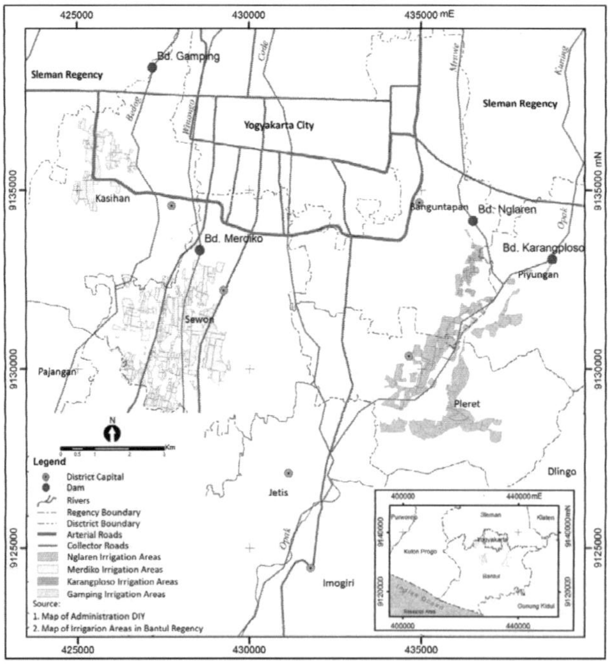

Figure 1. Irrigation trash assessment location map.

They were irrigation areas whose service was adjacent to Yogyakarta City, such as the district of Kasihan, Sewon and Banguntapan [22, 36]. Thus, this region experiences considerable environmental pressures [32] including land conversion, increasing population density, and domestic and commercial activities. The implication was that settlements and commercial activities were getting closer or closer to irrigation canals and agricultural land.

The data in this research was collected through questionnaires and interviews with farmers in 4 irrigation areas. In one of the irrigation areas, data collection was followed by direct waste measurement (figure 2). This research was conducted from December 2015 to April 2016.

In each irrigation area, 40 farmers were selected as respondents. They were members of the Group of Water-User Association (GWUAs). The number was based on input from GWUAs board. These farmers were those who were willing and had an opportunity to become the respondents. The questionnaires were used to collect information from farmers in the form of farmers' observation on the trash found, the time of trash emerging, and the impact of the trash on agricultural business. Interviews were also conducted with farmers to get a complete picture of the condition of irrigation, land and agricultural business.

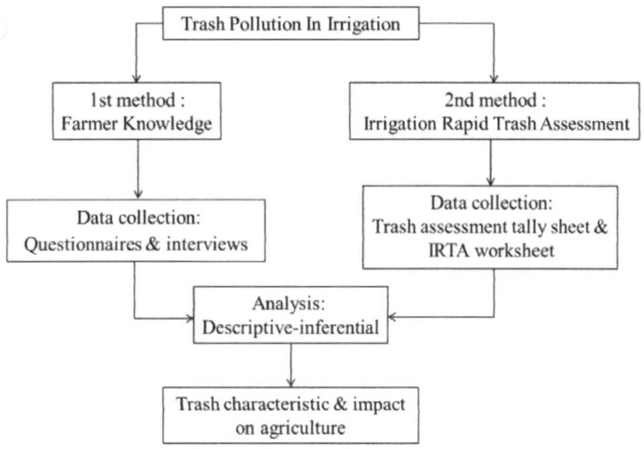

Figure 2. Research design flowchart.

There have not been any established methods developed for the assessment of trash in irrigation. The method used in this study for the assessment of waste in irrigation is called Irrigation Rapid Trash Assessment (IRTA), an adoption and modification of the California Water Board (CWB) method [6] (Table 1).

The California Water Board (CWB) method is a river trash assessment method that has the closest characteristics for the assessment of trash in irrigation, that is, water flows in the canal almost all the time. However, there is a difference in the flow aspect between unsteady flow in the river and continuous and steady flow in irrigation.

Similar with the CWB method, trash assessment using the IRTA method was carried out in two sheets of paper work, namely trash assessment tally sheet and IRTA worksheet. Trash assessment tally sheet combined sheet of California Water Board (CWB) [6] and Southern California Coastal Water Research Project (SCCWRP) [31], which contained ten trash groups, namely plastic, biohazard, construction, miscellaneous, metal, large, toxic, biodegradable, glass, and fabric and cloth, with an addition of agriculture residues into biodegradable groups. Calculation of trash in this research was based on trash unit found. The IRTA worksheet contained a thorough assessment against the parameters and condition categories of the data of trash found in irrigation. A total of six parameters, combining the parameters of illegal dumping and illegal littering, were assessed in four categories of condition: optimal, sub optimal, marginal and poor. Each subcategory was subdivided into five levels, except the poor with six levels and a continuous value of 0 to 20.

Table 1. Modified rapid trash assessment methods for rivers to irrigation.

Rapid Trash Assessment [6]		Irrigation Rapid Trash Assessment	
Trash Assessment Parameter	Condition Category	Trash Assessment Parameter	Condition Category
Level of trash	Visual observation of the presence and estimation of river filth. Value <10 up to >100 pieces.	Level of trash	Visual observation of the presence and estimation of river filth. Value <10 up to >100 pieces.
Actual number of trash items found	Amount of trash items per 100 ft stream	Actual number of trash items found	Trash items amount per 25 meters of irrigation canal
Threat to aquatic life	The trash types and nature that was hazardous for aquatic life	Disturbance in the operation and maintenance of irrigation, and crops	Trash type and nature that disrupts irrigation operation and maintenance, and rice crops
Threat to human health	Type, level of hazard and trash content that are hazardous to human health	Threat to human health and aesthetic	Type, level of hazard and trash contents that are hazardous to human health (especially farmers and irrigation operators), and aesthetic
Illegal dumping	Indication and evidence of illegal dumping	Illegal dumping	Indication and evidence of illegal dumping in borders or irrigation canals
Illegal littering	Indication and evidence of illegal littering	Illegal littering	Indication and evidence of illegal littering in borders or irrigation canals
Accumulation of trash	Indication and evidence of trash accumulation caused by river water flow	Accumulation of trash	Indication and evidence of trash accumulation due to irrigation water flow

Measurements with IRTA were carried out three times at weir, primary, terrain and tertiary canals with a gap of 3 days between measurements. The data was analysed descriptively to show the type, category, amount, and distribution of waste.

3. Result and discussion

3.1. Characteristic of trash pollution in irrigation

3.1.1. Trash characteristics in irrigation based on farmer observation. Observation of trash in irrigation had become part of the daily activities of farmers from the time of cultivation, planting, maintenance and harvesting. In the whole cycle of rice cultivation, they interacted with irrigation, land and crop for at least 3 months in each planting season. So, it was possible for them to know the condition of trash contamination in the irrigation canals and in rice fields.

Based on the research, all respondents stated that they had found trash in irrigation canals and their land. Meanwhile, observation on each irrigation network found different values with smaller trends for locations far from the farming lands (table 2).

Table 2. Outreach of farmers' observation on trash in irrigation area.

Description	Farmer answer	
	Frequency	Percentage (%)
In general, did you find trash in land and irrigation canals?	160	100
Did you find trash in		
a. your land?	160	100
b. tertiary canals?	160	100
c. secondary canals?	135	84.4
d. primary canals?	62	38.8

There are at least two things that can be explained from this condition. First, the farmer's observation focused more on the location close to the land they were working on, and did not pay much attention to the condition of canals far from their location. Second, it also indicated the ability of farmers' individual contribution to solve the problem of trash in irrigation which focused on the land they had, then gradually they would expand their contribution in areas which were not within their responsibility.

The trash which the farmers said existing in the irrigation canals and the rice field reached up to 23 types (figure 3). Frequency of the type of trash mentioned by the farmers was related to the following four factors: often found, arousing emotion, hazardous to health and safety of the farmers, and disrupting water flow. Plastic bags, or any types of plastic items, were the ones the farmers mostly mentioned because this type of trash was found throughout the irrigation network reaching up to the fields. Being easily carried away by water flow, these plastic items were driven away from their disposal sites.

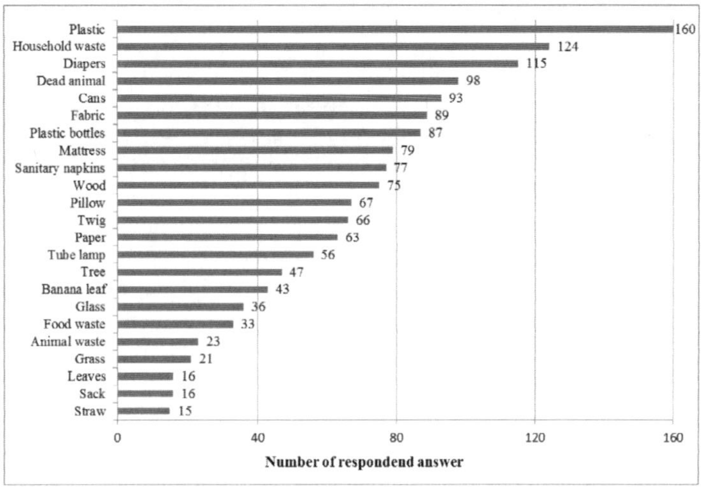

Figure 3. Type of trash based on farmers' observation, n=160.

Household trash, diapers and sanitary napkins were the types of trash that evoked emotion. Household waste, which the farmer declared as plastic bags filled with trash, evoked emotion due to the amount of trash in plastic bag units, and people who disposed the trash. When torn apart, the trash in the plastic bags would come out and spread. As a result, the number of the trash was large in unit although the number of bags was relatively small. According to the farmers, such domestic waste was disposed by workers who brought two bags– one briefcase and one plastic bag filled with trash – on their way to work. Diapers and sanitary napkins were supposed to essentially be disposed with certain treatment. Farmers considered this waste irritating because when they filled the canal and eventually were stranded in the field. They considered that such behaviour looked down at them because the trash made the farmers' environment dirty. Trash such as dead animals, lamp tubes and glass were considered hazardous to the health and safety of the farmers. Household trash, tree branches and plants were stated as trash that disrupted the water flow.

The time of trash emergence, which was divided into high and low-grade groups, in irrigation canals and rice fields referred to the terminology of water flow, time and farmer's activity (table 3). Based on the terminology of water flow, influenced by the season and water discharge, the terminology of time was expressed according to the time of disposal of trash into irrigation, while the terminology of the farmer's activity was related to activities inside and outside cultivation.

Table 3. Time of trash emergence in irrigation canals and rice fields.

Description	A lot of trash	A little trash
Terminology of water flow	Rainy season	Dry season
	Every water flow	Water does not flow
	Canal flushing	Canal does not work
Terminology of time	Morning	Afternoon
Terminology of farmer's activity	No collective work	After collective work
	Post harvest	The beginning of planting season

In the water flow terminology, it appeared that the trash polluting the canals and rice fields was influenced by the flow of irrigation water. The water flow could come from the operation of irrigation gates, or river flow in certain seasons that swept away the trash and it entered the irrigation system. From this condition, the tertiary canals and rice fields became locations that received a greater trash load from secondary and primary canals.

The morning and afternoon were the times related to when the trash was dumped into irrigation canals and the farmers found the trash in irrigation. Morning was the time when most of the households were cleaning the house, including throwing trash, and it was also the time when the farmers began their activities in the fields. These two conditions met at one time: in the morning. The afternoon was not the time for the households to dispose their trash, and farmers were no longer had activities in the fields.

Farmers' activities influenced the cleanliness of irrigation canals. When the harvest finished, farmers usually spent some time without doing any activities in the fields, including cleaning the irrigation canals. As a result, more trash stayed longer in the irrigation canals. At the time when planting season began, the farmers prepared the land and cleaned the irrigation canals so that the water could flow smoothly into the rice fields.

3.1.2. Trash characteristics in irrigation based on IRTA. As many as 22 locations, which consisted of weir, primary, secondary and tertiary canals, were visited to confirm the height level of trash present in the irrigation system in Gamping Irrigation Area. On average, covering the entire locations and repetition, 52.2% of the total trash consisted of plastic items (Figure 4). Biodegradable (17.91%) and miscellaneous (12.39%) items were also commonly found. The biodegradable items which was most found were food waste, yard waste and agriculture residues.

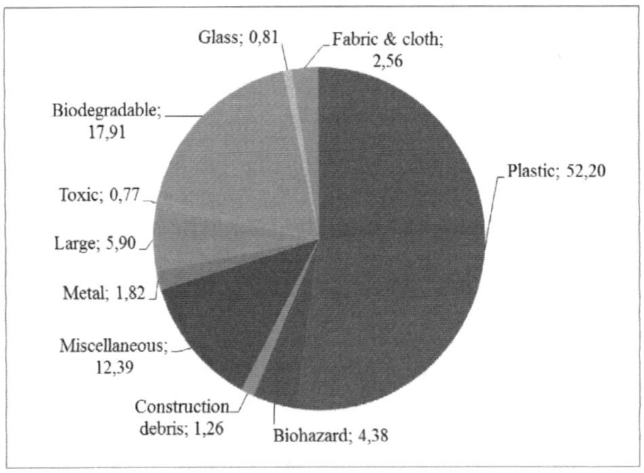

Figure 4. Composition of trash (%), by category, per 25 meters reach for all sites.

Most sites contained less than 200 pieces of trash, while several sites contained many more pieces, up to a maximum of 419 pieces. In the observation locations of the secondary canals, it was recorded that the average number of trash was 196 pieces, 2.7 times of those in the primary canals, and almost twice of those in the tertiary canals (Table 4).

Table 4. Comparison of the number of pieces of trash in each observation location.

Observation location	Number of pieces of trash		
	The lowest	The highest	Average
Weir	11	61	31
Primary canals	45	121	72
Secondary canals	120	419	196
Tertiary canals	69	141	104

The distribution of the number of trash pieces from the weir to the tertiary canals reflected how much stream flow affected the accumulation of trash. In accordance with the irrigation stream management, the stream in the primary canals would be greater than that in the secondary and tertiary canals [4]. Therefore, although in the border line of the primary canals a number of illegal dumping points were found, where trash could be strongly possible to enter the canals, the trash in the primary canals would flow to the downstream. Another cause was when there was an accumulation of trash in the primary canals, people would relieve the blockage and flow the trash to the downstream. Accordingly, trash accumulation would, in fact, happen in the tertiary canals and the farming land.

The IRTA result showed, however, that the highest number of trash units was found in the secondary canals. Such a condition happened because of water management pattern in the secondary canals which was balanced and without any sluicegates, trash from the primary canals entered the secondary canals, a small number of secondary canals, settlement and commercial activities around the secondary canals were higher than those around the primary and tertiary canals. Water management to the tertiary canals was done through off take doors, so that trash from the secondary canals spread out to the tertiary canals.

Based on irrigation network system, it was visible that the percentage of the type of trash that could easily be carried away by the water flow had increased in the location near the downstream (figure 5). It became an indication that tertiary canals and farming land bore a high risk of accommodating plastic trash from the canals located in the upper area, which was worsened by trash directly disposed to the tertiary canals. In the tertiary canals, trashes in the form of pieces and the ones that were still wrapped in plastic bags, as well as old and new trash, were found.

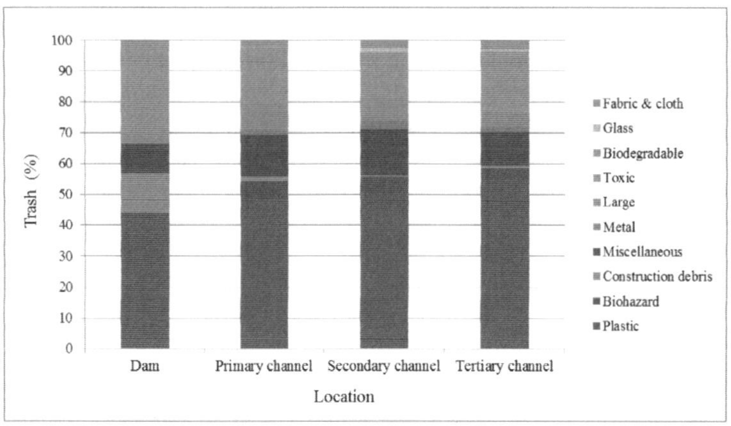

Figure 5. Percentage of trash based on category in each observation location.

In this research, plastic trash in various sizes filled with trash was categorised as "big" trash because the width of irrigation canals varied, starting from primary canals which was 2 meters wide to tertiary canals which was 0.4 meter wide. Therefore, a 0.25 m^2 plastic sheet was enough to clog half of the canal. It was different from what was defined as garbage bag of trash according to CWB [6], which was a plastic bag size 100x120 cm^2, because the object of the observation was river.

3.1.3. IRTA score. The average total Irrigation Rapid Trash Assessment (IRTA) score was 56, ranging from 38 to 110 and none had score of 120 (Figure 6). Lower IRTA score indicated the high level of trash. A high IRTA score, overall or in a specific category, represented that the location had a small amount of trash [6]. IRTA score was built from six parameters, each of which had a maximum score of 20. With an average IRTA score of 56, Gamping Irrigation Area fell under the category of marginal, which means that the level, amount, disturbance, and source of trash that entered the irrigation canals could result unsustainable irrigation system and need a serious handling.

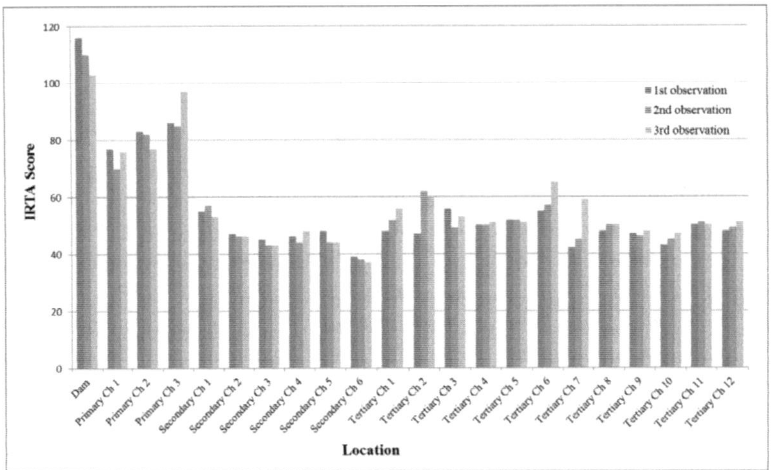

Figure 6. IRTA score in all observation locations and the repetition.

Based on IRTA, other parameters that became the causes of score obtained and impacts on irrigation management could be seen. In irrigation system, there are five important pillars that are used as reference, namely water supply, infrastructure, institutions, irrigation management, and human resources [1]. The pillar of irrigation management in the form of operation and maintenance (O&M) at all levels becomes important to ensure the management and distribution of water. Pilar of human resources includes self and all capability possessed by each individual in her or his role in irrigation. Trash in irrigation seems to create disturbance for the pillars of operation and maintenance of irrigation as well as human resources in the form of health disorder (figure 7). The types of trash that disturbed the operation and maintenance of irrigation included mattress, pillow, wood, garbage bag of trash, furniture, concrete, and tree, whereas the types of trash that were hazardous to the farmers included diapers, animal waste, dead animals, sanitari napkins, tube lamps, metal, and glass.

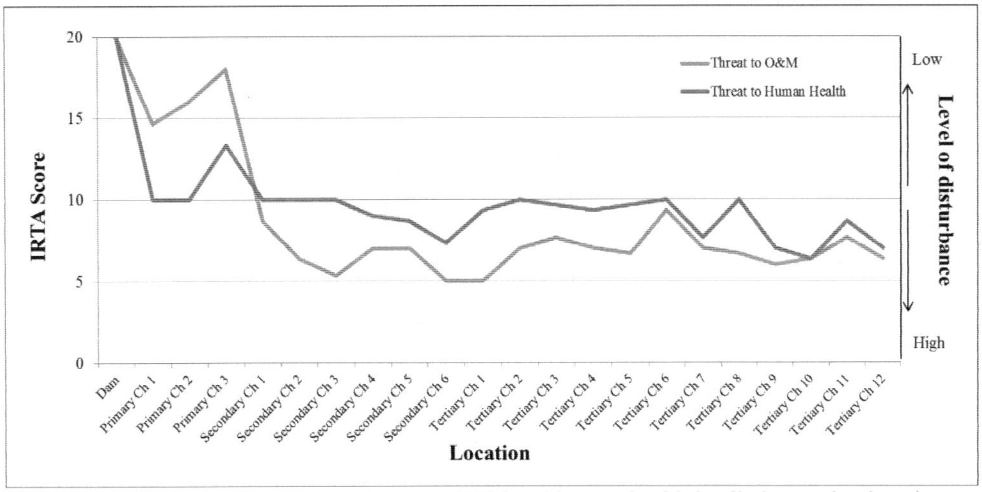

Figure 7. IRTA score in parameter threat to O&M and human health in all observation locations.

Increase of disturbance was higher in locations where the infrastructure dimension was getting smaller, which was in secondary and tertiary canals. Meanwhile, degradation in human health, particularly the farmers', increased in the irrigation canals where farmers did their activities. Both disturbances simultaneously would cause degrading performance of irrigation as well as degradation of the farmers' health.

3.2. Impacts of trash in irrigation on farmers

The impacts of trash in irrigation were identified to cause problems in operation and maintenance of irrigation, quality of irrigation water, and social relation. Operation and maintenance of irrigation were two activities that were interrelated. Operation of irrigation included water delivery schedule and maintenance included activities to keep the operation process running effectively through maintaining physical facility and cleanness of the canals.

The operation of irrigation disrupted by trash had something to do with the physical presence of big trash and high accumulation of trash which disrupted water flow, overflow of water over the canals, and increased delivery time. Considering the effectiveness of irrigation operation, the incrementing volume of trash in irrigation canals accordingly had increased irrigation maintenance in the form of irrigation canal cleaning. In reality, people in charge of irrigation, Water User Association (WUAs), and farmers had not yet been able to improve irrigation maintenance to overcome trash problems. Periodical irrigation maintenance had to be done more often as in three days, or even less, irrigation canals would be polluted by trash again.

Trash disposed in water bodies would degrade the quality of water physically, chemically [27, 30] and bacteriological [24, 25]. Although the study was conducted in rivers, since the type of trash disposed in the irrigation canals was generally the same, it could accordingly be predicted that the trash disposed in the irrigation canals would also degrade the quality of irrigation water. Poor quality of water used in agriculture affects the land and crop [2, 23, 38] which eventually will degrade the productivity of land and crop as well as the quality of the products generated. Therefore, the aspect of food security and safety will be disrupted.

Social turmoil also appeared in this condition. Complaints from the farmers to the irrigation officers about the late delivery of water, improper volume of water, and delivering water and trash, at the same time were conveyed. The farmers' trust to the irrigation officers would decline and the further impact was that the rules of cultivation pattern and water distribution that had been set were not thoroughly obeyed. There were also closed conflicts among the farmers in the form of bad expression and prejudice, especially by the farmers in the downstream against those in the upstream. When irrigation water flew in the tertiary canals in the downstream and carried along trash, farmers in the downstream would suspect that the community members in the upstream disposed their trash in the irrigation canals and farmers in the upstream did not clean the irrigation canals. Such a closed conflict could potentially become an open conflict among the farmers and between the farmers and the community members if it was not solved.

3.3. Discussion

The handling of trash pollution in irrigation canals should become an important concern as it involves multi stakeholders [28, 34]. Water resources institutions, including those which managed irrigation, agriculture, waste, and the environment, as multi stakeholders, were supposed to respond the changes happening in the community, rivers, irrigation, and farming land. The persistent behaviour of the community members even got more unfriendly against the water resources [14, 19, 22], ineffective institutional performance, lack of funding, environmental policies that were not jointly [34] made, all of which aspects have to be improved.

With a minimal level of trash services in Bantul [7] and community behaviour that still made water bodies trash disposal site [18, 19], irrigation canals, including irrigation borders located close to residential and commercial areas, would continuously became location of waste disposal. Sources of trash in irrigation were also from rivers that brought the flow of trash to irrigation canals.

Trash entering irrigation canals was not only that which disturbed the operation and maintenance of irrigation, but also hazardous trash that affected the health of the farmers. The loss they had suffered due to polluted irrigation was their decreasing time for working on their farming land as their time was consumed for cleaning the irrigation canals beyond the schedule set for irrigation maintenance, increasing healthcare expenses for suffering from wounds, infection, and disease contracting, as well as estrangement of social cohesion. Trash in irrigation had also caused degradation of water quality which affected the quality and quantity of their agricultural products. These losses would eventually press the economic life of the farmers and push them away from welfare. In addition, non-economic loss, which was social loss, also happened and for farmers, it was a loss of such a precious thing.

4. Conclusion

Irrigation as the supporting factor of food production had received pressure with the increasing pollution in rivers and directly in irrigation. Trash was one of the pollutants that had not been properly identified in irrigation, although in reality it was indicated to be in a state that needed to be seriously handled by all stakeholders. Trash spread out in all irrigation canals and farming land and tended to spread more widely to the downstream. Time of occurrence of the trash in irrigation canals had reflected that irrigation was exposed to trash disposal almost at any given time.

Trash in irrigation was dominated by plastic items up to 52.2%. The percentage of plastic trash was increasing in the downstream, which was in the tertiary canals. With the use of IRTA, biohazard trash, particularly diapers, was found in all observation locations. IRTA score in Gamping Irrigation Area was 56, meaning that it was at marginal state. Based on IRTA method, it was identified that trash disrupted the operation and maintenance of irrigation as well as the health of the farmers. Another impact of trash pollution in irrigation was disrupted social relation among the farmers and between the farmers and the community members, although it was still in closed conflicts.

Some impacts of trash on agricultural had indeed been identified. However, to know how much the impact of trash on agriculture is will need further research on the disturbance on the technical aspects of the operation and maintenance of irrigation, health of farmers, social relation between farmers and the community, and how to develop garbage-water governance. The government and the community can also prepare several ways and short-term activities to overcome trash in irrigation and prepare a more comprehensive strategy to manage water resources and agriculture in a sustainable manner.

References

[1] Arif S S Sastrohardjono S Subekti E Prabowo A Soekrasno S Sidharti T S Soetiarso L Mustofa A Sulaeman D Basuki R Ruzziyatno R Putra V R S Wicaksono Y R Yuliati N and Jahid I 2014 *Principles of Irrigation Modernisation in Indonesia* (Jakarta: The Ministry of Public Work and Housing).

[2] Ayers R S and Westcot D W 1976 *Irrigation and Drainage Paper 29: Water Quality for Agriculture* (Rome: Food And Agriculture Organization of The United Nations).

[3] Boateng S Amoako P Appiah D O Poku A F and Garsonu E K 2016 Comparative analysis of households solid waste management in rural and urban Ghana *Journal of Environmental and Public Health* JEPH (2016) 5780258.

[4] Bos M G Kselik R A L Allen R G and Molden D J 2009 *Water Requirements for Irrigation and The Environment* (Dordrecht: Springer Science + Business Media B.V.).

[5] Cahyaningsih A and Harsoyo B 2010 Spatial distribution of water pollution level in Citarum Watershed *Jurnal Sains & Teknologi Modifikasi Cuaca* 11(2): 1-9.

[6] California Water Board 2007 *A Rapid Trash Assessment Method Appied to Waters of The San Francisco Bay Region: Trash Measurement in Streams* (San Francisco: California Water Board).

[7] Development Planning Agency of Bantul 2011 *White Papers on Sanitation of Bantul Regency* (Bantul: Development Planning Agency of Bantul).

[8] Gogoi L 2013 Municipal solid waste disposal: A case study in Guwahati city to mitigate the man made disaster, *IOSR Journal of Humanities and Social Science (IOSR-JHSS)*, 9(3): 55-60.

[9] Hendrawan D 2005 Water quality of rivers and ponds in Jakarta *Makara Teknologi* 9(1): 13-9.

[10] Hill M K 2010 *Understanding Environmental Pollution* 3rd ed (New York: Cambridge University Press).

[11] Jambeck J R Geyer R Wilcox C Siegler T R Perryman M Andrady A Narayan R and Law KL 2015 Plastic waste inputs from land into the ocean *Science* 347(6223): 768-71.

[12] James L Smith J Reddy M Pochop L O and Lewis R W 1991 Design of turbulent fountain irrigation trash screens *Irrigation and Drainage Systems* 5: 267-75.

[13] Jati W R 2013 Involution of "River Purification Policy" in Yogyakarta *Masyarakat, Kebudayaan dan Politik* 26(4): 217-26.

[14] Jatnika L and Rahardyan B 2015 Application of contingent valuation method in sanitation quality improvement efforts at Cikapundung River in Bandung City *Majalah Ilmiah Globe* 17(1): 59-66.

[15] Kemper W Bondaurant J A and Trout T J 1986 Irrigation trash screen pay *Journal of Soil and Water Conservation* 41: 17-20.

[16] Machado Lake Trash TMDL Jurisdictional Group 2008 *Trash Monitoring & Reporting Plan: Machado Lake Trash TMDL* (California: The California Regional Water Quality Control Board).

[17] Midbust J Mori M Richter P and Vosti B 2014 *Reducing Plastic Debris in the Los Angeles and San Gabriel River Watersheds* (Santa Barbara: Bren School of Environmental Science & Management, University of California).

[18] Ministry of Health 2008 *Research on Basic Health 2007: A National Report of 2007* (Jakarta: Research and Development on Health Agency, The Ministry of Health of the Republic of Indonesia).

[19] Ministry of Health 2013 *Reseach on Basic Health 2013* (Jakarta: Research and Development on Health Agency, The Ministry of Health of the Republic of Indonesia).

[20] Ministry of Environment 2011 *Indonesia Environmental Status 2010* (Jakarta: The Ministry of Environment of the Republic of Indonesia).

[21] Ministry of Environment 2013 *Environmental pillars of Indonesia* (Jakarta: The Ministry of Environment of the Republic of Indonesia).

[22] Ministry of Public Work and Housing 2015 *Regulation of the Minister of Public Work and Housing of the Republic of Indonesia No. 14/PRT/M/2015 on Criteria and Stipulation of Irrigation Area* (Jakarta: Ministry of Public Work and Housing).

[23] Misaghi F Delgosha F Razzaghmanesh M and Myers B 2017 Introducing a water quality index for assessing water for irrigation purposes: A case study of the Ghezel Ozan River *Science of the Total Environment* 589: 107-16.

[24] Nartey V K Hayford E K and Ametsi S K 2012 Assessment of the impact of solid waste dumpsites on some surface water systems in the Accra Metropolitan Area, Ghana *Journal of Water Resource and Protection* 4: 605-15.

[25] Nkwocha E E Pat-Mbano E C and Nnaji A O 2011 Effect of solid waste dump on river water quality: A paradigm in a Nigeran tropical environment, *I.J.S.N.* 2(3): 501-7.

[26] Oktiawan W and Amalia S 2012 The impacts of drainage system condition, trash and water waste on the quality of the environment (A study case of Kelurahan Kuningan in North Semarang District) *Jurnal Presipitasi* 9(1): 41-50.

[27] Olabode A D and Lawrence A 2014 Environmental impact of indiscriminate waste disposal on river channel in part of Akoko-Region, Ondo State, Nigeria *International Journal of Innovation and Scientific Research* 5(2): 162-68.

[28] Özerol G Bressers H and Coenen F 2012 Irrigated agriculture and environmental sustainability: An alignment perspective *Environmental Science & Policy* 23: 57-67.

[29] Putra M S E and Rahardyan B 2017 *Problem Identification of Waste in Drainage Channels at Coblong Sub-District, Bandung*, accessible date 20 Juli 2017, http://publikasi.ftsl.itb.ac.id/assets/repositori/2013_10_19/2/1_2_15307033_berkas.pdf.

[30] Singh C R and Dey M 2014 Surface water quality with respect to municipal solid waste disposal within the Imphal Municipality Area, Manipur *International Journal of Scientific and Research Publications* 4(2): 1-4.

[31] Southern California Coastal Water Research Project (SCCWRP) 2016 *Southern California Bight 2013 Regional Monitoring Program: Volume III. Trash and Marine Debris. SCCWRP Technical Report 973.* Costa Mesa, CA. ftp://ftp.sccwrp.org/pub/download/DOCUMENTS/TechnicalReports/928_B13_Debris.pdf.

[32] Sudrajat 2016 Farmers commitment in maintaining wetted land ownership status in peri-urban area of Yogyakarta *IJG* 48(1): 91-101.

[33] Thevenon F Carroll C and Sousa J (eds.) 2014 *Plastic debris in the ocean: The characterization of marine plastics and their environmental impacts, situation analysis report* (Switzerland: Gland).

[34] Tropp H Jiménez A and Le-Deunff H 2017 Water integrity: From concept to practice *Freshwater Governance for the 21st Century* ed Karar E (Switzerland: Springer Open).

[35] United Nations Environment Programme 2016 *Marine Plastic Debris and Microplastics - Global Lessons and Research to Inspire Action and Guide Policy Change* (Nairobi: United Nations Environment Programme (UNEP)).

[36] Water Resources Agency 2016 *Inventory of Irrigation Network of Bantul Regency Authority* (Bantul: Water Resources Agency).

[37] Weston Solutions 2013 *Trash and Litter Investigation Special Study: Final Report* (California: Weston Solutions).

[38] World Health Organization 1989 *WHO Technical Report Series n. 778: Health Guidelines for The Use of Wastewater in Agriculture and Aquaculture* (Geneva: World Health Organization (WHO).

IOP Conf. Series: Earth and Environmental Science **148** (2018) 012029 doi:10.1088/1755-1315/148/1/012029

Baseflow index assessment and master recession curve analysis for karst water management in Kakap Spring, Gunung Sewu

H Fatchurohman[1], T N Adji[1], E Haryono[1], P Wijayanti[1]

[1]Karst Research Group, Faculty of Geography, Universitas Gadjah Mada, Indonesia

hendy.fatchurohman@gmail.com

Abstract Karst terrain occurs in combination of high solubility rock and well developed secondary porosity. Over the time, groundwater resources have not been well managed including karst aquifers. Karst aquifers formed in a very complex hydrological system. Developed in fracture media and soluble rocks have led karst aquifers into various porosity types and aquifer properties. Karst spring hydrograph is an essential element for water resource management. The form of karst spring hydrograph reflects the aquifer characteristics. The shapes of flood discharge hydrographs represent aquifer responses to recharge and contain information about the interior condition of karst drainage basin. Every year, Gunung Sewu karst area is suffering to severe water scarcity. The development of sub-terrain drainage networks lead into the minimum surface water resources. Kakap Spring is perennial gravity spring that located adjacently to the border of Gunung Sewu and the alluvial formation of Baturetno. Kakap spring play vital role regarding water supply in Giriwoyo sub-district as the spring fulfill most of the water needs in Giriwoyo sub-district. Kakap Spring utilized by the local authorities as the main source for pipeline water and distributed to the households. Water level data series obtained using automatic water level data logger and then correlated with manual discharge measurement to generate stage-discharge rating curve. The stage-discharge rating curve formula for Kakap Spring calculated as y = 14,504e8,9763x with r2 value =0.8582. From the MRC result, flow regimes formula determined as + 400 (1-0,005t) + 700 (1-0,01t)., indicated that the aquifer dominated by turbulent flow regime. From the MRC formula, the degree of karstification in Kakap Spring classified at eighth scale. The average baseflow index in Kakap Spring calculated using recession curve analysis with the BFI index= 0,7485.

1. Introduction

Karst is an international terminology that widely known to describe a distinctive landscape type controlled by dissolution processes. Karst landform characterized by extensive subterranean drainages, caves, and closed depressions. Karst terrain is particularly developed in soluble rock such as carbonate rocks, evaporates and quartzite [8] [16] [15]. Approximately, more than 20 % of the world's ice free land occupied by karst rock. Despite lack of surface water availability, karst areas contain abundant groundwater resources. Karst groundwater resources reside as underground rivers beneath the earth surface or emerge as karst springs. About 20-25% of the world's population depends on karst water resources. Most of the biggest springs in the world are located in karst area. Therefore, karst aquifers play an important role for human water supply and one of the most significant aquifer formations. [7] [15] [18].

Gunung Sewu karst area is one of the largest karst areas in Indonesia with distinctive sinusoidal conical hills. Several karst springs emerge with various discharge rate and characters. Karst springs play an important role as potable water sources in Gunung Sewu karst area that considerably lack of surface water resources [1] [3]. Every year, Gunung Sewu karst area is suffering a severe water scarcity. Through the sinkholes and epikarst layer, the surface waters enter the saturated zone as allogeneic and autogenic systems. The development of sub-terrain drainage networks leads into the minimum surface water resources. The most dependable surface water resources in karst area are karst springs and doline ponds/telaga [17] [2].

Kakap spring is located in Giriwoyo sub-district, Wonogiri Regency, Central Java. Kakap Spring located in the north-eastern part of Gunung Sewu (figure 1). Kakap Spring is perennial gravity spring that located adjacently to the border of Gunung Sewu and the alluvial formation of Baturetno. Kakap spring play significant role regarding water supply in Giriwoyo sub-district as the spring fulfill most of the water needs in Giriwoyo sub-district. Kakap Spring utilized by the local authorities as the primary source for pipeline water and distributed to the households in 4 villages. Kakap Spring also utilized for irrigation purposes for 340 hectares agricultural area [4].

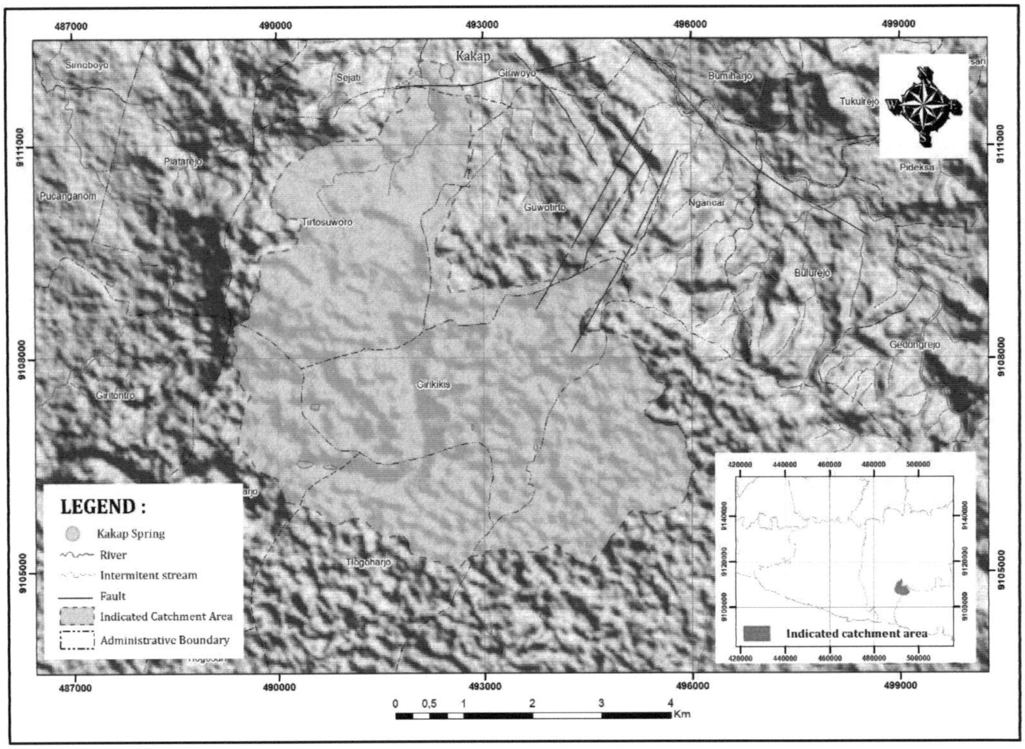

Figure 1. The sitemap of Kakap Spring.

Karst aquifer receives, stores, and transmits more water compared to all surface streams. Karst springs are known as the world's largest springs with the highest discharge rates compared to springs from other aquifer types. Among the largest water sources in the world, karst springs ensure that their importance for water supplies is widespread [19]. Karst spring hydrograph is an essential element for water resource management. The form of karst spring hydrograph reflects the aquifer characteristics. The shapes of flood discharge hydrographs represent aquifer responses to recharge and contain information about the interior condition of karst drainage basin. Hydrograph is generally more stable during the recession period and represents hydraulic and geometrical characteristics of the aquifers [14]. Besides, recession curve of karst spring hydrograph describes the significant progress when the maximum discharge recorded until the next consecutive pulse. Each part of the single flood hydrograph curve will represent the contribution of different porosity types [10]. Karst spring hydrograph analysis especially focused on recession curve data possibly provides valuable information in order to identify some characteristics of the karst aquifer and water resources management [9] [15] [24].

Recession curve analysis provides essential information to interpret the hydrological significance and characterize specific parameters of the springs. The main advantage of analyzing spring recession curves is that the drainage mechanisms in the aquifer and its quantitative parameters will be able to be estimated [5] [20]. Recession curve analysis plays an important role in determining the geological and storage characteristics of the karst aquifer. However, the rising limb of the hydrograph must have been analyzed to determine the karstification degree of the aquifer [22]. Karstification degree is very useful in the application of karst river basin management and deciding protection zone for karst aquifer. Groundwater management and protection zone in karstic aquifer can be based on the determination of karstification degree. According on those issues, this study focused on spring hydrograph analysis to provide general information of the aquifer systems in karst drainage basin within Gunung Sewu karst area. The objectives of this study were to determine the characteristics of recession curves of Kakap spring; Assess the karstification degree and the Baseflow Index of Kakap Spring as for karst water resources management.

2. The Methods

The main material of this research is hydrograph data series that measured using HOBO automatic water level logger. The HOBO water level data logger sensor measure the water pressure with operation range from 0 to 400 kPa (0 to 58 psi); approximately 0 to 30.6 and fully compensated with barometric pressure and logged with 30 minutes interval. Manual discharge measurements conducted using area velocity method, in this case surface float. After several discharge data obtained from numerous measurements, stage-discharg rating curve generated to determine linear regression between water level and stream discharge. The rating curve's regression function was used to convert the measured water level data into hydrograph series data. After hydrograph series created, the master recession curve (MRC) constructed. MRC construction was done by assembling several recession curves using semiautomatic tool called RC 4.0 that included in HYDROOFFICE software. MRC is also the main input to determine the karstification degree.

The determination of karstification degree in Kakap Spring is adapted form [20]. [20], introduced a methodology for karstification degree determination using a 10-degree scale, from the 1st degree for the lowest karstification, and 10th degree for best developed karstic behavior. The number of flow sub-regimes that presence in MRC used as the main consideration in determining the karstification degree. Typical recessional equations used to describe recessional types that characterized by the existence of different flow sub-regimes.

The baseflow separation [11] introduced automatic technique for hydrograph separation using digital filtering method. Separation of total discharge into two components, baseflow and direct runoff performed using recursive digital filtering as expressed in Eq.1.

$$b_{k}=\frac{(1-BFImax)aqb(i-1)+(1-a)BFImax\,qi}{1-aBFImax} \quad (1)$$

This filter parameter describes baseflow recession function that expressed as an exponential equation, indicating recession periods without groundwater recharge. BFImax value that based on preliminary results have been suggested to minimize the subjective influence.

3. Result and Discussion

Kakap Spring located in the north-eastern flank of Gunung Sewu. Kakap Spring is perennial gravity spring that located adjacently to the border of Gunung Sewu and the alluvial formation of Baturetno. Kakap spring play very important role regarding water supply in Giriwoyo sub-district as the spring fulfil most of the water needs in Giriwoyo sub-district. The study at Kakap Spring considered as initial research since this location has not been studied before, especially for hydrograph monitoring. The water level observation at Kakap spring that used for this research as obtained during March 2014-April 2015. However, the water level data series that used for the analysis in this study only from March 2014-April 2015. A couple of automatic water level data logger and barometric pressure were installed with 30 minutes interval. Stage-discharge rating curve of Kakap spring generated from eleven manual discharge measurements. The water level-discharge correlation of Kakap Spring expressed in Eq.2.

$$y = 14{,}504e^{8{,}9763x} \tag{2}$$

Where y is discharge (liters/second) and x is water level (meters). The stage-discharge rating curve of Kakap Spring can be seen in figure 2.

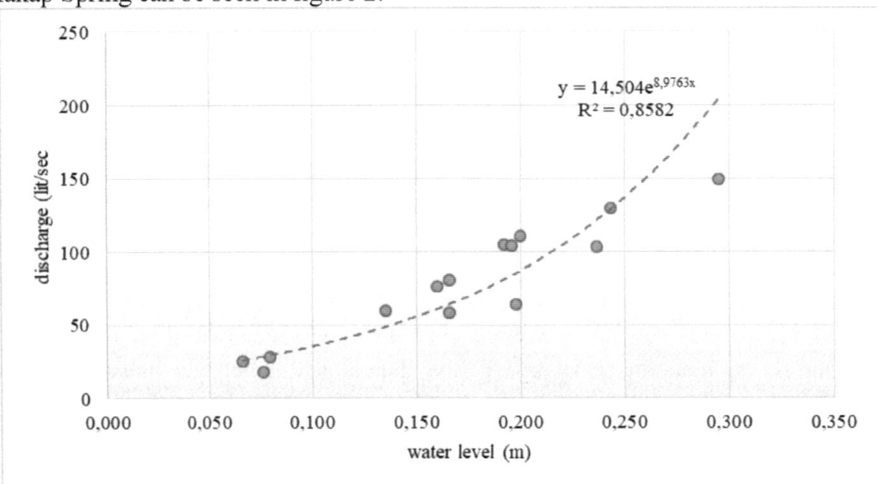

Figure 2. Stage-discharge rating curve of Kakap Spring.

After the stage-discharge rating curve equation determined, the water level data series converted to hydrograph data. Based on the hydrograph result, major recessional period at Kakap Spring occurred from March-December. The minimum discharge recorded at 38,239 liters/second. Intensive flood periods started in December and the discharge periodically increase as the rainy season has begun. The maximum discharge at Kakap spring recorded at 3314, 24 liters/second, five times larger than the average discharge rate at 600,087 liters/second. Based on the hydrograph data series, 42 flood events were recorded with various magnitude. The hydrograph data series of Kakap Spring shown in figure 3.

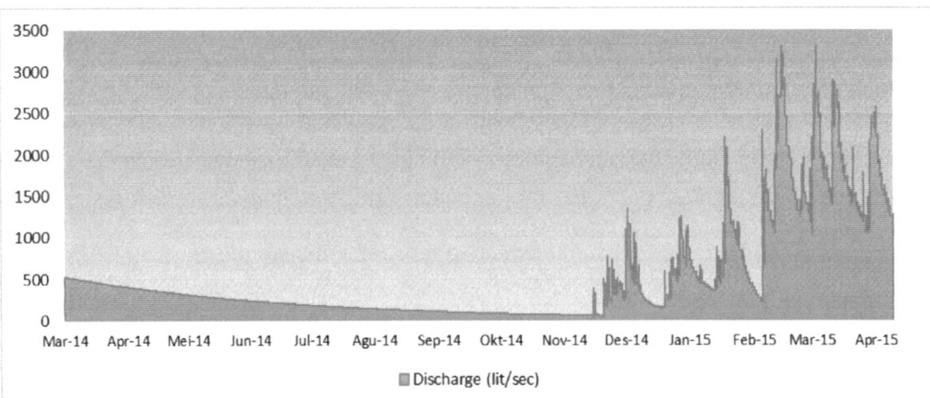

Figure 3. The hydrograph of Kakap Spring.

3.1. Master recession curve of Kakap Spring

The discharge fluctuation of Kakap Spring is relatively high, especially in the rainy season. A long recessional period occurred from March-December with very minimum precipitation events. Basically, several small precipitation events were occurred during that time but didn't give significant

IOP Conf. Series: Earth and Environmental Science **148** (2018) 012029 doi:10.1088/1755-1315/148/1/012029

impacts to the hydrograph. Therefore, the hydrograph were adjusted with the removal of linear segments and insignificant flood pulses. The huge flood period started in December and periodically increase as the season entered the peak of rainy season.

Based on automatic recession curves filters, 26 flood events were identified and 23 events analyzed. The selection of ideal recession curve based recession length and the inclination of the recession limb. Subsequently, each recession curve was calibrated with the appropriate flow model, based on the shape and inclination of the recession limb. The shape of each recession curve is unique and differ representing the aquifer response to the precipitation intensity. After each recession curve has calibrated, all the recession curves subsequently combined to generate Master Recession Curve. The shape of calibrated MRC in Kakap Spring shows the combination of two turbulent flow regimes and one laminar flow regime as it presented in figure 4. The advantage of using MRC for hydrograph analysis is because of its ability to easily identified flow regimes. Flow regimes identification from MRC is relatively more accurate than from combined single recession curves. The determination of karstification degree from flow regimes identification is a powerful tool for scientific adjustment before formulating management actions in karst area [12].

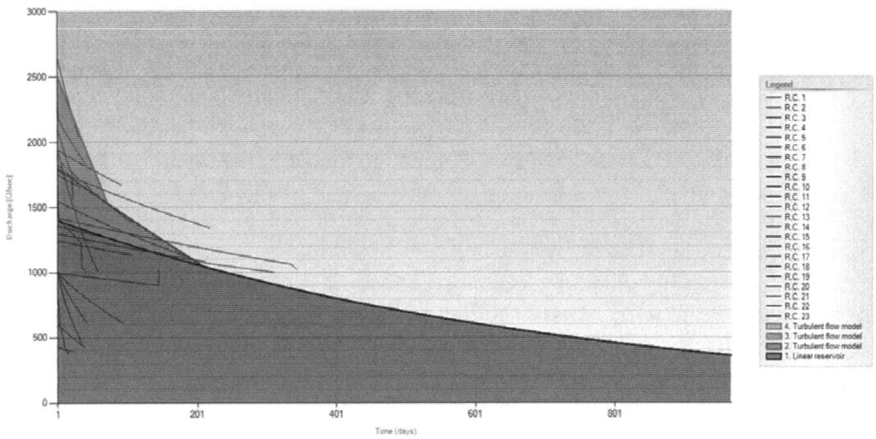

Figure 4. Master Recession Curve of Kakap Spring.

These results indicate that the aquifer feeding Kakap's stream is complex and formed by the combination of pores, micro-fractures, fissures, and conduits. However, based on the MRC shape this mixed type aquifer also strongly influenced by the conduit channel. This indication convinced by the fact that the falling limb steepness is relatively high with short periods to baseflow condition. Rapid discharge responses in Kakap Spring also indicate the domination of concentrated flow in the conduits as the stream turns turbid several 2-3 hours after the rain events. Example of various discharge conditions in low flow period and flooding period presented in figure 5.

From the hydrograph presented in figure 4 , it can be seen that most of the flood pulses shape is relatively steep, either in rising limb or falling limb. This condition indicates a short residence time of quick flow in the aquifer [6]. However, discharge response to rain events also affected by several factors. [23] stated that after infiltrating through the catchment's surface-soil, the rainwater recharges two portions of the catchment. The storage-rate of water in particular portion of the spring catchment depends on its permeability and the volume of water stored in particular portion of the spring depends on its storage capacity. Based on this justification, it is concluded that the maximum volume of water in shorter time will be stored in the portion with highest permeability and capacity. Conversely, the portion with highest permeability will deplete earlier in comparison to catchment portions with low permeability.

Figure 5. Kakap Spring at various discharge rate and turbidity.

Based on those findings, it is clear that the role of conduit channel feeding the stream is significant, especially during the rainy season. Fast depletion time in falling limb indicates that the aquifer with high permeability play an essential role feeding the stream during rainy season with quick flow characteristics. Based on the MRC of Kakap Spring, falling lime of quick flow period occurred at the first five hours after the peak flow. Subsequently, the recession period took four more hours to reach the low flow condition. The presence of more turbulent flow in the aquifer proofs the domination of developed karstic networks in the aquifer formation.

3.2. Karstification degree of Kakap Spring

The karstification degree assessment at Kakap Spring attributed from twenty-three individual recession events. Based on the MRC result of Kakap spring, one laminar sub-regime and two turbulent sub-regimes were identified. The superposition equation of Kakap spring represented in Eq.3.

$$Q_t = 1400e^{-0,0014t} + 400\,(1\text{-}0,005t) + 700\,(1\text{-}0,01t) \tag{3}$$

According to the Eq.2, two turbulent flow sub-regimes were identified with a higher β2 value that indicates a high karstification degree. Based on the superimposed equation of flow regimes, Kakap Spring's karstification degree classified at eight scales. At this scale, the aquifer considered as highly karstified with a strong domination of conduit channels. The domination of conduit channels represents rapid groundwater transmissions within the aquifer. This type of turbulent flow through conduit channels possibly causes a linear segment in the recession limb [13]. This condition implemented in the Eq.2 that shows the domination of linear equations, representing turbulent flow regimes. A higher value of β2 at Kakap Spring represents more rapid aquifer drainage that can be associated with conduits draining from the saturation zone.

A rapid recharge responses at Kakap spring also indicates the domination of conduit channels. Short time to peak with the average time two up to three hours and the following turbid water flows that recharges indicates rapid water transmissions within conduit channels. Besides controlled by the internal structures of conduit channels, the response time of karst aquifers also depends on several factors such as the size of catchment area, carrying capacity, and the contribution of allogenic recharge and internal runoff [25]. According to [21] water flow in karst aquifer considered as turbulent flow when the value of transmissibility > 5 m/day. While when the k value less than 1 m/day the water flow considered as laminar flow. Rapid temperatures and chemical compounds response to storm pulses also indicates the so called "conduit flow springs". This type of spring is characterized by highly karstic limestone aquifers that dominated by open cave systems [25]. According to the automatic temperatures recording at Kakap Spring, it shows rapid temperatures response that also indicates the domination of conduit channels within the aquifer. A short time to peak also represents a quick aquifer response to storm events. These quick stream responses respectively followed by rapid temperature responses as it shown in figure 6.

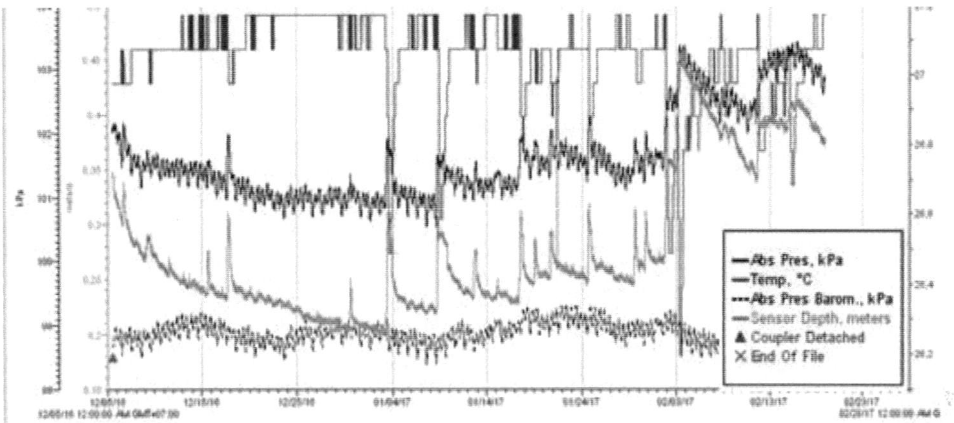

Figure 6. Rapid temperature responses to discharge and storm events at Kakap Spring.

3.3. Baseflow separation

The main purpose of baseflow separation is to determine the percentage of flow component supplying the spring. Two flow components that separated were direct flow (conduit-fissure) and baseflow (diffuse). According to BFI index calculation, the baseflow index of Kakap Spring varies from 0.1401 to 0.9419 with the average value 0.49. The average value of BFImax used as the main input for baseflow separation as it already described in Eq.1. Baseflow recession constant for Kakap Spring calculated from five major food. The value of baseflow recession constant in Kakap Spring varies from 0.9913 to 0.9995 with average value 0,9914. The base flow separation result shows the average baseflow percentage during March 2014-April 2015 is 81.20 % from the total run-off (figure 7).

The ratio of baseflow during rainy season in some certain month was relatively smaller compared to the baseflow presentation in the dry season. However, the ratio of baseflow gently increased in the end of rainy season. The average baseflow ratio in December 2014 was recorded at 69.54 % that indicates the role of conduit flow supplying the spring. The ratio of diffuse flow during flood is relatively low compared to the normal flow condition. In the first recorded flood (November 17, 2014), baseflow ratio decreased up to 17.68 %. This fact indicates that the conduit flow dominates the total flow through the large open channel and contributes directly to the spring.

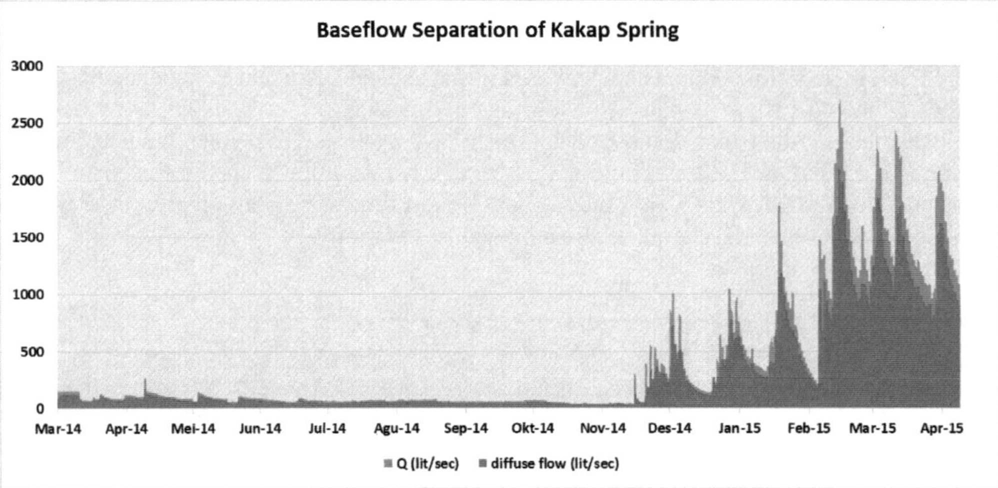

Figure 7. Baseflow separation of Kakap Spring.

4. Conclusion

The karstification degree of Kakap Spring is classified at the 8th scale, which means that the aquifer considered as highly karstified with a strong domination of conduit channels. The domination of conduit channels represents rapid groundwater transmissions within the aquifer. Low baseflow ratio during flood events also indicates the domination of conduit channel in term of its contribution to the total run-off. Those results show that Kakap Spring catchment area play a big role as water reservoir and must be considered as conservation area. However, according to the base flow separation result, the percentage of base flow is apparently too high for karst aquifer. The default BFI max value that suitable for karst aquifer is 0,25. While in this study, the average BFI value was 0,49. According to the base flow separation, the default BFI max value for karst aquifer is more appropriate than the average BFI max that derived from the study location.

5. Acknowledgement

This research is part of the first author's master thesis entitled **"Characterization and Management of Karst Drainage System based on Karst Spring and Underground River Hydrograph Analysis in Gunung Sewu Karst Area"**. Authors gratefully acknowledge the research funding and support from "Research Grant for Post Graduate School (Hibah Penelitian Pasca Sarjana), titled "Model Tingkat Perkembangan Pelorongan Akuifer Karst untuk Identifikasi Kapasitas Penyerapan Karbon" with the referrence number : SK-254/UN.1.PIII?DIT-LIT/LT/2017. Authors would also like to thank all parties involved in the research preparation, field survey, and evaluation.

References

[1] Adji TN Bahtiar IY 2016 Rainfall–discharge relationship and karst flow components analysis for karst aquifer characterization in Petoyan Spring, Java, Indonesia. *Environmental Earth Sciences,* 75:735.

[2] Adji TN Haryono E Fatchurohman H Oktama R 2016 Diffuse flow characteristics and their relation to hydrochemistry conditions in the Petoyan Spring, Gunungsewu Karst, Java, Indonesia. *Geosciences Journal, Vol. 20, No. 3, p. 381–390, June 2016.*

[3] Adji T N 2012 Wet Season Hydrochemistry of Bribin River in Gunung Sewu Karst, Indonesia, Environmental Earth Sciences Journal. Springer, 2012, *vol. 67:1563–1572 pp.*

[4] Ahmadi A N 2015 Optimalisasi Pemanfaatan Air Mataair Karst Kakap terhadap Pemenuhan Kebutuhan Air bagi Penduduk di Kecamatan Giriwoyo, Kabupaten Wonogiri. *Undergraduate Thesis.* Surakarta :UNS.

[5] Amit H Lyakhovsky V Katz A Strasinsky A and Burg A 2002 Interpretation of Spring Recession Curves. *Groundwater vol 40 no* 5, 543-551.

[6] Atkinson T C 1977 Diffuse flow and conduit flow in limestone terrain in the Mendip Hills, Somerset (Great Brittain) *Journal of Hydrology vol. 35*, 93-110.

[7] Bakalowicz M 2005 Karst groundwater: a challenge for new resources *Hydrogeol J*, 13 :148-160.

[8] Bögli A 1980 *Karst Hydrology and Physical Speleology* Berlin Springer-Verlag.

[9] Bonacci O 1993 Karst spring hydrograps as indicators of karst aquifers. *Hydrological Sciences-Journal-des Sciences Hydrologiques* 38,1,2, 51-62.

[10] Civita M V 2008 An improved method for delineating source protection zone for karst springs based on analysis of recession curve data *Hydrogeology Journal* 16: 855-869.

[11] Eckhardt K 2005 How to construct recursive digitalfilters for baseflow separation. *Hydrol. Process.*19, 507– 515.

[12] Fatchurohman H Cahyadi A Suprayogi S 2016 Hydrograph Analysis for Sustainable Karst Aquifer Protection and Management Geomedia Volume 14 Nomor 1 Mei 2016.

[13] Fiorillo F 2011 Tank-reservoir drainage as a simulation of the recession limb of karst spring hydrographs *Hydrogeology Journal* 1009-1019.

[14] Fiorillo F 2014 The Recession of Spring Hydrographs, Focused on Karst Aquifers. *Water Resource Management*, 1781-1805.

[15] Ford D & Williams P 2007 *Karst Hydrology and Geomorphology* West Sussex: John Wiley & Sons Ltd.

[16] Gillieson D 1996 *Caves: Processes, Development andManagemant.* Oxford: Blackwell Publisers.

[17] Haryono E 2000 Some properties of epikarst drainage system in Gunung Kidul Regency, Yogyakarta, Indonesia. *The Indonesian Journal of Geography, vol. 32, No. 79-80.*

[18] Haryono E Danardono Mulatsih S Putro ST and Adji, TN 2016 The Nature of Carbon Flux in Gunungsewu Karst, Java-Indonesia, Acta Carsologica 45/1, 173–185, Postojna.

[19] Kresic, N 2013 Water in Karst; Management, Vulnerability, and Restoration. New York: McGraw-Hill.

[20] Malik P and Vojtkova S 2012 Use of recession-curve analysis for estimation of karstification degree and its application in assessing overflow/underflow conditions in closely spaced karstic springs . *Environ Earth Sci. 2012* 65, 2245–2257.

[21] Milanovic P T 1981 *Karst Hydrogeology.* Colorado: Water Resources Publications.

[22] Rashed K A 2012 Assessing degree of karstification : a new method of classifying karst aquifers. *Sixteenth International Water Technology Conference* (pp. 1-9). Istanbul: IWTC.

[23] Vashisht A K and Bam B 2013 Formulating the spring discharge-function for the recession period by analyzing its recession curve: A case study of the Ranichauri spring (India). *Journal of Earth System Science .122,* No. 5 1313–1323.

[24] White W B 1988 *Geomorphology and Hydrology of Karst Terrains* New York Oxford University Press.

[25] White W B 2002 Karst hydrology: recent developments and open questions. *Engineering Geology* 65 , 85–105.

Evaluation of spatial plan in controlling stream flow rate in Wakung Watershed, Pemalang, Central Java, Indonesia

Y Anwar[1], I Setyasih[1], M A Setiawan[2] and N Christanto[2]

[1]Faculty of Teacher Training and Education, Mulawarman University, Indonesia
[2]Faculty of Geography, Gadjah Mada University, Indonesia

yaskinul.anwar@fkip.unmul.ac.id

Abstract. Evaluation study for such a regional spatial plan (RTRW) in Indonesia has not been evaluated for its effectiveness in controlling the surface run off that contributed to streamflow. This necessity can be accomplishsed by applying a modeling approach, such as Soil Water Assessment Tool (SWAT). The objectives of this research are 1) to simulate the streamflow of Wakung watershed based on actual landuse, 2) to predict streamflow of Wakung watershed based on RTRW, and 3) to evaluate the effectiveness of the RTRW of Pemalang District in controling streamflow rate at Wakung Watershed. ArcSWAT model was used to determine the erosion rate prediction. The model was then calibrated by using SWATCUP. Model performance were tested by using R^2 and E_{NS}. The calibration and validation results showed that R^2 and E_{NS} (monthly) > 0.5. The result of SWAT simulation in Wakung sub-watershed reaching 161 - 4950 m^3/s/years for W-A scenario (actual landuse and weather data of 2013), for scenario W-R (RTRW and weather data of 2013), 330 – 4919 m^3/s/year. The comparison between actual and spatial plan land use data for stream flow is showing that the W-A scenario is lower than the W-R scenario in 19 sub watersheds. This is because there are many plans for adding land use for urban and intensive horticulture land in areas with steep slopes (> 25%). This condition is caused by the demands of fulfilling the needs of settlement and food for people in the Wakung watershed.

1. Introduction

The Watershed Management Plan can't be implemented optimally, as watershed planning compiled by the Watershed Management Bureau (BPDAS) must be supplemented by the District/City Spatial Plan (RTRW). The challenge is that the watershed area is not always coincident with the administrative area so as to allow synergies between agencies cannot run harmoniously and produce no real technical action [19]. For example, making a grand design of a watershed can have a different direction of environmental function with a land allocation, which is arranged in the Regional Spatial Plan (RTRW). Therefore, the evaluation of the suitability of the regional functions in the RTRW and the watershed functions is necessary to be studied.

Mismatching land use allocation with watershed functions may result in a decrease in watershed support capacity [17]. The decreases are characterized by flood, landslide, erosion, sedimentation and drought. Adjustment of the spatial plan covering the watershed should consider the environmental carrying capacity [17] [22]. Conditions like this occur in the Wakung watershed.

Wakung watershed is a part of the Comal watershed system that falls into the priority watershed (critical) category since the start of the current decision [11] [12] [13] [14]. Although many techniques of conservation have been conducted in the Comal watershed but still cannot reduce the criticality of the watershed. This matter is critical of Wakung watershed not only because of the human factor and also physical factor of the watershed, which naturally is vulnerable to experience criticality.

In implementing spatial planning in watershed priority needs monitoring and evaluation [13]. Monitoring and evaluation conducted for streamflow data. One of the streamflow monitoring techniques usually uses the model because of limited resources and available funds. One of the model for predicting streamflow is the Soil and Water Assessment Tool (SWAT). SWAT is a model developed by the United States Department of Agriculture [3] [7], it is an effective tool to model streamflow in a variety of watersheds [5] [8] [10] [16] [22] [23].

The objective of the study is to simulated and predicted of Wakung watershed based on actual and spatial plan land use. Result of this simulation is intended to see streamflow changes. These changes can provide an overview to evaluate the effectiveness of spatial plan of Pemalang District in controlling streamflow rate at Wakung watershed.

2. Methods and Material

2.1 SWAT model description

ArcSWAT (Arc GIS Soil and Water Assessment Tool) is a software based on ArcGIS geographic information system (GIS) as an additional extension of ArcGIS software based on GUI (Graphical User Interface) by using SWAT model (Soil and Water Assessment Tool) [2] [15]. This program is issued by the Texas Water Resource Institute, College Station, Texas, USA. SWAT is designed to predict the effect of land management for streamflow, erosion, sedimentation, and pollution in a complex relationship to a watershed including land type, land use, planting calendar or land management. For modeling purposes, the ArcSWAT program makes users easy to divide the watershed into several sub-basins and the HRUs (Hydrological Response Units) [2]. HRUs describe spatial heterogeneity in terms of land cover, soil type and slope within a watershed.

SWAT offers two methods for estimating surface runoff: The Soil Conservation Service (SCS) Curve Number (CN) procedure and the Green and Ampt infiltration method [15]. Using daily or sub-daily rainfall amounts, SWAT simulates surface runoff, volumes and peak runoff, rates for each HRU [15]. In this study, the SCS curve number method was used to estimate surface runoff volumes because of the unavailability of sub-daily data for the Green and Ampt method.

2.2 Preparation and model input

SWAT is a comprehensive model that requires diverse information in the process of running. Input required in SWAT model is land use (actual and based on spatial plan), digital elevation model (DEM), soil characteristics, climate (precipitation, temperature, solar radiation, wind speed and relative humidity), and cropping calendar. The availability of data input will influence the results of research, more complete of sustainable data input such as climatology data will provide simulation results that are almost close to reality in the field.

The DEM data input is derived from topographic map (scale 1: 25.000), while the existing land use input is derived from the interpretation and field check of the image from google earth and land use based on spatial plan using spatial plan (RTRW) map Pemalang District. Soil input in the Wakung watershed is derived from landform (figure 1), due to the unavailability of detailed soil data in the Wakung watershed. All three inputs above are the main inputs as the formation of unit analysis or known as hydrologic response units (HRUs). The availability of climate data in the Wakung watershed is only rainfall data, so other climate data use climate data from the global weather already provided by SWAT.

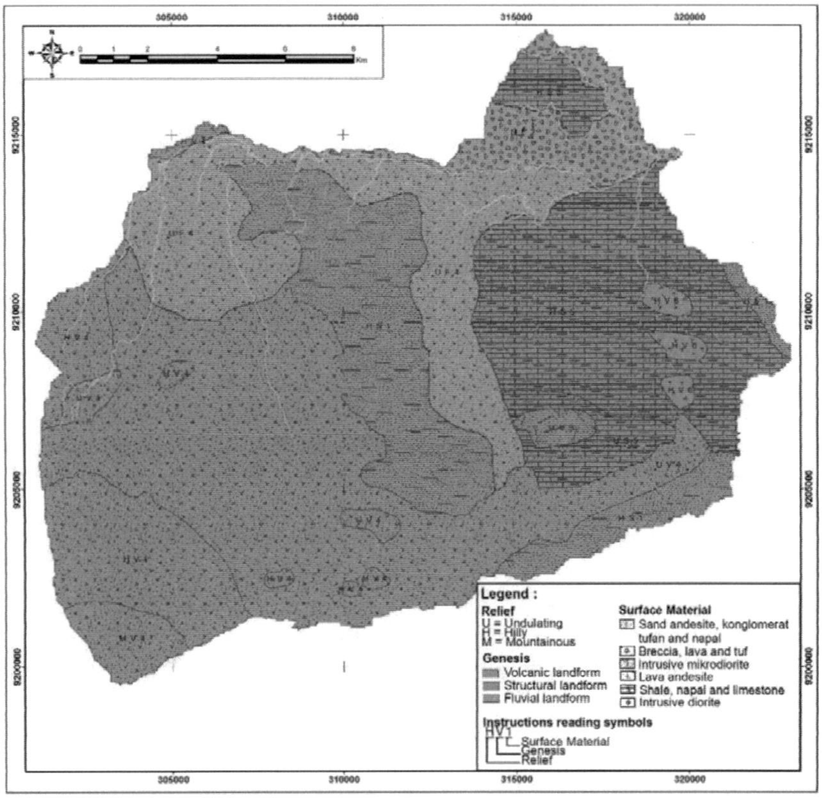

Figure 1. Landform of Wakung Watershed.

2.3 Scenario running model

SWAT scenario running model to predict the rate of streamflow in Wakung watershed consists of two scenarios; (1) running SWAT model based on actual land use (W-A) (2) running SWAT model based on land use based on spatial plan (RTRW) (W-R). Limitations of climate data in the Wakung watershed make the model unable to make predictions with a homemade weather generator (WGN), so running model only uses actual climate (2013 climate date).

2.4 Model calibration and validation

The calibration process was using SWAT-CUP software (Soil and Water Assessment Tool-Calibration and Uncertainty Programs). Model calibration performed by comparing daily streamflow of Sub watershed out from Nambo dam outlet, with SWAT simulation result from June 2009 to May 2010. While validation use daily streamflow data June 2012 - May 2013.

Statistical analysis used in calibration and validation using coefficient of determination (R^2) and Nash-Sutcliffe coefficient of Efficiency (NSE). R^2 and NSE values ranged from 0 to 1. The values of R^2 and NSE close to 1 indicate a close relationship between the simulated data and the observed data. The equations R^2 and NSE are determined by equation 1 and 2.

$$R^2 = \frac{[\sum_i^n (Q_{Obs,i} - \overline{Q}_{Obs,i})\,(Q_{Cal,i} - \overline{Q}_{Cal,i})^2]}{\sum((Q_{Obs,i} - \overline{Q}_{Obs,i}))^2 \sum((Q_{Cal,i} - \overline{Q}_{Cal,i}))^2} \quad (1)$$

$$NSE = 1 - \left(\frac{\sum_{i=1}^{n}\left(Q_{Obs,i}-Q_{Cal,i}\right)^2}{\sum_{i=1}^{n}\left(Q_{Obs,i}-\overline{Q}_{Cal,i}\right)^2}\right) \quad (2)$$

Description:

$Q_{Obs,i}$ / O_i = streamflow observation (m^3/s)

$Q_{Cal,i}$ / P_i = streamflow simulation (m^3/s)

$\overline{Q}_{Obs,i}$ = average of streamflow observation (m^3/s)

$\overline{Q}_{Cal,i}$ = average of streamflow simulation (m^3/s)

Classified the results of the NSE-based simulation into 5 groups [6] as in Table 1.

Table 1. Statistical criterion of NSE.

No	Statistical criterion	NSE
1	Very good	0.75 ≤ NSE ≥ 1.00
2	Good	0.65 ≤ NSE ≥ 0.75
3	Satisfactory	0.50 ≤ NSE ≥ 0.65
4	Acceptable	0.40 ≤ NSE ≥ 0.50
5	Un-satisfactory	NSE ≥ 0.40

3. Result and Discussion

3.1. Unit analysis of SWAT model

The SWAT model divides the watershed into sub-watersheds and HRUs as the unit of analysis. Based on the SWAT analysis Wakung Watershed divided into s35 sub-watersheds shown in figure 2 (a), 345 HRUs based on actual land use and 336 HRUs based on spatial plan (RTRW) land use. HRUs are the result of SWAT model by overlaying land use, soil and slope maps in figure 2 (b). The HRUs for the Wakung Watershed have been generalized with the value thresholds of land use, soil and slopes are 25 Ha (Scale 1: 100,000).

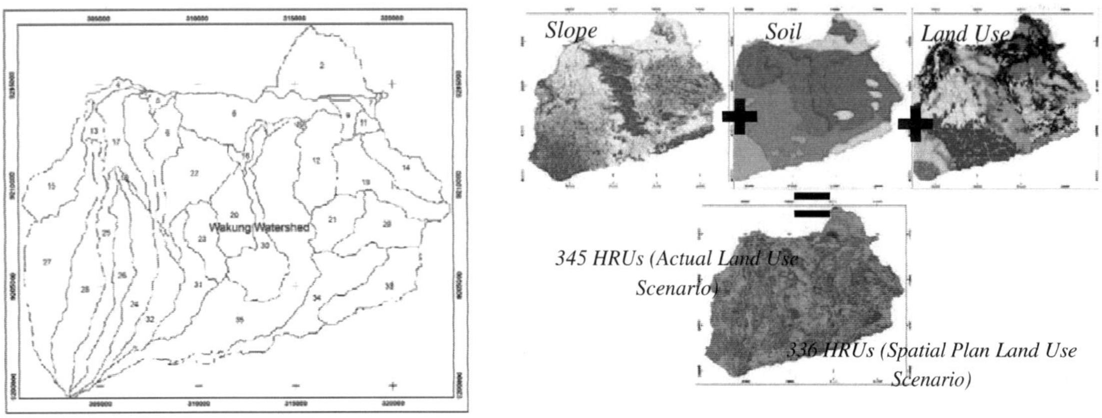

(a) (b)

Figure 2. (a) Sub-watershed of Wakung Watershed and (b) HRUs of Wakung Watershed.

3.2. Calibration and validation of SWAT Model

3.2.1. Parameter Sensitivity Test. During SWAT Model simulation, some parameters are enhanced through calibration process automatically with SUFI-2 (SWAT CUP). The results of the SUFI-2 algorithm are used to determine the fit value of each parameter and how sensitive each parameter. The parameter sensitivity is used as the key parameter of whether the parameter is sufficiently influential to the streamflow, if the parameter is altered the fit-value and parameter sensitivity in the Wakung Watershed uses simulated streamflow comparison with the daily short-observation streamflow from the Nambo dam June 2009 - May 2010, due to limited availability of streamflow data in the Wakung watershed. Based on the results of the running algorithm SUFI-2 (SWAT CUP) with 100 iterations can be obtained the value of fitted-value parameter as in table 2.

Table 2. Fitted, Minimum, Maximum and Sensitivity Value of Streamflow in Wakung Watershed.

Parameter_Name	*Fitted Value*	*Min value*	*Max value*	*t – Stat*	*P - value*
CN2.mgt	0.12	0	2	-1.04	0.30
ALPHA_BF.gw	0.59	0	1	-0.23	0.82
GW_DELAY.gw	29.54	20	120	0.84	0.40
GWQMN.gw	0.93	0	2	-0.51	0.61
GW_REVAP.gw	0.03	0.02	0.2	0.43	0.67
RCHRG_DP.gw	0.20	0	1	0.62	0.54
ESCO.bsn	0.90	0	1	-0.09	0.92
EPCO.bsn	0.33	0	1	-1.49	0.92
SURLAG.bsn	0.31	0	10	-1.16	0.25
SOL_AWC.sol	0.26	0	2	-0.50	0.14
SOL_K.sol	1.77	0	2	0.59	0.56

In addition to fitted value analysis of eleven parameters with SUFI-2 SWATCUP with minimum range and maximum value as in Table 2, also obtained the sensitivity value of eleven parameters. The most sensitive parameters are EPCO, SURLAG, CN2 and GW_DELAY. Greater the t-stat value indicates if the parameter is more sensitive, this can be proved by the value of p-value which is closer to zero, which means the level of significant t-stat significance. This value indicates that change of the value of four parameters gives quite a lot of change of simulation result of streamflow in Wakung watershed.

3.2.2. Performance calibration and validation. Calibration using simulation and observation streamflow daily and monthly dates from June 2009 to May 2010.

Table 3. Performance of Calibration and Validation.

Performance Calibration of Streamflow June 2009 – May 2010	R^2	wr^2	E_{NS}
Daily	0.61	0.36	0.35
Monthly	0.77	0.42	0.50
Performance Validation of Streamflow June 2012 – May 2013			
Daily	0.61	0.24	0.37
Monthly	0.89	0.33	0.51

Validation using simulation and observation streamflow daily and monthly dates from June 2012 to May 2013. Selection of this time range is chosen because of limitations continuous daily observation streamflow dates without a vacuum of instantaneous streamflow observation from the weir guard. Table 3 shows the performance of calibration and validation streamflow in Wakung Watershed and the value of R^2 calibration and daily validation entered in satisfactory category with wr^2 0.36 for calibration and 0.24 for validation. This weighted value of R^2 indicates that the range of observation and simulation values whose value is almost close is 36% for calibration and 24% for validation. The monthly R^2 value shows good correlation with R^2 weighting 0.42 for calibration and 0.33 for validation.

Monthly statistical tests show better results than daily. This is because monthly statistical test has lower value of streamflow less than daily statistical test for simulation of streamflow at peak of rainy season. However, when looked from the value of wr^2 for calibration and validation (daily and monthly) have low value, because many values are not equal of streamflow observation and simulation, especially at peak of rainy season (figure 3).

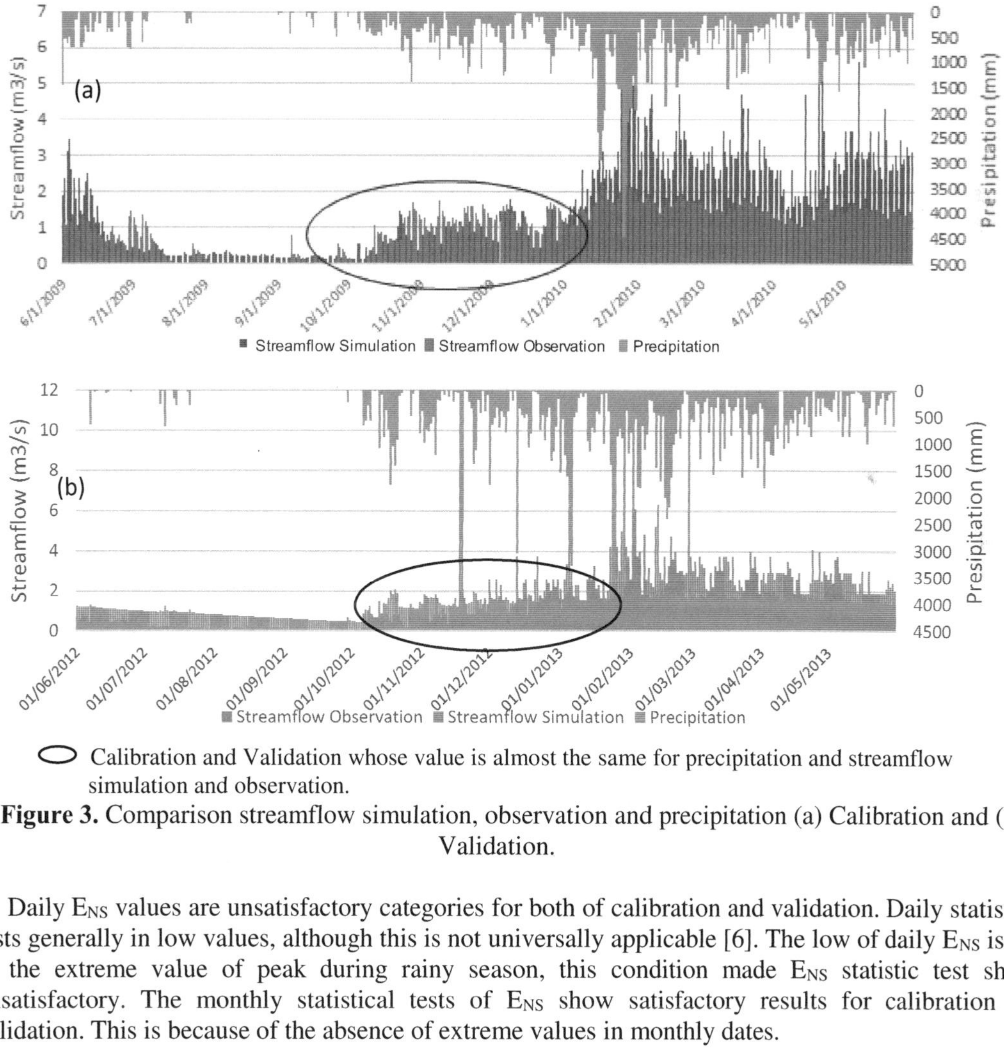

Figure 3. Comparison streamflow simulation, observation and precipitation (a) Calibration and (b) Validation.

Daily E_{NS} values are unsatisfactory categories for both of calibration and validation. Daily statistical tests generally in low values, although this is not universally applicable [6]. The low of daily E_{NS} is due to the extreme value of peak during rainy season, this condition made E_{NS} statistic test shows unsatisfactory. The monthly statistical tests of E_{NS} show satisfactory results for calibration and validation. This is because of the absence of extreme values in monthly dates.

IOP Conf. Series: Earth and Environmental Science **148** (2018) 012030 doi:10.1088/1755-1315/148/1/012030

3.3. Streamflow in Wakung Watershed

Running SWAT model in this research using two scenarios; 1) running SWAT model using actual land use (W-A), and 2) running model SWAT using spatial plan land use (W-R). Based on W-A Scenario, streamflow rate in Wakung watersheds has ranged from 161 - 4950 m^3/s /year (figure 4). The highest streamflow value was in sub-watershed 21st (4950 m^3/s /year) with 8.8% of the slopes> 40% and the use of pine forests, and 8% of sub-watershed on slopes of 25% - 40%. The smallest streamflow rate in 3rd sub-watershed (161 m^3/s /year) with slopes 0 - 8 and land use in the form of paddy fields.

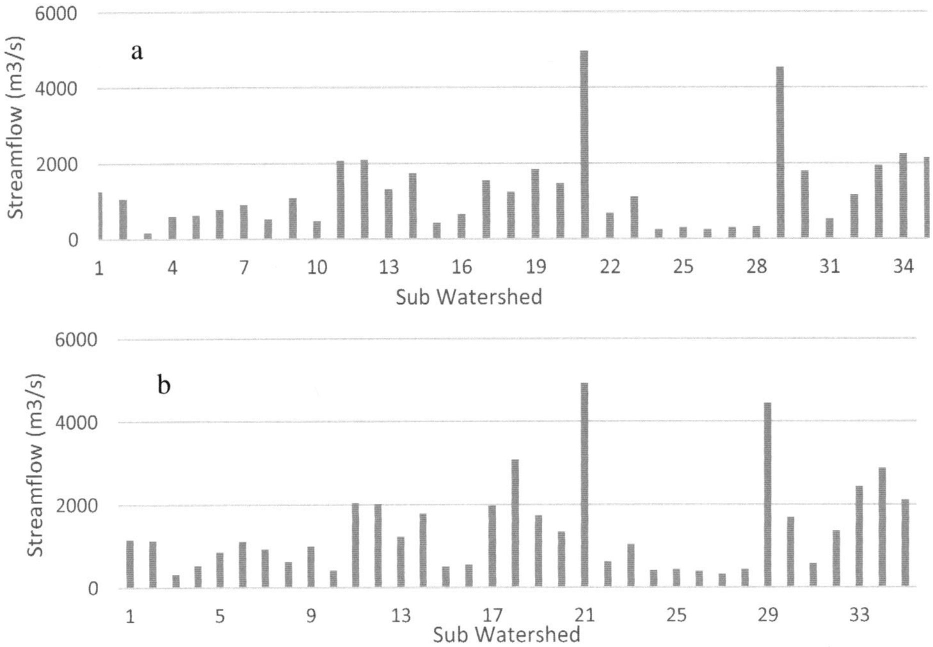

Figure 4. Streamflow Graphs (a) W-A Scenario (b) W-R Scenario.

The result of running model SWAT scenario W-R has streamflow range 330 – 4919 m^3/s/year (figure 4). The highest and lowest streamflow is still same in 21st sub-watershed (4919 m^3/s/year) and 3rd sub-watershed (330 m^3/s/year).

3.4. Evaluation Spatial Plan Land Use (RTRW) to controling streamflow

To evaluate spatial plan land use to control streamflow use comparison between streamflow from actual land use and spatial plan land use. The comparison of two streamflow is showed in figure 5. Spatial plan land use can control and reduce streamflow in 16 sub-watershed and can't control and increase streamflow in 19 sub-watershed. Streamflow reductions ranged of 16 sub watershed are 116 to 31 m^3 /s/ year. Streamflow reductions occurring in sub-watersheds that planned for forest areas or for horticultural areas with perennials plant. Land areas was planned for forests such as sub-watershed 20, 22, 30 and 35, and while for horticultural area with perennials plant are located in sub watershed 22, 12, 19, 21, 22, 23, 29 and 35. The highest reduction streamflow was in sub-watershed 20 (116 m^3/s/year) with forest area planning of 1039.5 Ha.

IOP Conf. Series: Earth and Environmental Science **148** (2018) 012030 doi:10.1088/1755-1315/148/1/012030

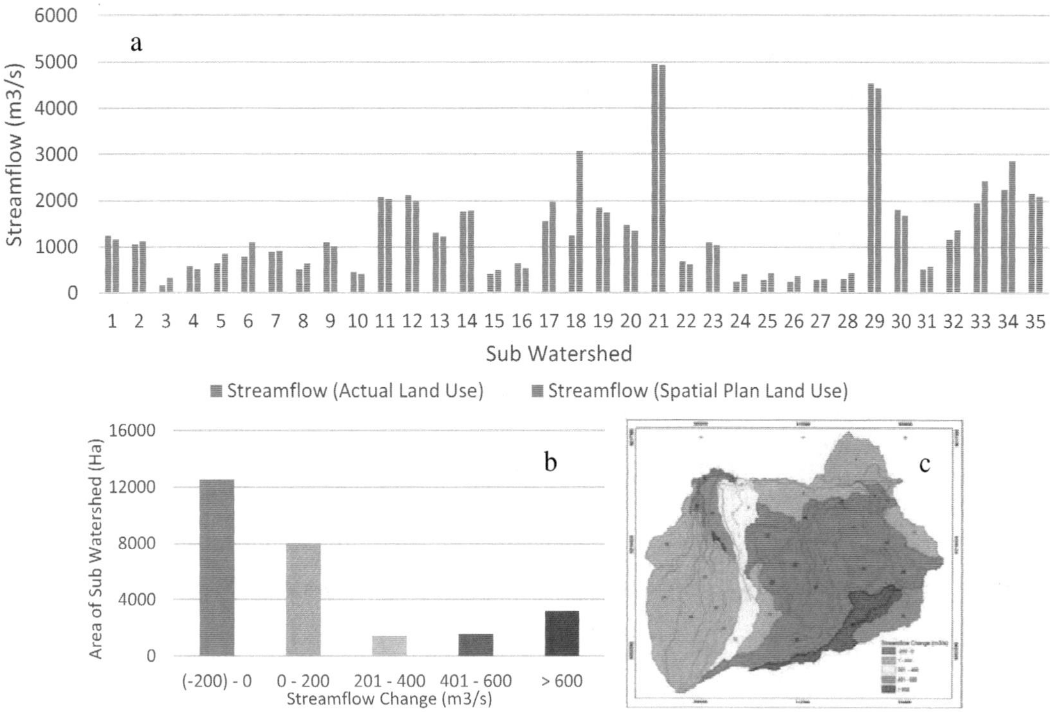

Figure 5. (a) Comparison Streamflow Change, (b) Sub Watershed Area of Streamflow Change (c) Distribution of Streamflow Change in Wakung Watershed.

The increasing streamflow ranged from 21 - 1841 $m^3/s/year$. This is due to increase in residential area planning and intensive agriculture such as seasonal horticultural agriculture or paddy and corn agriculture. The highest increase of streamflow occurred in sub-watershed 18 (1841 $m^3/s/year$) with almost all of the sub watershed allocated for settlements (URHD). While the lowest increasing streamflow occurred in sub DAS 27 (21 $m^3/s/year$). This is because the land allocates in the sub-watershed is not too much change with the actual land use.

This condition is caused by the amount of land allocated for the fulfillment of housing and food needs. So much of the land with a steep slope is used for agricultural and settlement areas. as a result of this condition made increasing of streamflow in 19 sub watersheds. For more details of land use planning in Wakung watershed can be seen in table 4. Based on table 4, the increase of streamflow is due largely to the addition of settlements almost across the watershed, especially in sub-district town centers, such as Belik, Pulosari, Randudongkal, and Moga. The most extensive addition of settlements occurred in Randudongkal and Belik.

Table 4. Comparison Actual and Spatial Plan Land Use in Wakung Watershed.

Clasification of Land Use	Actual Land Use	Area (Ha)	Spatial Plan Land Use	Area (Ha)
Open space	*Barren*	29.4		
Horticulture	*Cabbage*	3750.5	*Cabbage*	3561.6
	Coconut	37.8	*Coconut*	131.2
	Corn	170.2	*Corn*	287.4
	Vineyard	470.5	*Vineyard*	692.0
	Peppers	23.9		
	Poplar	3356.8	*Poplar*	4388.9
	Pineapple	1081.6	*Pineapple*	770.4
Forest	*Oaks*	239.9	*Forest-Evergreen*	3454.1
	Pine	5286.3	*Pine*	5108.3
Paddy	*Rice*	5413.3	*Rice*	4162.6
Brush	*Range-Brush*	3035.3		
Urban	*URLD (Urban Low Density)*	2054.7	*URLD (Urban Low Density)*	1543.7
	URMD (Urban Middle Density)	112.0		
	URHD (Urban High Density)	349.0	*URHD (Urban High Density)*	1345.5
Water	*Water*	34.7		
	Total	25445.7		25445.7

This streamflow change when viewed based on the width of the sub-basin area, implemented of land use planning will be able to reduce the streamflow in the 16 sub watersheds (0 - 200 m^3/s/year) of 12275.17 Ha (47% of Wakung watershed). while for increasing streamflow from 1 - 200 m^3/s/year occurred in 12 sub watersheds with 7771.96 Ha (30% of Wakung watershed) and for streamflow increase > 200 m^3/s/year occurred in 7 sub watersheds with 5398.62 Ha (23% of Wakung watershed).

4. Conclussion

SWAT was applied to simulate actual land use and spatial plan land use to determine streamflow changes in Wakung watershed. The comparison between actual and spatial plan land use data for streamflow showing that the W-A scenario is lower than the W-R scenario in 19 sub watersheds. This is because there are many plans for adding land use for urban and intensive horticulture land in areas with steep slopes (> 25%). This condition is caused by the demands of fulfilling the needs of settlement and food for people in the Wakung watershed.

References

[1] Arsyad S 1989 *Konservasi Tanah dan Air* (Bogor: Institut Pertanian Bogor).

[2] Arnold J G Kiniry J R Srinivasan R Williams J R Haney E B and Neitsch S L 2009 *Soil and Water Assessment Tool Input/Output File Documentation Version 2009.* Grassland, Soil and Water Research Laboratory – Agriculture Reseach Service Blackland Resech Center – Texas Agrilife Reseach.

[3] Arnold J G Srinivasan R Muttiah R S and Allen P M 1999 Continental scale simulation of the hydrologic balance. *J. Amer. Water Resour. Assoc.*, 35, 1037–1051.

[4] Arnold J G Srinivasan R Muttiah R S and Williams J R 1998: Large area hydrologic

modelling and assessment. Part I: Model development. *J. Amer. Water Resour. Assoc.*, 34, 73–89.

[5] Bouraoui F S Benabdallah A and Bidoglio G 2005: Application of the SWAT model the Medjerda river basin (Tunisia). *Phys. Chem. Earth*, 30, 497–507.

[6] da Silva M G Netto A O A Neves R J J Vasco A N Almeida C and Faccioli G G 2015 Sensitivity Analysis and Calibration of Hydrological Modeling of the Watershed Northeast Brazil, *Journal of Environmental Protection, 837-850.*

[7] Grizzetti B Bouraoui F and Marsily G 2005 Modelling nitrogen pressure in river basins: A comparison between a statistical approach and the physically based SWAT model, *Physics Chem. Earth 30(8-10): 508-517.*

[8] Jurgen S Karim C A Raghavan S and Hong Y 2008 Estimation of freshwater availability in the West African sub-continent using the SWAT hydrologic model. *J. Hydrol., 352, 30–49.*

[9] Krause P Boyle D P and Base F 2005 Comparison of different efficiency criteria for hydrological model assessment, *Advances in Geosciences 5, 89–97.*

[10] Lu E Takle E S and Manoj J 2009 The relationships between climatic and hydrological changes in the Upper Mississippi river basin: A SWAT and multi-GCM study. *J. Hydrometeor*,11, 437–451.

[11] Ministerial decree of Forestry and Plantation. 058/Kpts-II/1970, On the determination of priority order of the Watershed

[12] Ministerial decree of Forestry and Plantation. 124/Kpts-II/1984, On the determination of priority order of the Watershed

[13] Ministerial decree of Forestry and Plantation. 284/Kpts-II/1999, On the determination of priority order of the Watershed

[14] Ministerial decree of Forestry and Plantation. 328/Kpts-II/2009, On the determination of priority order of the Watershed

[15] Neitsch S L Arnold J G Kiniry J R and Williams J R 2011 *Soil and Water Assessment Tool Theoretical documentation Version 2009*. Grassland, Soil and Water Research Laboratory – Agricultur Reseach Sevice Blackland Resech Center – Texas Agrilife Reseach.

[16] Pisinaras V Petalas C Gikas G D Gemitzi A and Tsihrintzis V A 2010 Hydrological and water quality modeling in a medium-sized basin using the Soil and Water Assessment Tool (SWAT). *Desalination*, 250, 274–286.

[17] PP. No. 37 Year 2012, Watershed Management.

[18] PerMen. PU NO. 16/PRT/M/2009, the Guidelines for Spatial Planning of the Regency.

[19] Setiawan M A 2012 Integrated Soil Erosion Risk Manangement in the Upper Serayu Watershed, Wonosobo District Central Java Province, Indonesia. *Dissertation.* Innsbruck.

[20] Srinivasan R Arnold J G and Jones C A 1998 Hydrologic modeling of the United States with the Soil and Water Assessment Tool. *Int. J. Water Resour. Dev.*, 4, 315–325.

[21] Law No. 7 Year 2004, Water Resources.

[22] Law No. 26 Year 2007, Spatial Planning.

[23] Zhang X Srinivasan R Debele B and Hao F 2008 Runoff simulation of the headwaters of the Yellow River using the SWAT model with three snowmelt algorithms. *J. Amer. Water Resour. Assoc.*, 44, 48–61.

IOP Conf. Series: Earth and Environmental Science **148** (2018) 012031 doi:10.1088/1755-1315/148/1/012031

Optimization of waste transportation route at waste transfers point in Lowokwaru District, Malang City

S Hariyani[1] **and C Meidiana**[1]

[1]Urban and Regional Planning Department, Faculty of Engineering, University of Brawijaya, Malang, East Java, Indonesia

septianahariyani@ub.ac.id

Abstract. Increasing population led to the emergence of the urban infrastructure services issue including waste problems especially waste transportation system. Data in 2016 shows that the amount of waste in Malang was 659.21 tons / day. The amount of waste transported to landfill only reached 464.74 tons / day. This indicates that not all waste can be transported to the landfill Supiturang because Level of Service (LoS) reached 70.49%. This study aims to determine the effectiveness of waste transportation system and determine the fastest route from waste transfers point in Lowokwaru district to the landfill Supiturang. The data collection method in this research were 1) primary survey by interview officials from the Sanitation and Gardening Agency which questions related to the condition of the waste transportation system in waste transfer point, 2) Secondary survey related to data of waste transportation system in Malang City i.e the amount of waste generation in waste transfer point, number of garbage trucks and other data related to the garbage transportation system. To determine the fastest route analyzed by network analyst using ArcGIS software. The results of network analyst show that not all routes are already using the fastest route to the landfill Supiturang.

1. Introduction

Municipal solid waste collection (MSWC) is a solid waste management process consisting of generation, collection, transfer, treatment and final disposal [1, 2, 3, 4, 5]. The costly aspect of the process municipal solid waste was collection and transportation, because of the massive use of vehicles and labour intensity of the work [6]. About 85 % proportion of total cost for solid waste was waste collection [1, 2, 3].

A large number of population and the diversity of activities in major cities in Indonesia has resulted in problems in terms of urban infrastructure services, such as waste issues. One of the problems in waste management is waste transportation. According to Law No. 18 of 2008, waste transportation is part of waste management [7]. Waste transportation is defined as an activity of carrying waste from source and or from waste transfer point (TPS) to final solid waste disposal site (SWDS) in landfill (TPA).

[8] explain that waste transport activities are one of the important components and require careful calculation where the target is to optimize the transport time required in the system. The process of costs incurred in the transport system. Inefficient routes will cause longer distance, thus higher cost for fuel. In determining the route can affect the required haul time to the landfill. Inefficient route selection can also result in addition; the transportation process will also affect the surrounding environmental conditions, associated with exhaust emissions of waste transporting vehicles. Good domestic waste transport system services with optimal routes will reduce the adverse impacts of these activities on the environment [9]. The most significant reason for implementing network analysis and transport route planning is because businesses are interested in determining the best route to minimize cost and time [10]. Business in waste transportation system considers waste transportation route much since the length of waste transportation route is proportional to the expense of related cost of vehicle fuel and vice versa.

Based on the data from the Department of Hygiene and Gardening Malang City in 2016, the waste generation in Malang was 659.21 tons / day, while the amount of waste transported to the landfill was only 464.74 ton / day. This amount is assumed to be the same for 2017 since there is no significant improvement of waste management in Malang City. The amount indicates that not all waste is transported to final SWDS in Supiturang landfill. The value of level of service (LoS) of waste transportation is 70.49%, below the effective rate which should be 100% where there is no waste remain in waste transfer point. Lowokwaru is a district with a high number of educational facilities, evidenced by the number of universities both public and private, such as Brawijaya University, Malang State University, Muhammadiyah University of Malang, and others. Consequently, the number of the universities leads to positive population growth followed by more the activities. This can result in the amount of volume of waste generated is also greater.

The condition of transfer points in Lowokwaru district, Malang also show problem in waste transportation system. Delay waste pick up, overload truck capacity, and lack of human resources capacity are typical problems exist in transfer points. In addition, waste is still often seen in the waste transfer point at the time of transfer stations must be in a state of free waste. It shows that the level of effectiveness of transportation in waste transfer point Lowokwaru district does not reach 100%. According to [11] target method, source, or process can be used to assess the effectiveness. From target's point of view, which is transporting waste goal, waste transport can be said to be effective if there is no more waste after the last waste pick up schedule [11].

Waste management, including proper waste transport can improve efficiency and effectiveness. The efficiency of this waste transport can be seen from several factors such as transportation facilities, carrier fleet routes and accessibility. The right transport system and route selection will improve accessibility leading to decrease of cost and time consuming waste transportation. Conditions in some transfer stations in Lowokwaru district, Malang City indicate that transportation performance is not optimal so that the need for evaluation. Therefore, this research is more focused on route optimization, so it could be obtained the shortest and fastest route.

2. Research Methods

The study was conducted on waste transfer point which is located in Lowokwaru District in Malang City. The study area could be seen in Figure 1. The samples of waste transfer point in this research are 12 waste transfer points managed by the Department of Sanitation and Gardening of Malang City (Dinas Kebersihan dan Pertamanan/DKP) and also managed by Market Office (Dinas Pasar) of Malang City.

There were waste transfer point 1) Tunggulwulung, 2) Merjosari, 3)Tlogomas, 4) Dinoyo and Ketawanggede, 5) Sumbersari, 6) Menjing, 7) Tunjungsekar, 8) Borobudur, 9) Tawangmangu, 10) TPS Tasikmadu I, 11) TPS Tasikmadu II, and 12) Joyogrand. Scope of material discussed in this research is the process of waste transport system from waste transfer point to landfill. The scope of the study could be seen in figure 2.

IOP Conf. Series: Earth and Environmental Science **148** (2018) 012031 doi:10.1088/1755-1315/148/1/012031

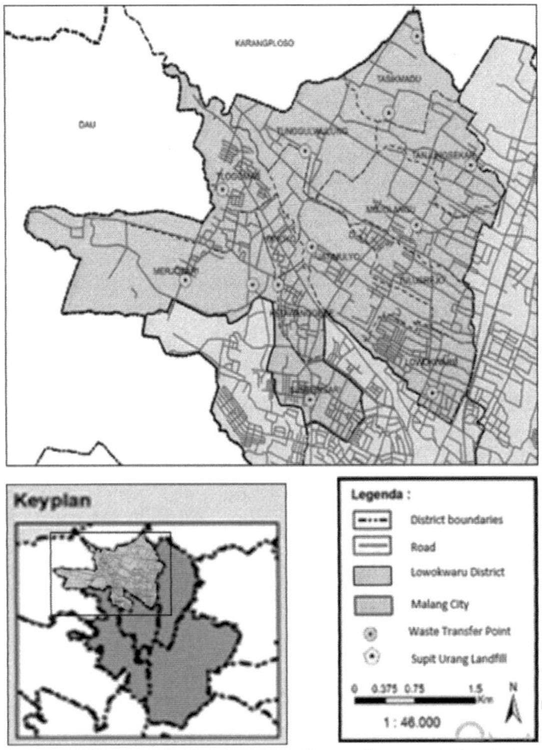

Figure 1. Study area Lowokwaru District.

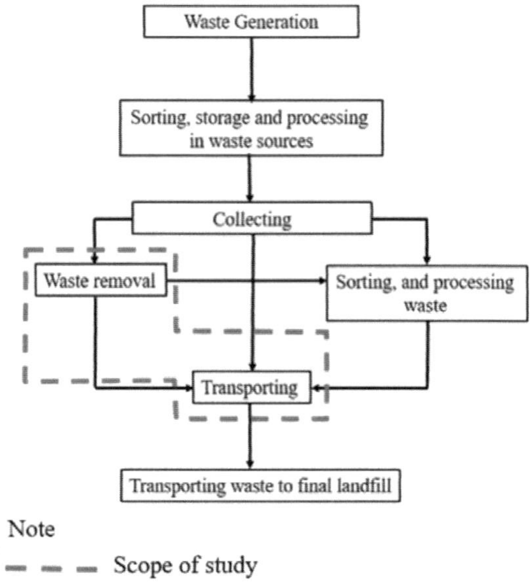

Figure 2. Operational techniques of waste management.

Observation is a method of collecting data through direct observation or review conducted carefully and directly in the field or research location. Methodology research implied that the observation is a systematic observation and recording of a symptom that appears on the object of research [12, 13, 14].

Observations in this study were conducted to obtain the required data related to waste transportation system in waste transfer points in Lowokwaru district. Observations include the condition of the waste transfer point, the condition of equipment and means of transporting waste and the condition of roads traversed by waste trucks. Moreover, observation is also done by following the waste truck from waste transfer point to landfill to measure the average speed of the waste truck. The average speed data is used in network analyst calculations for the fastest route determination of waste transfer point in Lowokwaru district to landfill.

Network analyst was developed by Gallardo et al (2015) based on different vehicle speeds of the day. Network Analyst has the ability to calculate the shortest travel time between two locations by determining start time [15]. In 2003, Echols developed GIS application to find out the fastest route between two destinations [16]. Some previous research determined the optimal route of solid waste collection using GIS analysis tools [17, 18].

The use of Network analyst in this research is intended to find out the optimal route can be taken from the existing transfer points in Lowokwaru district to Supiturang landfill. The first step is to fill the road attributes of vehicle speed and also the length of each road segment. Vehicle speed is obtained by following waste truck from waste transfer points in Lowokwaru district to Supiturang landfill. The average speed is calculated afterwards and inserted into the attribute speed in each road segment. The length of road segment can be obtained from the shp file of Malang City. There are two steps to be taken to determine the fastest route which is creating a network file dataset and analysing the fastest route on the network analyst. Network analyst in this research is part of analysis in ArcGIS 10.1 for student.

IOP Conf. Series: Earth and Environmental Science **148** (2018) 012031 doi:10.1088/1755-1315/148/1/012031

3. Result and Discussion

Please follow these instructions as carefully as possible so all articles within a conference have the same style to the title page. This paragraph follows a section title so it should not be indented.

3.1. Fleet routes

Fleet routes are made to facilitate workers and means of transporting waste can be used effectively and efficiently. The assessment of the optimal transport route is done based on the distance and the fastest time from the waste transfer point to Supiturang landfill. The route of waste transportation in Malang City has been determined by the Office of Sanitation and Gardening of Malang City (Dinas Kebersihan dan Pertamanan/DKP). The transportation routes describe the waste transportation from waste transfer point to the final waste disposal site (landfill) through certain streets (St.). The following section shows all routes for waste transportation in the district.

a. Tunggulwulung Waste transfer point
Tunggulwulung Waste transfer point -Candi Panggung St.–Soekarno Hatta St.– Mayjen panjaitan St.–Bogor St.–Veteran St.–Bendungan Sutami St– (Galunggung St. –Raya Langsep St.–Mergan St.–Bandulan St.–Mulyorejo St.–Supiturang landfill.

b. Merjosari Waste transfer point
Merjosari Waste transfer point–Sigura-gura St.-Bendungan Sutami St.–(Galunggung St.–Raya Langsep St.–Mergan St.–Bandulan St.–Mulyorejo St.–Supiturang landfill.

c. Tlogomas Waste transfer point
Tlogomas Waste transfer point–Merjosari St.–Sumbersari St.–Bendungan Sutami St.–(Galunggung St.–Raya Langsep St.–Mergan St.–Bandulan St.–Mulyorejo St.–Supiturang landfill.

d. Dinoyo and Ketawanggede Waste transfer point
Dinoyo dan Ketawanggede Waste transfer point–Gajayana St.–MT Haryono St.–Mayjen panjaitan St.–Bogor St.–Veteran St–Bendungan Sutami St.(Galunggung St.–Raya Langsep St.–Mergan St.–Bandulan St.–Mulyorejo St.–Supiturang landfill.

e. Sumbersari Waste transfer point
Sumbersari Waste transfer point-Mayjen panjaitan St.–Bogor St.–Veteran St.–Bendungan Sutami St.–(Galunggung St.–Raya Langsep St.–Mergan St.–Bandulan St.–Mulyorejo St.–Supiturang landfill.

f. Menjing Waste transfer point
Menjing Waste transfer point–Pisang Kipas St.–Soekarno Hatta St.–Mayjen panjaitan St.–Bogor St.–Veteran St.–Bendungan Sutami St.–(Galunggung St.–Raya Langsep St.–Mergan St.–Bandulan St.–Mulyorejo St.–Supiturang landfill.

g. Tunjungsekar Waste transfer point
Tunjungsekar Waste transfer–Ikan Piranha Atas St.–Ikan Tombro St.-Soekarno Hatta St.–Mayjen panjaitan St.–Bogor St.–Veteran St.–Bendungan Sutami St.–(Galunggung St.–Raya Langsep St.–Mergan St.–Bandulan St.–Mulyorejo St.– Supiturang landfill.

h. Borobudur Waste transfer point
Borobudur Waste transfer point-Soekarno Hatta St.–Mayjen panjaitan St.–(Ijen St.–Jakarta St.–GedeSt.–Wilis St.–Raya Dieng St.–Raya Langsep St.–Mergan St.–Bandulan St.–Mulyorejo St.–Supiturang landfill.

i. Tawangmangu Waste transfer point
Tawangmangu Waste transfer point–Tawangmangu St.–Sarangan St.–(Letjen Suparman St.–Letjen Sutoyo St.–Buring St.–Ijen St.–Wilis St.–Raya Dieng St.–Raya Langsep St.–Mergan St.–Bandulan St.–Mulyorejo St.–Supiturang landfill.

j. Tasikmadu I Waste transfer point
Tasikmadu I Waste transfer–Kyai Yusuf St.–Ikan Gurame St.–Sudimoro St.–Soekarno Hatta St.–Gajayana St.–Bendungan Sutami St.–(Galunggung St.–Raya Langsep St.–Mergan St.–Bandulan St.–Mulyorejo St.–Supiturang landfill.

k. Tasikmadu II Waste transfer point
   Tasikmadu II Waste transfer point–Bulutangkis St.–Renang St.-Kyai Yusuf St.–Ikan Gurame St.–Sudimoro St.–Soekarno Hatta St.–Gajayana St.–(Bendungan Sutami St.–Galunggung St.–Raya Langsep St.–Mergan St.–Bandulan St.–Mulyorejo St.– Supiturang landfill.
l. Joyogrand Waste transfer point
   Joyogrand Waste transfer point–Merjosari St.–Sumbersari St.-Bendungan Sutami St.–(Galunggung St.–Raya Langsep St. – Mergan St.–Bandulan St.–Mulyorejo St.–Supiturang landfill.

3.2. Transport System Effectiveness

According to [11], there are three approaches that can be used to measure effectiveness: target approach, source approach, and process approach. In this study the researchers used a target approach to determine the effectiveness of waste transport system in Lowokwaru district. The purpose of transporting waste is to transport all the waste from the waste transfer point to the landfill. If there is still waste in the waste transfer point it means that the waste transport is not effective and if the waste is transported all to landfill has the meaning of effective waste transport. Here is the waste condition at Lowokwaru waste transfer point shown in table 1.

Table 1. Effectiveness of Lowokwaru Waste Tranfer Point.

No.	(1)	(2)	(3)	(4)	(5)	(6)
1.	Tunggulwulung	12	12	2	Waste left over	Ineffective
2.	Merjosari	26	26	3	Waste left over	Ineffective
3.	Tlogomas	24	24	-	No Waste	Effective
4.	Dinoyo & Ketawanggede	28	28	2	No Waste	Effective
5.	Sumbersari	24	24	-	No Waste	Effective
6.	Menjing	25	25	-	No Waste	Effective
7.	Tunjungsekar	22	22	-	No Waste	Effective
8.	Borobudur	50	50	-	No Waste	Effective
9.	Tawangmangu	30	30	-	No Waste	Effective
10.	Tasik Madu I	12	12	-	Waste left over	Ineffective
11.	Tasik Madu II	6	6	1	Waste left over	Ineffective
12.	Joyogrand	16	16	2	Waste left over	Ineffective

Note: 1=Waste Transfer point 2=Number of Waste Carts
3= Waste Generation (m^3) 4=The Rest of The Waste (Waste carts)
5=Waste Conditions at Waste transfer point 6=Effectiveness

Based on table 1, there are seven waste transfer points (Tlogomas, Dinoyo & Ketawanggede, Sumbersari, Menjing, Tunjungsekar, Borobudur, and Tawangmangu) which show how effective it is because in this transfer depo there is no waste. If all waste is transported to landfill, it indicates that waste transport is effective. In the other hand, five waste transfer points that still have the rest of the waste after the last trip of waste transport. The five waste transfer points are Tunggulwulung, Merjosari, Tasik Madu I, Tasik Madu II, and Joyogrand. If there is still waste in the waste transfer point it means that the waste transport is not effective. Therefore, there should be a replacement of vehicles with greater tamping power or by increasing the number of trucks to avoid waste in the waste

IOP Conf. Series: Earth and Environmental Science **148** (2018) 012031 doi:10.1088/1755-1315/148/1/012031

transfer point.

3.3. Network analyst

The use of Network analyst in this study is intended to find out the fastest route that can be taken from the existing waste transfer point in Lowokwaru district to Supiturang landfill. The fastest process is obtained from the accumulated travel time of each road segment that can be reached from waste transfer points in Lowokwaru district to Supiturang landfill. While the travel time of each road segment obtained from the division between the lengths of each road segment with the speed of the vehicle, the average speed of the vehicle obtained from the interview with waste truck drivers.

In this study, in determining a route assisted by Network Analyst ArcGIS 10.1 for student. Network analyst will automatically generate the fastest route that can be passed from the point of the initial location to the point of destination location. The following is the fastest route analysis result done by Network Analyst on ArcGis from each waste transfer point to Supiturang landfill. The fastest routes describe the waste transportation from waste transfer point to the final waste disposal site (landfill) through certain streets (St.). The following section shows all the fastest waste transportation in the district.

a. Tunggulwulung waste transfer point
Tunggulwulung waste transfer point–Arumba St.–Akordion St.–Candi Panggung St–Soekarno Hatta St.–Mayjen Panjaitan St.–Bogor St–(Jakarta St.–Gede St.–Simpang Wilis St.–Raya Dieng St.–Raya Langsep St.–Bandulan St.–Raya Mulyorejo St.–Rawisari St.–Supiturang landfill).
b. Merjosari waste transfer point
Merjosari waste transfer point–Bendunag Sigura-gura St.–Bendungan Sutami St.–(Galunggung St.–Raya Langsep St.–Bandulan St.–Raya Mulyorejo St.–Rawisari St.–Supiturang landfill).
c. Tlogomas waste transfer point
Tlogomas waste transfer point–Kecubung St.–Tlogo Wulan St.–Tlogo Indah St.-Joyo Utomo St.-Joyo Tambaksari St.–Sumbersari St.–Bendungan Sutami St.–(Galunggung St.–Raya Langsep St.–Bandulan St.–Raya Mulyorejo St.–Rawisari St.–Supiturang landfill).
d. Dinoyo and Ketawanggede waste transfer point
Dinoyo and Ketawanggede waste transfer point-Joyo Tambaksari St.–Sumbersari St.–Bendungan Sutami St.–(Galunggung St.–Raya Langsep St.–Bandulan St.–Raya Mulyorejo St.–Rawisari St.–Supiturang landfill).
e. Sumbersari waste transfer point
Sumbersari waste transfer point–Sumbersari St.–Bendungan Sutami St.–(Galunggung St.–Raya Langsep St.–Bandulan St.–Raya Mulyorejo St.–Rawisari St.–Supiturang landfill).
f. Menjing waste transfer point
Menjing waste transfer point–Pisang Kipas St.–Soekarno Hatta St.–Mayjen Panjaitan St.–Bogor St.–(Jakarta St.–Gede St.–Simpang Wilis St.–Raya Dieng St.–Raya Langsep St.–Bandulan St.–Raya Mulyorejo St.–Rawisari St.–landfill Supiturang).
g. Tunjungsekar waste transfer point
Tunjungsekar waste transfer point–Ikan Piranha Atas St.–Sudimoro St.-Soekarno Hatta St.–Mayjen Panjaitan St.–Bogor St.–(Jakarta St.–Gede St.–Simpang Wilis St.–Raya Dieng St.–Raya Langsep St.–Bandulan St.–Raya Mulyorejo St.–Rawisari St.–Supiturang landfill).
h. Borobudur waste transfer point
Borobudur waste transfer point–Borobudur St.-Soekarno Hatta St.–Mayjen Panjaitan St.–Bogor St.–(waste transfer point Gede St.–Simpang Wilis St.–Raya Dieng St.–Raya Langsep St.–Bandulan St.–Raya Mulyorejo St.–Rawisari St.–Supiturang landfill).
i. Tawangmangu waste transfer point
Tawangmangu waste transfer point–Sarangan St.– (Letjen Sutoyo St.–Jaksa Agung Suprapto St.–Brigjen slamet Riyadi St.–Bromo St.–Semeru St.–Ijen St.–Wilis St.- Raya Dieng St.–Raya Langsep St.–Bandulan St.–Raya Mulyorejo St.–Rawisari St.–Supiturang landfill).

j. Tasikmadu I waste transfer point
Tasikmadu I waste transfer point–KH Yusuf St.–Ikan Gurami St.–Sudimoro St.- Soekarno Hatta St.–Mayjen Panjaitan St.–Bogor St.–(Jakarta St.–Gede St.–Simpang Wilis St.–Raya Dieng St.–Raya Langsep St.–Bandulan St.–Raya Mulyorejo St.–Rawisari St.–Supiturang landfill).

k. Tasikmadu II waste transfer point
Tasikmadu II waste transfer point–Bulutangkis St.–Atletik St.–KH Yusuf St.–Ikan Gurami St.–Sudimoro St.- Soekarno Hatta St.–Mayjen Panjaitan St.–Bogor St.– (Jakarta St.–Gede St.–Simpang Wilis St.–Raya Dieng St.–Raya Langsep St.–Bandulan St.–Raya Mulyorejo St.–Rawisari St.–Supiturang landfill).

l. Joyogrand waste transfer point
Joyogrand waste transfer point–Perum Joyogrand St.–Joyo Sari St.-Joyo Utomo St.-Joyo Tambaksari St.–Sumbersari St.–Bendungan Sutami St.–(Galunggung St.–Raya Langsep St.–Bandulan St.–Raya Mulyorejo St.–Rawisari St.–Supiturang) landfill.

3.4. Comparison of existing route with the fastest route

Based on the fastest route from network analyst result there is difference of distance and also travel time between existing routes. Table 2 show the difference between the distance and travel time between the existing route and the fastest route based on network analyst. Network analyst in this study is intended to find out the fastest route that can be taken from the existing waste transfer point in Lowokwaru district to Supiturang landfill. The fastest process is obtained in table 2.

Table 2. Differences Existing Route with the Fastest Route Based Network Analyst.

No	Waste Transfer point	Existing Route (minute)	Fastest route (minute)	Existing Distance (Km)	Fastest Distance (Km)
1	Tunggulwulung	32.8	31.2	14.24	13.82
2	Merjosari	23.3	23.3	9.88	9.88
3	Tlogomas	26.2	26.2	11.44	11.44
4	Dinoyo and Ketawanggede	29.9	24.4	12.00	9.10
5	Sumbersari	17.2	17.2	7.25	7.25
6	Menjing	28.4	26.7	12.04	11.62
7	Tunjungsekar	32.7	31.8	13.23	12.81
8	Borobudur	29.6	29.1	12.84	12.73
9	Tawangmangu	25.4	25.4	11,07	11.0
10	Tasik Madu I	34,6	32.4	14.24	13.89
11	Tasik Madu II	39,2	37.1	16.31	15.96
12	Joyogrand	27.1	27.1	11.74	11.74

Based on table 2 it can be seen the difference of travel time and also the distance between the existing route with the fastest route. There was also the travel time and distance is the same with network analysis, there were Merjosari, Tlogomas, Sumbersari, Tawangmangu and Joyogrand waste transfer point. The difference of travel time and the largest distance is found in Dinoyo and Ketawanggede waste transfer point with the difference of travel time reaches 5.4 minutes and 2.9 km. The difference in travel time and mileage (distance) indicates that the determination of the existing route is still inefficient.

4. Conclusion

The existing transportation route is a route that has been determined by the Department of Hygiene and Gardening of Malang City (Dinas Kebersihan dan Pertamanan) as the responsible of the process of transporting waste Malang City. The value of effectiveness of waste transportation system in waste transfer point in Lowokwaru district is 48% effective. While as many as 52% is still not effective due to the remaining waste in the waste transfer point. The comparisons between existing route with fastest route based Network Analyst obtained the shortest and fastest route as follows 1): six waste transfer points have the fastest route in travel time (minute) and 2) seven waste transfer points have the fastest route in travel distance (km).

5. Acknowledgments

The authors would like to thank to Rakih Wicaksono for helping to collect practical data and improve the quality of the work.

References

[1] Apaydin O 2004 Geographical information system (GIS) supported optimization of solid waste management alternatives in Trabzon City, *Ph.D. Thesis*, Yildiz Technical University, Science Institute, Environmental Engineering, Istanbul.
[2] Apaydin O Arslankaya E Avsar Y and Gonullu M T 2004 GIS supported optimization of solid waste collection: Trabzon case study, *Sigma*, 4, 249-254.
[3] Apaydin O 2005 A GIS Supported Optimization Application of Solid Waste Collectio System in Trabzon City, *Ekoloji*, 54, 1-6.
[4] Jaydeep L Mandla R V and Xuan Z 2017 Solid waste collection/transport optimization and vegetation land cover estimation using Geographic Information System (GIS): A case study of a proposed smart-city, *Sustainable Cities and Society 35*, 336-349
[5] Gallardo A Carlos M Peris M Colomer F J 2015 Methodology to design a municipal solid waste pre-collection system. A case study. *Waste Manage.* 36, 1–11.
[6] Amponsah S K Salhi S 2004 The investigation of a class of capacitated arc routine problems: the collection of garbage in developing countries. *Waste Manage*. 24 (7), 711–721.
[7] Undang-Undang Republik Indonesia No.18 tahun 2008 tentang Pengelolaan Sampah
[8] Damanhuri E and Padmi T 2010 *Diktat Kuliah Pengelolaan Sampah* Institut Teknologi Bandung, Bandung.
[9] Clifford T 2008 *Waste Collection Optimisation Tools for Waste Managers* United Kingdom: Indecon.Co, Ltd
[10] Memon M A 2010 Integrated solid waste management based on the 3R Approach. *Journal of Material Cycles and Waste Management*, 12(1), 30-40. doi: 10.1007/s10163-009-0274-0
[11] Sudarto 2008 *Keefektifan Pengelolaan Sampah Kota dalam Meningkatkan Kualitas Lingkungan di Kecamatan Kajen Kabupaten Pekalongan* Universitas Sebelas maret Surakarta.
[12] Sutrisno H 1987 *Metodologi Research Jilid 2* Yogyakarta. Andi Offset.
[13] Nasution 2003 *Metode Research: Penelitian Ilmiah* Jakarta: PT. Bumi Aksara.
[14] Prastowo A 2010 *Menguasai Teknik-teknik Koleksi Data Penelitian Kualitatif* Yogyakarta: DIVA Press.
[15] Winyoopradist S Siangsuebchart S1999. Network Analysis for variable traffic speed. *Paper presented at the ESRI User Conference.*
[16] Echols N 2003 Finding the Fastest Route. Paper presented at the ESRI User Conference, URL: http://gis. esri. com/library/user-conf/proc03/p0320.pdf, Redlands, California.
[17] Abousaeidi M Fauzi R and Muhamad.R 2016 Geographic Information System (GIS) modelling approach to determine the fastest delivery routes. *Saudi Journal of Biological Science*. 23, 555-564.
[18] Malakahmad A Bakri M P Mokhtar M R M and Khalil N 2014 Solid Waste Collection Routes Optimalization via GIS techniques in Ipoh City Malaysia, *Procedia Engineering 77*, 20-772

IOP Conf. Series: Earth and Environmental Science **148** (2018) 012032 doi:10.1088/1755-1315/148/1/012032

The impact of human activities in the Wulan Delta Estuary, Indonesia

L N Fadlillah[1], Sunarto[2], M Widyastuti[2] and M A Marfai[2]

[1]Faculty of Geography, Universitas Gadjah Mada, Indonesia
[2]Department of Environmental Geography, Faculty of Geography, Universitas Gadjah Mada, Indonesia

lintang.nur.f@mail.ugm.ac.id

Abstract. The increasing of human population in the watershed and the coastal area and the need of life exert pressure in the delta that provides various resources. Wulan Delta is one of active Delta in Central Java, Indonesia. It has been experienced multiple pressures because of natural factors and human factors. In order to provide the scientific solution and to analyze the impact of human intervention in delta, we collected several pieces of evidence based on secondary data and primary data. The secondary data is water quality data on sites 6 and 7, meanwhile the secondary data is the water quality data in site 1 to 5. This paper present a review and problems identification in Wulan Delta, based on hydrological condition, land use, and human activities in the delta. Meanwhile, the human intervention in the land which is land use exchange leads to several problems such as the land use changes, high sediment load, and water degradation. Almost 80% of Delta has been transformed into the fish pond by local communities.

1. Introduction

Delta system is significantly threatened by various kind of drivers, such as aquaculture, agriculture, fish catching, transportation, climate change, and another human activities [2]. The delta and estuaries region is a dynamic area. It represents transactional zone between land and sea, characterized by high sediments and rich of bio-diversities [1]. Due to that factors, this region has become critical studies for multidisciplinary studies [11]. Delta is part of watershed system [15]. The watershed system which is connected by rivers plays significant role as a link in sediment transport and nutrient transport [22]. The watershed characteristic in term of land use, topography, hydrology in the upstream area will affect the downstream area [8].

Indonesia is an archipelago which has 5,8 million km of shoreline (KEP.18/MEN/2011) which consist of small islands and big islands. The increasing of human population in the coastal area and global climate change influence the productivity and environmental health in the delta system. The decreasing of water quality condition in the delta is a response of physical processes that altered by hydrodynamics processes in the deltaic environment [4]. Hydrodynamics is important in deltaic environment due to its contribution to sediment transport processes, nutrient transports, and mixing processes.

The Wulan Delta is one delta in Java Island, Indonesia. This delta situated in Demak and Jepara City, Indonesia, between latitude 6°43'30" S and 6°46'30" S and longitude 110°32'0" E to 110°36'0" E. The area covers 31.75 km2. The delta is located in the Serang Watershed's estuary. Wulan Delta has been developing until recently [21]. The development of Wulan Delta has been noticed since 1920 [16] and noted by several research until now. Not only disturbed by huge sediment load, but this location is also affected by human intervention in the form aquaculture and settlement in the delta. It has potential to harm the water environment and the delta morphology. The research aims to find evidence of the impact altered by human intervention in the delta. This review is beneficial to predict the future condition and formulate the policy regarding delta management [6].

2. Method

The present study was carried out using primary and secondary data sources. The information of delta shoreline and hydrological data were obtained from governmental and non-governmental organization. The water quality data was collected from governmental and field survey. The data obtained from the government were Temperature, pH, Biological Oxygen Demand (BOD), Dissolved Oxygen (DO), and Total Suspended Solid (TSS) data in 2014 and 2016 (Site 6 and 7). Primary data of water quality in Wulan Delta were collected in 2016 (Site 1 to 5). The primary data collected was TSS, DO, and Nitrat (NO_3). The sampling was conducted twice in May and August 2016. The sample were taken at the highest high water level (HHWL) in each month and taken in ebb tides and flood tides. There were 5 sampling sites to represent the estuary (figure 1). The water quality data in the Wulan Delta were compared to Water Quality Standards (WQSs) for Estuary.

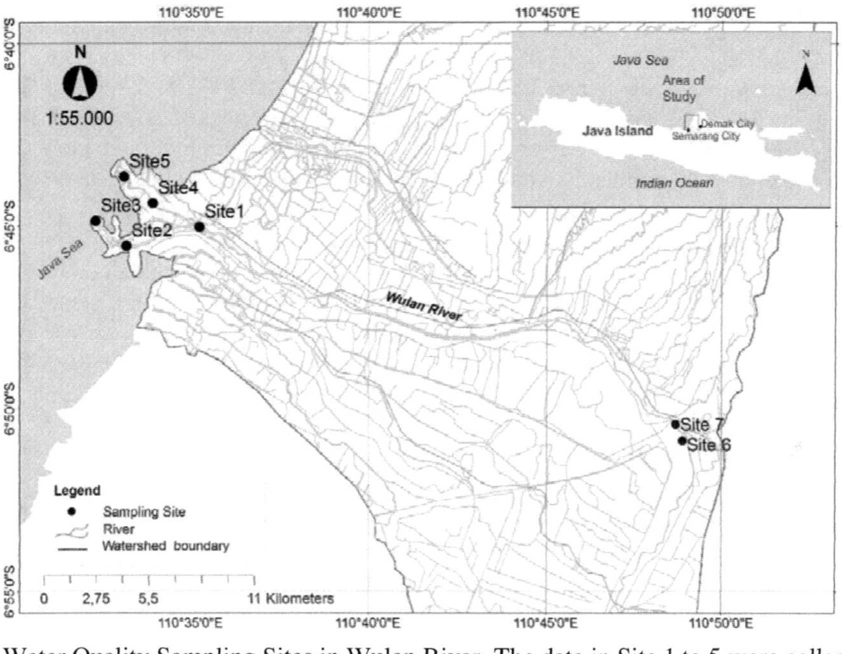

Figure 1. Water Quality Sampling Sites in Wulan River. The data in Site 1 to 5 were collected in 2016 during Field Survey. Meanwhile, the data for Site 6 and 7 were a secondary data which collected in 2014 and 2016.

Data were also collected from the local people in the Menco hamlet or sub-village, Berahan Wetan and Berahan Kulon Village, Wedung District, Demak, Indonesia using Focus Group Discussion (FGD) method. The informants for FGD are consisting of 10 people. All of them ate fish farmers whose own fish pond in Wulan Delta. The FGD data were used to validate the accretion and erosion processes and gain information regarding waste disposal system in the Delta Wulan's aquaculture. The land use data were gained from base map of Peta Rupa Bumi Indonesia in 2008 and Satellite Images of IKONOS in 2016. The satellite interpretation was used to investigate the changing of land use between the years of 2008 to

2016 in a simple way. The accretion and erosion data were collected from several publications conducted in Wulan Delta.

3. Result and Discussion

3.1. Hydrological characteristic of Wulan Delta

The Wulan River is bifurcated into 2 channel, namely Wulan Baru (New Wulan) and Wulan Lama (Old Wulan). According to the history, the Wulan Lama formed first, and the Wulan Baru formed latter due to dredging projects. The Serang River runs from the South to The North Coast of Java Island over a length of 48,62 km. The Wulan Baru is approximately 6 km long and the Wulan Lama is 5 km long. The average rainfall in Wulan Delta is 2187 mm/year. The mean water discharge in 2016 is 102 m^3s^{-1}. Based on Schmidt-Fergusson climate classification, this area climate is moderate.

3.2. Land use change in Wulan Delta

Along with the increasing of daily needs, the delta nowadays was changed into fish pond. Figure 2 shows the difference of land use in Wulan Delta in 2008 and 2016. The major use of delta was fish pond. Most of saltwork land, rice field, and crop transformed into fish pond. The fish pond expands from 21,862,251.84 m^2 in 2008 to 24,946,439.35 m^2 in 2016. The fish pond percentage expansion increase from 82.94 % (2008) to 91.55 % (2016) of major use in Wulan Delta. Table 1 and table 2 indicate the difference of the total land between the years of 2008 to 2016. There was sedimentation in the West Part of Wulan Delta (figure 2).

The evidence shows that the wetland increase of 888,787.8 m^2 in 2016 due to sediment accretion. The delta development rate before 1994 was 0.000338 km^2/year [21]. However, the delta development rate in 2000-2011 was 0.085 km^2/year and in 2011-2015 was amounted up to 0.06 km^2/year [18]. The delta development allegedly affects its surrounding. The morphological changes in delta are assumed as one the main factor of fish pond loss in Jepara and Demak due to the changing pattern of longshore current in the north coast.

Table 1. Land Use in 2008.

Land Use	Area (m^2)
Fish Pond	21,862,251.84
Bush	1,166,519.20
Salt work land	52,074.26
Salt Marsh	856,661.12
Farm	77,972.13
Settlement	279,492.47
Grass	363,689.83
Swamp	608,678.60
Rice Field	648,953.53
Crop/moor	441,905.14
Total	26,358,198.11

Table 2. Land Use in 2016.

Land use	Area (m^2)
Fish Pond	24,946,439.35
Bush	1,166,519.20
Salt marsh	412,872.92
Farm	77,972.13
Settlement	279,492.47
Grass	363,689.83
Total	27,246,985.91

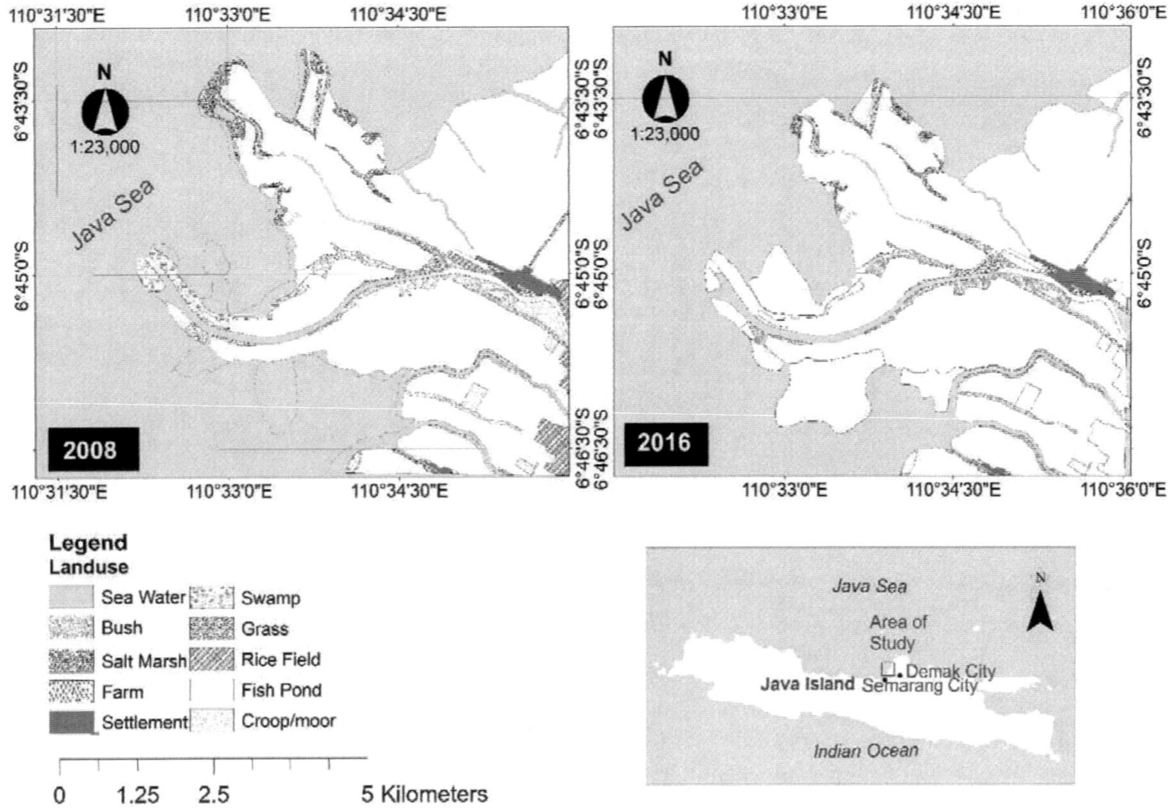

Figure 2. The Land use change in Wulan Delta in 2008 and 2016. It also shown that there was an accretion process in the delta in the West part of the delta.

3.3. Possibility of water quality degradation

The aquaculture in Wulan Delta is supplied by Wulan River and The North Coast. The Wulan River flowed into the fish pond into channels which connected the river and the sea directly. The tidal cycle which occurs in everyday potentially bring nutrient as well as pollutants in the estuary. It may affect the aquaculture and the benthic ecosystem in the delta estuary. The estuary receives threats not only from the deltaic system, but also from the upstream area such as agriculture and domestic use. Several parameters are sensitive to aquaculture, i.e. BOD, DO, and TSS. We compare the laboratory and direct measurement for several parameters mentioned to Water Quality Standards (WQSs) for Estuaries and Aquaculture according to Decree of The Minister of State for The Environment of Republic of Indonesia Number 51/MENKLH/2004 for marine biotic. The WQSs is important for preserving aquatic ecosystem and human health and establish the standard for aquatic ecosystem protection [13].

Dissolved Oxygen (DO) is fundamental in the coastal water as well as in the other water bodies and its used as indicators of pollution in the aquatic areas [19]. The decreasing of DO may happen due to biological processes such as photosynthesis and respiration of the benthic organism or algae. Several studies have shown that the limit of DO for benthic organism is 2mg/L [17]. The Indonesian Water Quality Standard for DO in the estuary is 5 mg/L. The average DO in Wulan Delta in 2016 is 6.82 mg/L.

DO is sensitive parameter towards temperature. In high temperature, the respiration and photosynthesis of aquatic organism was increase, which makes decreasing of DO concentration in the water column.

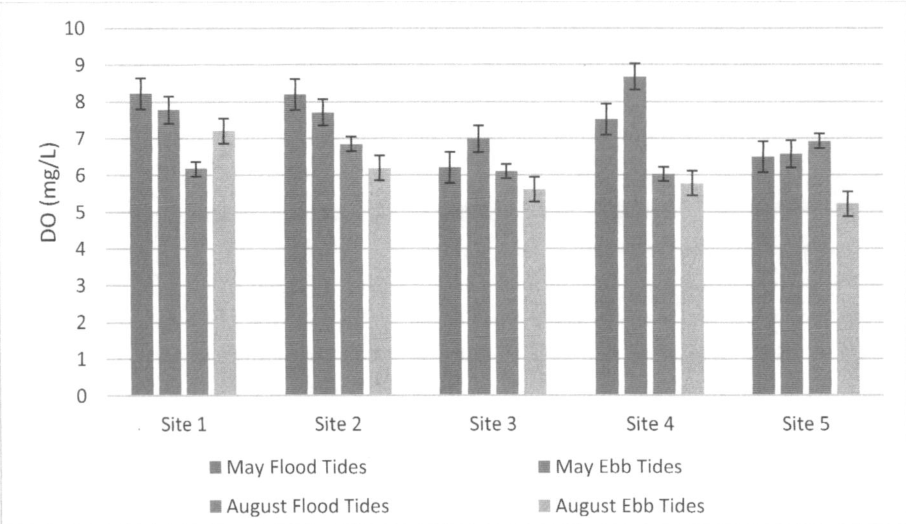

Figure 3. The Dissolved Oxygen Concentration during Field Survey in 2016.

The water quality standard of TSS for aquaculture is < 25 mg/L, while TSS between 25-80 mg/L is considered to have bad impact on the aquaculture. Average TSS concentration in Wulan Rivers is 22,4 mg/L in 2016 during the field survey. Yet, form the secondary data shows high concentration of TSS in April 2016, both in the upstream station and downstream area (table 3). This condition may occur due to flood or high water discharge. TSS concentration commonly equivalent to the water discharge. The higher the water discharge, the higher TSS concentration [9]. High concentration on TSS will increase turbidity and prevent light to penetrate in the water column and disturb the photosynthesis process [5]. High concentration of TSS happens due to high intensity of erosion and land degradation in the upstream area, or the river bank erosion. In general, TSS in the river mouth sites is higher than any sites in the delta estuary.

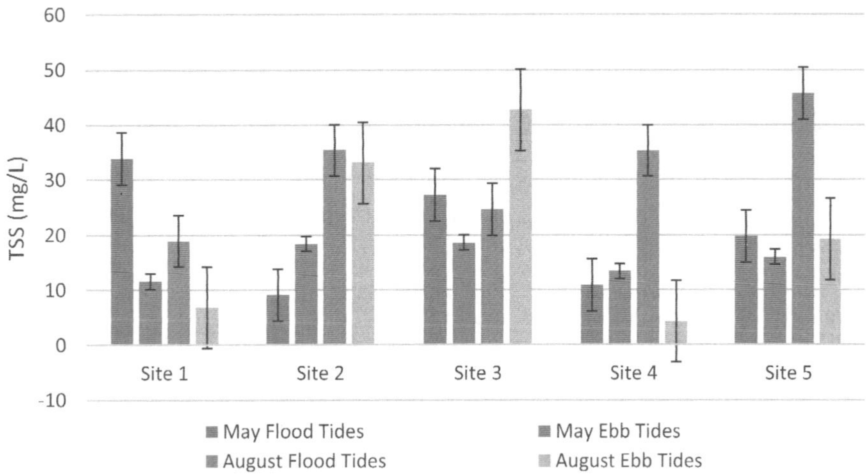

Figure 4. The TSS concentration during Field Survey in 2016.

Biochemical Oxygen Demand (BOD) is useful parameter to trace the wastewater effects in aquatic ecology. This parameter provides information about the potential of microbial respiration to breakdown the organic material present in the water [19]. Indeed, from the BOD data (**Table 3** and **4**). The BOD Shows high concentration which is > 3 mg/L. According to Indonesian Water Quality Standards it shows that in the Site 6 and 7 (Fig), the BOD was above the WQSs, the indication of high organic contamination from agriculture or industrial activities. in the upstream area [10]. Site 6 and 7 was far from the estuary, so that the aquaculture may not be the reason for high contamination. Yet, the BOD was not collected in the Site 1 to 5.

Table 3. Water Quality in Wulan River (Site 6).

	Temperatur	**pH**	**TSS**	**BOD_5**	**DO**	**Total Coliform**
May 2014	30.8	8	26	8.256	7.07	450
August 2014	30.6	8.2	21	11.45	5.61	4500
April 2016	28.2	7.8	2275	14.95	2.16	92000
August2016	30	8.1	24	3.28	6.31	450

Table 4. Water Quality in Wulan River (Site 7).

	Temperatur	**pH**	**TSS**	**BOD_5**	**DO**	**Total Coliform**
May 2014	31.9	7.8	54	71.96	5.49	680
August 2014	31.5	7.6	38	66.02	2.81	26000
April 2016	29	7.8	2592	21.45	1.19	93000
August2016	30	7.7	58	16.86	6.05	680

The main source of ammonia in the water column includes the decomposition of organic material from mangroves, shrimp, or fish in the bottom of the pond [7]. The concentration of Ammonia (NH_3^--N) in the Delta Estuary are still in a good condition, there was no data exceed the WQSs. Water bodies which have DO> 5 usually do not have high concentration of Ammonia. This means that the aquaculture didn't give bad impact to water bodies in Estuaries. However, the Total Maximum Daily Load is better method to understand the limitation of river to accommodate the pollution load which enters the water bodies.

Table 5. The Average Ammonia Concentration (mg/L).

Sampling Site	May Flood Tides	May Ebb Tides	August Flood Tides	August Ebb Tides
Site 1	0,0000	0,0275	≤0,0094	≤0,0094
Site 2	≤0,0094	≤0,0094	≤0,0094	≤0,0094
Site 3	≤0,0094	≤0,0094	≤0,0094	≤0,0094
Site 4	≤0,0094	0,0200	≤0,0094	≤0,0094
Site 5	≤0,0094	≤0,0094	≤0,0094	≤0,0094

3.4. The role of local society to preserve the delta

In the new land formed by sedimentation in delta, the aquaculture gradually becomes dominant. Most of the delta was transform into fish pond. The local said that during the raining season, flood comes along with huge sediment and formed new land which can be used as fish pond. Fish pond waste may induce water degradation. The waste was carried out by tides back to the pond and affects the fish farming [14]. The unplanned fish pond may destroy the environment and ecosystem [12].

Contrastingly, the water quality in Delta Wulan Estuary in general didn't show the bad impact of fish farming. The water quality in the fish pond areas deteriorates because of excessive feeding of food and in several countries use antibiotics as medicine [12]. Meanwhile, in Wulan Delta most of the fish pond is managed in traditional way. Farmers build his muddy pond in the delta and allow water from the river and the sea to enter the pond naturally and rely on natural food. During ebb and flood tides, the tides flush the pollution in the fish pond. They build the channels from the fish pond directly to the sea. This type of fish farming is beneficial for environment, yet it gives low productivity in the fish production [20].

4. Conclusion

This study presents the evidence of human intervention impact in the Wulan Delta. It shows that due to complicated reason, the Wulan Delta has been experiencing land use change from many land use type into fish pond. The human intervention in delta may alter water quality degradation due to aquaculture, domestic, and industrial. However, the water quality data shows there were no significant impacts of aquaculture to the decreasing of water body. The traditional aquaculture management which conducted based on natural food and tides system was suitable for the delta ecosystem and can be used as a key to preserve the deltaic environment. Based on this simple research, we can estimate the delta development in the future. The Estuaries monitoring are needed to control the pollution or sediment load in the Serang Watershed.

Acknowledgement

This project is supported by the Ministry of Research, Technology, and Higher Education, Indonesia as part of the grant for Master Leading to PhD Project with the contact number: 4550/UN1-P.III/LT/DIT-LIT/2016. The authors express their gratitude to reviewers for their advices.

References

[1] Bao J and Gao S 2016 Environmental Characteristics and Land-use Pattern Changes of the Old Huanghe River Delta, Eastern China, in the Sixteenth to Twentieth Centuries. *Sustain Sci.* 11: 695-709.

[2] Chea R Grenouillet G Lek S 2016 Evidence of Water Quality Degradation in Lower Mekong Basin Revealed by Self-Organizing Map. *PLOS ONE.* 11(1). DOI:10.1371/journal.pone.0145527.

[3] Decree of The Minister of State for The Environment of Republic of Indonesia Number 51/MENKLH/2004 for Marine Biotics Water Quality Standards.

[4] DiLorenzo J L Filadelfo R J Surak C R Litwack H S Gunawardana V K and Najarian T O 2004 Tidal Variability in the Water Quality of an Urbanized Estuary. *Estuaries.* 27(5): 851-860.

[5] Effendi H 2003 *Telaah Kualitas Air.* Yogyakarta: Kanisius.

[6] Etcheber H Schmidt S Sottolichio A Maneux E Chabaux G Escalier J M Wennekes H Derriennic H Schmeltz M Quéméner L Repecaud M Woerther P and Castaing P 2011 Monitoring Water Quality in Estuarine Environments: Lessons from the MAGEST Monitoring Program in The Gironde Fluvial-Estuarine System. *Hydrology and Earth System Sciences.* 15: 831-840.

[7] Garno Y S 2004 Pengembangan Budidaya Udang dan Potensi Pencemarannya pada Perairan Pesisir. *Jurnal teknik Lingkungan. P3TL-BPPT.* 5(3): 187-192.

[8] Gaspar R Marques L Pinto L Baeta A Pereira L Martins I Marques J C Neto J M 2017 Origin here, Impact There- The need of integrated Management for River basin and Coastal Areas. *Ecological Indicators 72*: 794-802.

[9] Gray A B Warrick J A Pasternack G B Watson E B and Goni M A 2014 Suspended Sediment Behavior in Coastal Dry Summer Subtropical Catchment: Effects of Hydrological Preconditions. *Geomorphology.* 214: 485-501.

[10] Huong and Hoa 2012 Aquaculture and Agricultural Production in the Mekong Delta and Its Effects on Nutrient Pollution of Soil and Water. Book Chapter in Renaud F. dan Kuenser C. 2012. *The Mekong Delta System: Interdisiplinary Analyses of a River Delta*. Dordrecht: Springer. DOI 10.1007/987-94-007-3962-8. Chapter 14:363-393.

[11] IPCC 2014 IPCC WG II AR 5 *Climate Change 2014: Impacts, Adaptation, and Vulnerability* Cambridge University Press: Cambridge.

[12] Islam S M D and Bhuiyan M A H 2016 Impact Scenarios of Shrimp Farming in Coastal Region of Bangladesh: An Approach of An Ecological Model for Sustainable Management. *Aquaculture Int.* 24:1163-1190.

[13] Kwak J I Nam S H and An Y J 2017 Water Quality Standards for The Protection of Human Health and Aquatic Ecosystem in Korea: Current State and Future Perspective. *Environ Sci Pollut Res.* DOI.10.1007/s11356-017-8923-7.

[14] Kon K Kawakubo N Aoki J Tongnunui P Hayashizaki K dan Kurokura H 2009 Effect of Shrimp Farming Organic Waste on Food Availability for Deposit feeder Crabs in A Mangrove Estuary, Based on Stable Isotope Analysis *Fish Sci.* 75:715-722.

[15] Renaud F and Kuenser C 2012 *The Mekong Delta System: Interdisiplinary Analyses of a River Delta.* Dordrecht: Springer.

[16] Ruswanto M 1996 Delta Wulan Jepara terus Meluas. *Media Riset Akuntansi Trisakti.* 14: 16-19.

[17] Sany S B T Hashim R Rezayi M Salleh A and Safari O 2014 A review Strategies to Monitor Water and Sediment Quality for Sustanability Assessment of Marine Environment. *Environ Sci Pollut Res.* 21:813-833.

[18] Septiangga B 2017 Analisis Multitemporal Morfodinamika Wilayah Kepesisiran Muara Delta Wulan dan Sekitarnya Tahun 1995-2015. *Skripsi.* Yogyakarta: Universitas Gadjah Mada.

[19] Sheldon J E and Alber M 2011 Recommended Indicators of Esuaries Water Quality for Goergia. *Proceedings of the 2011 Georgia water Resources Conference*, held April 11-13, 2011, at the University of Georgia.

[20] Soliman N F and Yacout D M M 2016 Aquaculture in Egypt: Status, Constraints, and Potentuals. *Aquacult Int.* 24: 1201-1227.

[21] Sunarto 2004 Perubahan Fenomena Geomorfik Daerah Kepesisiran di Sekeliling Gunung Api Muria Jawa Tengah (Kajian Paleogeomorfologi). *Disertasi*. Yogyakarta: Universitas Gadjah Mada.

[22] Wang S Fu B Liang W Liu W Wang Y 2017 Driving Forces of Changes in the Water and Sediment Relationship in the Yellow River *Science of the Total Environment* 576:453-461.